The Aerobiological Pathway of Microorganisms

The Aerobiological Pathway of Microorganisms

C. S. Cox
Chemical Defence Establishment
Porton Down, Salisbury

A Wiley–Interscience Publication

JOHN WILEY & SONS
Chichester · New York · Brisbane · Toronto · Singapore

Library of Congress Cataloging-in-Publication Data:
Cox, C. S.
The aerobiological pathway of microorganisms.

Includes index.
1. Air—Microbiology. 2. Aerosols. 3. Airborne infection.
I. Title.
QR101.C69 1987 576'.15 86-15710

ISBN 0 471 91170 4

British Library Cataloguing in Publication Data:

Cox, C. S.
The aerobiological pathway of microorganisms.
1. Microorganisms—Dispersal 2. Air—Microbiology
I. Title
576'.190961 QR101

ISBN 0 471 91170 4

Printed and bound in Great Britain.

Contents

Introduction

Aerobiology is the science of the aerial transport of microorganisms and other microscopic biological materials together with their transfer to the air, their deposition and the ensuing consequence for life forms including the microscopic entities themselves. The Aerobiological Pathway consists of the launch, aerial transport and subsequent deposition of such particulate matter: topics which play important roles for public health, laboratory workers, agriculture, microbiology, genetic engineering, industry, environmental studies, space exploration and the biological control of insects (Winkler, 1973; Edmonds and Benninghoff, 1973). Even so, few books deal specifically with aerobiology, possibly because it is a relatively new subject, is broadly-based and involves many different branches of science. The alternative of referring to original papers usually provides a confused picture to students as the experimental data can often appear to lack consistency and have no obvious patterns. One major difficulty stems from the large number of interdependent factors which affect the outcome of experiments. The problem is confounded further because the importance of some of these factors has been recognized only comparatively recently. Consequently previous results sometimes require re-evaluation in the light of this later research. For example, results of an experiment conducted in 1960 in air and explained only in terms of desiccation will not have taken account of the toxic action of oxygen which remained undiscovered until 1965.

The discovery of oxygen toxicity for some bacteria together with insights into repair and into dehydration–rehydration death mechanisms now provide a coherence of data not possible in 1969 when *An Introduction to Experimental Aerobiology* was published. An important illustration is that even though dehydration and rehydration processes are common virtually to all microbial aerosols, their elucidation during the 1970s was required before it became apparent that bacteria and viruses when airborne were inactivated by similar basic mechanisms. A little over a decade ago such causality was never promulgated since bacteria and viruses as aerosols appeared to be such different entities.

The aim of this textbook is to provide a background and explanation of the Aerobiological Pathway and of microbial survival and infectivity. It relates these to biohazards to be found in public health, laboratories, agriculture, microbiology, genetic engineering, industry, environmental studies and space exploration. It is pertinent also to the biological control of insects and pests. The first nine chapters deal with the more physical aspects of microbiological aerosols, whereas the remaining eight chapters cover the more biological ones. Chapter 1 considers physical properties of aerosols appertaining to aerobiology while Chapter 2 discusses laboratory techniques specifically for handling microbiological aerosols. Chapter 3 covers aerosol sampling — a function of particular concern — and Chapter 4 describes aerosol monitoring methods. Chapter 5 deals with the difficulties of determining the sizes of aerosol particles and discusses different approaches to the problem. In Chapter 6 measures for controlling and containing biohazardous aerosols are presented; Chapter 7 considers techniques and certain difficulties associated with work in the field. The aerial transport of microbial aerosols both indoors and outdoors are topics for Chapter 8 while the take-off and landing of aerosol particles forms the subjects for Chapter 9. The next chapter is the first of those concerned with viability and infectivity of microbial aerosols. It deals specifically with dehydration–rehydration death mechanisms and the role in these of temperature. These subjects are covered first because movements of water molecules occur whenever microorganisms are airborne and, therefore, represent one of the most fundamental processes of microbial aerobiology. Chapter 11 is concerned with the toxic effects of oxygen on airborne bacteria and Chapter 12 describes the Open Air Factor — an atmospheric pollutant highly effective in killing many species of airborne microbes. Other environmental parameters such as pressure fluctuations and air ions are covered in Chapter 13. In Chapter 14 the important topic of repair mechanisms is considered while Chapter 15 deals with factors specifically involved with microbial infectivity. The application of catastrophe theory in Chapter 16 shows how the kinetics of viability/infectivity loss are consistent with known chemical reactions evoked in the denaturation of membranes. The final chapter gives some examples of how the events of previous chapters come together in the Aerobiological Pathway of microorganisms operating in practice.

REFERENCES

Edmonds, R. L. and Benninghoff, W. S. (1973). *Aerobiology and its Modern Applications*. Report No. 3, Aerobiology Component, U.S. Component of the International Biological Program.

Winkler, K. C. (1973). In *Fourth International Symposium on Aerobiology*, J. F. Ph. Hers and K. C. Winkler (Eds), Oosthoek, Utrecht, The Netherlands, pp. 1–11.

Chapter 1

Aerosol physical attributes

1.1 INTRODUCTION

Aerosols comprise matter finely divided and suspended in air and have a composition as varied as matter itself. Sensibly this state constitutes particles less than about 25 μm diameter while in the present context the lower limit is about 0.5 μm diameter and the upper one about 15 μm (for reasons which will become apparent from later chapters). Aerosol particles may be liquid or solid and contain microorganisms, organic and/or inorganic material. But in one very important aspect all aerosols are the same, that is, in their aerodynamic behaviour. Consequently microbial aerosols are subject to the same physical laws as other airborne particles. Unlike them though, microbial aerosols are subject as well to laws concerned with their special property of being living entities, i.e. being able to multiply in kind with the expenditure of energy and production of nucleic acid.

Airborne microbes need to be considered in terms of their physical attributes and their biological ones. Such constitutes the subject of Aerobiology while their Aerobiological Pathway is the take-off, aerial transport and their subsequent landing. This chapter deals with the more important aspects of their physical behaviour while later chapters are more concerned with the discovery, understanding and mathematical representation of biological features of the Aerobiological Pathway. That it is a difficult subject is most succinctly put by Phalen (1984): 'In short, when it comes to characterizing aerosols it may be said that Mother Nature is a witch'.

1.2 BROWNIAN MOTION

As for particles suspended in liquids, particles suspended in air are constantly bombarded by molecules of the surrounding medium. Some molecules are sufficiently energetic that their collision with a particle causes it to move slightly. Given the random motion of molecules the motion they impart to particles likewise is random while the high speed molecules of gases results in a high collision frequency. The net result is that particles rapidly 'vibrate' about a point which drifts randomly. This particular motion is termed Brownian motion and the intensity increases with temperature and with decreasing particle size. It may be expressed by the Einstein equation which in simplified form is:

$$\bar{x} = 5.0 \times 10^{-6} \sqrt{\frac{t}{r}} \tag{1.1}$$

where, $\bar{x}$ = root mean square of the particle displacement,

t = time (s),

r = particle radius (cm).

Dimmick (1969) shows that diffusion due to Brownian motion is less than gravitational settling (Section 1.4) for particles larger than about 1 µm diameter. He provides also the following useful argument. Given that the mean diameter of the human alveolus is 15×10^{-3} cm, then for a particle to travel a distance of 7×10^{-3} cm (i.e. alveolus radius) in a 2 s holding time, its displacement velocity would need to be larger than 3.5×10^{-3} cm/s. Particles less than about 0.1 µm or greater than about 1 µm diameter possess that displacement velocity. That is, particles less than 0.1 µm or greater than 1 µm diameter are likely to be retained by alveoli to an extent greater than for particles having sizes within that range. Alternatively, particles between 0.1 µm and 1µm diameter are those retained in alveoli with lowest probability. That this is so in practice is discussed later.

1.3 ELECTRICAL GRADIENT

Airborne particles on generation nearly always are charged unless steps are taken to neutralize them and the mean charge per particle, $\bar{q}$, generated by a two fluid atomizer is:

$$q = 5.6 \times d^{\frac{1}{2}} \text{ ions/particle (for } 10^{18} < N < 10^{20}) \tag{1.2}$$

$$\text{or, } \bar{q} = 8.2 \times 10^{-7} d^{\frac{3}{2}} \sqrt{N} \quad (\text{for } N < 10^{15})$$

where, d = particle diameter (µm)

N = number of ions of one charge (+ve or −ve) per ml of spray fluid.

For distilled water with ionic strength of $N \simeq 10^{14}$/ml, a 2 µm droplet of it would have about 24 charges, while on average the whole aerosol will have equal numbers of positively and negatively charged particles. For an aerosol with both positively and negatively charged particles and overall charge of zero the aerosol reaches equilibrium after some time. This charge equilibrium is known as the Boltzmann equilibrium for which the fraction (F) of particles of diameter (d) carrying (n) elementary units of charge (c) is,

$$F = \frac{1}{s} \exp\left[-\frac{(nc)^2}{4\pi\varepsilon_0 dkT}\right] \tag{1.3}$$

$$\text{where, } S = \sum_{-\infty}^{+\infty} \exp\left[-\frac{(nc)^2}{4\pi\varepsilon_0 dkT}\right]$$

ε_0 = dielectric constant of vacuum

k = Boltzmann constant

T = °K

The charge reached by a particle is approximately given by the equation,

$$Q_s = 2\pi\varepsilon_0 dV$$

where, d = sphere diameter

V = voltage (0.1V can be assumed).

When in an electric field, the force operating on a charged particle is:

$$F = nqE \tag{1.4}$$

where, n = number of electron charges on the particle
$q = 4.8 \times 10^{-10}$ Coulombs
E = electric field intensity (V/cm)

while the particle velocity (v) is slowed by a viscous drag due to the air and is:

$$v = \frac{10^4 nqE}{3\pi\eta dC} \text{ (cm/s)} \tag{1.5}$$

where, η = viscosity of air (poise, dyne $-$s/cm^2)
d = particle diameter (μm)
C = Cunningham slip correction factor (see equation 1.9).

Since particle mobility (Z) by definition is:

$$Z = V/E$$

$$\text{then, } Z = \frac{10^4 nq}{3\pi\eta dC} \tag{1.6}$$

Hence, in a field of 2000 V/cm, a 1 μm particle bearing 100 electron charges could attain a maximum velocity of about 2 cm/s, i.e. an electrical mobility of 10^{-3} cm^2/V/s. Electrical behaviour like this is invoked in electrostatic aerosol samplers and particle sizers.

Another facet appears in particle coagulation and particle deposition. For unipolar aerosols (i.e. all particles carry a like charge) repulsive forces occur between particles effectively preventing coagulation whereas for aerosols having mixed charge electrostatic attraction between particles causes increased coagulation rates compared to uncharged aerosols. However, the rate decreases fairly rapidly due to depletion of charged particles. Induced charging of particles in high electrostatic fields produces dipoles in particles and increases their coagulation rate and deposition probability. Coagulation rate is proportional to the square of the number of particles per unit volume of air.

Particle deposition can be markedly altered by charge. If equipment used to hold, transport or collect aerosols has appreciable charge physical loss of particles invariably is enhanced. But such deposition does not occur equally for all particle sizes, tending to be greatest for smallest particles owing to their higher electrical mobility. Therefore, particle size distributions can change through differential electrostatic precipitation. One should particularly avoid utilizing non-metallic tubing to connect to ungrounded metal chambers while plastic aerosol chambers are to be avoided if at all possible. Even then use of all metal systems does not always prevent severe problems due to enhanced electrostatic deposition. One pertinent example is that of dry powder microbiological aerosols at low RH, when being highly charged, particles induce charges of opposite sign in metal surfaces and are deposited by image Coulombic forces. When severe, one technique which helps to prevent it is to incorporate a radioactive isotope (e.g. C^{14}) into the dry powder to be disseminated (Cox *et al.*, 1970). Another is to work at RH values greater than about 70% RH when electrostatic phenomena are much reduced in their effects due to increased conductivity of surfaces and rapid dissipation of charge. In most work with microbiological aerosols effects due to their electrical charge tend to be overlooked except in the application of agricultural materials as sprays and aerosols.

1.4 GRAVITATIONAL FIELD

A particle in a parcel of still air will be acted upon by gravity and it will fall with a velocity dependent upon its mass. As particle rate of fall increases so too does the drag or viscous frictional force of the air. When the two forces are equal the particle attains its terminal (or final) velocity. The equation relating terminal velocity to particle size, mass, etc. is known as Stokes law and is derived by equating acceleration due to gravity to viscous drag forces. For a spherical particle,

$$v = \frac{\rho d^2 g C}{18\eta} \quad \text{cm/s} \tag{1.7}$$

where, ρ = particle density (g/cm^3)
d = particle diameter (cm)
g = acceleration due to gravity (cm/s^2)
η = viscosity of air (g/cm s^{-1})
C = Cunningham slip correction factor (see equation 1.9)

For ambient conditions, the equation becomes,

$$v = 3.2 \times 10^5 \rho d^2 \quad \text{cm/s} \tag{1.8}$$

and a spherical particle of diameter 10 μm of unit density would fall with a velocity of 3.2×10^{-1} cm/s. Such a prediction for aerosol particles of this size and greater is accurate to better than 1%. However, for smaller sizes the values are smaller than observed. The discrepancy occurs because as particle size decreases it approaches the mean free path of air molecules (about 6.5×10^{-6} cm at ambient conditions) and particles slip between them. A correction for this effect is called the Cunningham slip correction factor (C) where,

$$C = (1 + 2A\lambda/d) \tag{1.9}$$

and, λ = mean free path of air molecules

A = constant, approximately unity.

According to Dimmick (1969) a good approximation for ambient conditions is to add 0.08 μm to the actual diameter before computing velocity or to subtract 0.08 μm from particle diameter calculated from terminal velocity. For a particle of 0.5 μm or less, its terminal velocity is so low as to be difficult to measure and for all practical purposes can be considered zero.

In many papers concerned with settling velocity of microbiological particles a value of 1.1 g/cm^3 often is used for particle density, and is that most commonly occurring in the observed range of 0.9 to 1.3 g/cm^3 (Orr, 1966). One outcome of gravitational settling is that on storage in metal tanks, cylinders, spheres, etc., aerosols decrease in concentration with respect to time simply because particles reach the bottom of the container. This decrease in particle number of microbial aerosols is referred to as their 'physical decay' as opposed to 'biological decay' which reflects, specifically, loss of viability and/or infectivity. In the very early days of studying these aerosols a technique actually used was to have a cylindrical holding chamber which occasionally (on the basis of calculation) was inverted, as is an hour glass. Nowadays more sophisticated techniques are available as described in this and the following chapter. They each represent an approach to getting over the problem of loss of aerosol over extended time due to gravitational settling. None completely avoids the problem (except that of the microthread technique, q.v.) rather the effects of gravitational sedimentation are reduced. All invoke stirred-settling in one form or another, but even then larger particles of polydisperse aerosols are preferentially lost. The outcome is a particle size distribution which continuously changes with respect to time together with a shift to smaller particle sizes. This behaviour can be especially troublesome when measurements of polydisperse aerosols are required over extended time.

Physical decay of aerosols can be modelled most simply in terms of a first order decay process, viz.

$$\frac{N_t}{N_0} = \exp(-kt) \tag{1.10}$$

where, N_0 = number of particles at t = time = 0

N_t = number of particles at $t = t$

k = decay rate constant.

$$\text{Hence,} \quad \ln\left(\frac{N_t}{N_0}\right) = -kt \tag{1.11}$$

$$\text{or,} \quad \log_{10}\left(\frac{N_t}{N_0}\right) = \frac{-kt}{2.303} \tag{1.12}$$

A plot of the logarithm of the airborne fraction as a function of time will be (approximately) a straight line having negative slope equal to $-k$. The value of k is a function of particle sedimentation velocity and chamber dimensions.

A very convenient and useful concept is that of aerosol half-life, i.e. the time required for the aerosol concentration to become halved, or

$$\ln(0.5) = kt_{\frac{1}{2}}$$

$$t_{\frac{1}{2}} = \frac{0.69}{k} \tag{1.13}$$

According to Dimmick (1969) the value of k approximately is given by,

$$k = \frac{v}{H}$$

where, v = particle terminal velocity
H = height of a chamber with vertical walls or an effective height, otherwise.

For a chamber of height of 100 cm, neglecting Cunningham's slip correction factor, the half-life can be obtained from the nomograph of Figure 1.1. To allow for polydispersity the decay curve is calculated for each size class of the distribution normalized to its frequency and the composite curve derived by addition of the separate decay curves. Dimmick (1969) provides examples of this process which although relatively simple gives results close to those found in practice. Consequently, the method is proved and it is possible to conclude that aerosol particles sediment independently of one another provided excessive aerosol concentrations are avoided.

There have been various attempts to prolong the airborne state beyond that achievable with a stirred settling chamber. All were based on providing a rising column of air to counter particle fall under gravity but failed because isothermal and laminar flow conditions could not be maintained for long periods. However, the answer was simple and provided by Goldberg *et al.* (1958); it was to rotate the stirred settling chamber about a horizontal axis, i.e. the rotating drum. Such a system is somewhat analogous to the cylinder which is inverted as mentioned above. An analysis of the way rotating a drum enables the airborne state to be prolonged (e.g. half-life of 24 h) is provided by Dimmick and Wang (1969). Suffice to say here that the contained air rotates with the drum and the suspended particles achieve a circular path with respect to the drum axis. The result is that particles thereby experience a continuous acceleration which offsets that due to gravity. By proper selection of drum diameter and speed of rotation (e.g. 2 m diameter rotated at 3 rev/min), the acceleration forces approximately cancel and the particles remain airborne.

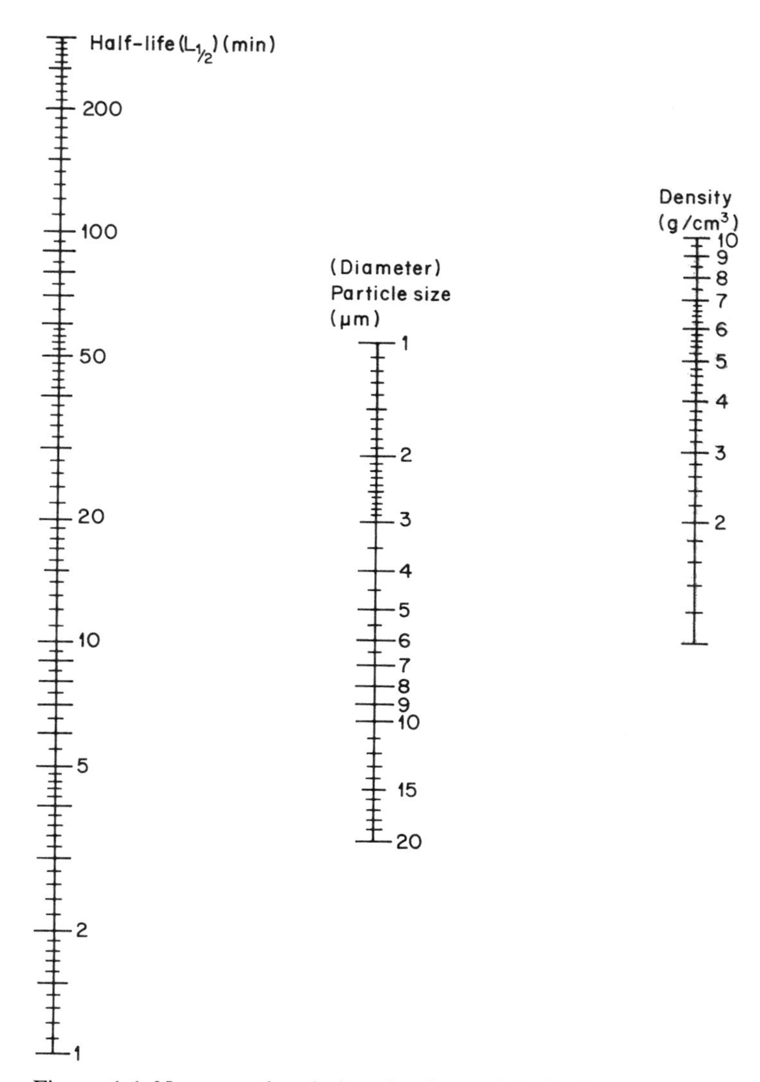

Figure 1.1 Nomograph relating density and half-life to particle size for aerosols undergoing stirred settling in a chamber of 100 cm effective height. A straight line drawn from appropriate density to observed half-life intersects at correct particle size. (From *An Introduction to Experimental Aerobiology*, R. L. Dimmick and Ann B. Akers (Eds), Wiley Interscience, 1969.) Reprinted with permission of the American Medical Association.

In practice, the radius of the circular path of a particle in a rotating drum is not constant but changes as it moves from the drum axis to the periphery. Likewise, the force of acceleration changes, becoming larger as the path radius increases. The net result is that particles eventually all drift to the walls

Table 1.1. Half-lives of monodisperse aerosol particles (density 1.1) in a drum rotating at 3 rev/min

Particle diameter (μm)	Half-life (minutes)
1	17 200
2	4308
3	1914
5	689
8	269
10	172
12	120

of the drum. If the drum rotation speed is reduced to try to prevent this occurring particle deposition still occurs but now because of gravitational settling.

Another factor is particle size which as in other stirred systems reduces aerosol half-life as it increases. The effect is quite marked as indicated by the calculated data presented in Table 1.1. That these half-lives are not attained in practice is thought to be owing to extent of sampling, temperature gradients, mechanical vibration and perhaps particle charge. Extent of sampling detracts because the removed sample volume of air has to be replaced by clean air causing a dilution of the aerosol in the drum, i.e. a depletion of the contained aerosol. Even so, under ideal conditions a half-life of 12 000 min has been observed (Dimmick and Wang, 1969).

1.5 INERTIAL FORCES

It is usual to think that when air carries particles they move together. Provided there is no change in direction (e.g. a straight pipe) this is approximately true. However, when a change in direction does occur (e.g. a curved pipe) the air molecules having low inertia follow the radius of the curve. In contrast, particles having much larger inertia try to continue in a straight line thereby crossing streamlines. Streamlines are hypothetical lines or threads of visible material which behave the same as air molecules; fine jets of smoke used for visualization of air flows are analogous.

Whether particles contained in an airstream flowing through a curved pipe will collide with the pipe wall or exit at the far end of that pipe is a question which arises frequently when working with aerosols. As a general rule when it is necessary to transport aerosols through pipes, the pipes should be straight and of relatively wide bore, e.g. at least a few centimetres. The air flow velocity should be sufficiently slow that laminar flow occurs (equation 1.19), i.e. minimum turbulence and mixing of parallel streamlines. Then, particle losses due to inertial forces will be minimal. When curved pipes are unavoidable

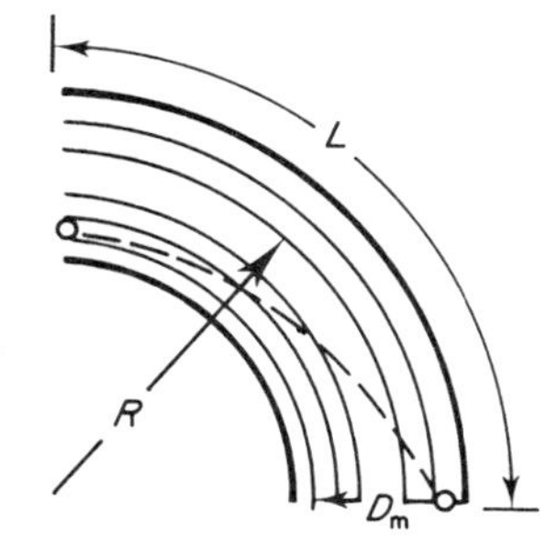

Figure 1.2 Particle crossing air streams in the bend of a pipe. D_m is the distance the particle has travelled across the streamlines. (From *An Introduction to Experimental Aerobiology*, R. L. Dimmick and Ann B. Akers (Eds), Wiley Interscience, 1969.) Reprinted by permission of John Wiley & Sons Inc.

losses are higher as illustrated by the following useful example taken from Dimmick (1969). Consider a particle of diameter 1 μm and unit density in a pipe 0.6 cm diameter with a curve of radius 5 cm and an air flow of 10 l/min (Figure 1.2),

Linear air velocity, $$V = \frac{F}{\pi(D/2)^2} \text{ cm/s} \tag{1.14}$$

where, F = flow rate (cm^3/s)

D = pipe diameter (cm)

is calculated as, $$V = \frac{10^4}{60} \times \frac{1}{\pi \times 0.3^2}$$

$$V \simeq 600 \text{ cm/s}.$$

Particle velocity, $$v = \frac{\rho d^2 A}{18\eta} \text{ cm/s} \tag{1.15}$$

which is the same as equation 1.7 except that gravitational acceleration (g) is replaced by (A), that is, any acceleration other than gravitational acceleration. Hence,

$$v = \frac{1 \times (1 \times 10^{-4})^2 A}{18 \times 18 \times 10^{-4}}$$

and, $$A = \frac{V^2}{R} = \frac{(600)^2}{5} \text{ cm/s}^2 \tag{1.16}$$

where, V = air velocity (cm/s)

R = radius of curvature of the pipe (cm)

Hence, $v = 0.22$ cm/s.

For a curve of length L, the time spent in the curve is,

$$t = \frac{L}{V} \tag{1.17}$$

Supposing the bend is a quarter of the circumference of a circle of radius R, then,

$$t = \frac{\frac{1}{4} \times 2\pi R}{V} = \frac{\frac{1}{2} \times 3.14 \times 5}{600} = 0.013\,\text{s}.$$

Therefore, during the 0.013 s that the particle is in the pipe it would continue in its original direction for that 0.013 s, i.e. a distance of $0.013 \times v$, i.e.

$$0.013 \times 0.22 = 0.0029\,\text{cm}.$$

The centripedal force acting on the particle moves it 0.0029 cm closer to the inside wall of the curved pipe, and, therefore, it would not collide. For a 10 μm particle, $v = 22$ cm/s and the distance travelled becomes 0.29 cm. Such a particle has a probability of collision with the pipe wall equal to this distance divided by pipe diameter, i.e.

$$p = \frac{0.29}{0.6} \simeq 0.5$$

that is, only about 50% of 10 μm particles entering such a curved pipe, on average, may be expected to exit. This situation would occur provided the air flow round the particle is not turbulent, otherwise the turbulence would affect the particle's trajectory. The nature of that air flow may be determined by calculation of the particle Reynolds number, $Re(p)$, from the equation,

$$Re(p) = \frac{\rho_a v r}{\eta} \tag{1.18}$$

where, ρ_a = air density (g/cm^3)

v = particle velocity (cm/s)

r = particle radius (cm)

η = viscosity of air (g/cm s^{-1})

and $Re(p)$ is a dimensionless number.

The ratio $(\eta/\rho_a) = 0.15$ and when vr approaches the same value the air flow round the particle will be found to be turbulent, i.e. when the value of $Re(p)$ approaches unity.

For the 10 μm particle considered above, $Re(p) = 0.1$ and the air flow round it will have minimal turbulence as the particle crosses streamlines. A 20 μm particle, in contrast, ($Re(p) > 1$), will experience turbulent flow. Although mathematically complex, techniques now exist for calculating the average trajectory of particles in turbulent flow which turns out to be not totally random.

Whether the air flow itself is likely to be turbulent or not can be deduced from the Reynolds number appertaining to that flow, viz.

$$Re = \frac{\rho_a V D}{\eta} \tag{1.19}$$

where, V = air velocity (cm/s)
D = pipe diameter (cm) } cf. equation (1.18)

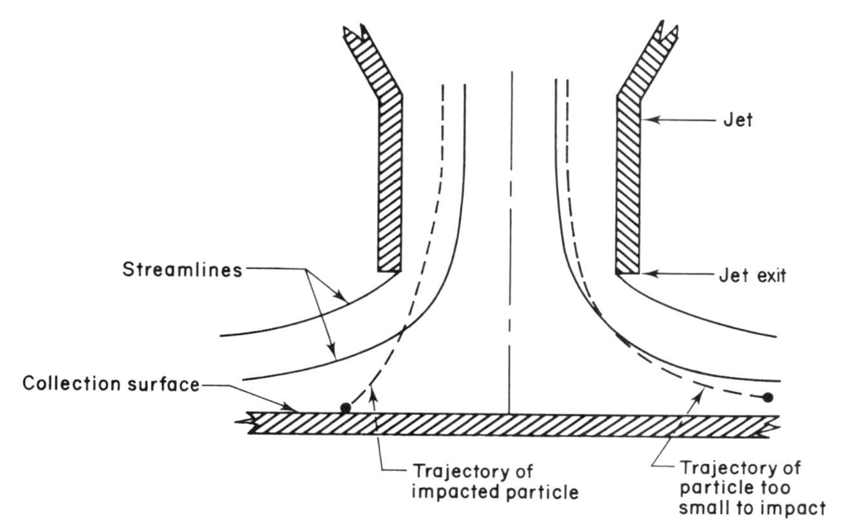

Figure 1.3 Streamlines and particle trajectories in an impaction jet. (After Rao, Ph.D. Thesis, *An Experimental Study of Inertial Impactors*, University of Minnesota.)

When *Re* is 2 to 3 × 10^3 or greater, turbulent air flow is probable; the value of *Re* in the above example is 2.4 × 10^3 and the air flow in the pipe would be turbulent. The effect of this turbulence also can be modelled and is to increase the probability of large particles negotiating the bend while that for small particles is slightly decreased. On exit a polydisperse aerosol has a size distribution changed from that at the inlet.

In practice, gentle bends are preferable to sharp ones but ideally only straight pipes should be used to carry aerosols. While calculations such as those described above aid in the design of aerosol systems, their performance in terms of what happens physically to aerosol particles should be determined experimentally. If all particles introduced into a system cannot be sensibly accounted for, there must always be doubt as to the validity of results achieved using it, e.g. what's 'special' about the missing particles?

Another area where inertial forces play a dominant role is that of aerosol sampling by such devices as impactors, impingers, cyclones and stacked sieve samplers. All rely on particle inertia for their particulate sampling action. Actual samplers are discussed in Chapter 3: of concern here is the principle of their action. This principle will be described in terms of inertial impaction occurring in impactors. Basically, an impactor consists of a tapered jet either rectangular or circular below which is an impaction plate (Figure 1.3). By applying low pressure, air is rapidly sucked through the jet and on approaching the impaction plate it changes direction. Entrained particles owing to their momentum and inertia cross the streamlines and move towards the impaction plate. The distance moved (D) will be the particle velocity (v) in the jet multiplied by time, viz.

$$D = vt.$$

This time is given by,

$$t = \frac{L}{V}$$

where, L = length of curved particle trajectory (analogous to equation 1.17)

V = air velocity

Hence, $v = \dfrac{DV}{L}$

but from equations 1.15 and 1.16,

$$v = \frac{\rho d^2(V^2/R)}{18\eta}$$

$$\text{so, } \frac{DV}{L} = \frac{\rho d^2(V^2/R)}{18\eta}$$

$$\text{or, } D_M = \frac{d^2V}{18\eta} \times \frac{L}{R} \tag{1.20}$$

for which L/R is dimensionless and equal to 1.56 (i.e. the ratio of $\frac{1}{4}$ circumference divided by radius).

The value of D_M is known as Sinclair's stopping distance and is the vertical distance (Figure 1.3) a particle travels after directional change. It has obvious application in determining the collision probability of a particle and a surface. In the figure the stopping distance of the large particle is greater than the diameter of the exit tube — the particle impacts. The value of D_M for the smaller particle is considerably smaller than the diameter of the exit tube — the particle does not impact.

Obviously there is a particular size for which the probability for impaction is 0.5 and equal to the probability of passage. This size is termed the characteristic diameter for that particular jet and its operating conditions (e.g. air flow rate).

Arguments like those above apply widely for aerosols and are developed further in Chapters 3 and 5 in relation to particle sizing techniques.

1.6 ELECTROMAGNETIC RADIATION

Electromagnetic radiation interacts with aerosol particles in four basic ways. These are reflection, refraction, absorption and scattering. Absorption is capture of photons by aerosol particles which are re-emitted as fluorescence, for example, or cause a temperature rise with heat lost by radiation and convection. Fluorescence, etc., can be used for particle identification as discussed in Chapter 4, as can radiation. Photophoresis arises when the heating causes one side of the particle to become hotter than the opposite one when the particle then moves photophoretically. Because transparent particles behave as lenses the opposite side to that of the electromagnetic radiation source will be hottest and such particles migrate towards it. In contrast, an opaque particle suspended in air having a temperature gradient will move down that gradient (i.e. from hot to cold) because more air molecules strike it from the warmer end of the gradient. Such is the principle of thermal precipitation aerosol samplers which are limited to particles of less than about 3 μm diameter because of a diminishing thermal effect as particle size increases.

Scattering of electromagnetic radiation by aerosol particles is a complex process with maximum scattered intensity at 0° and 180° to the incident radiation. When the particle size is much less than the wavelength Raleigh elastic scattering occurs and the scattered intensity between 0° and 180° changes smoothly. (Elastic scattering is when the incident and scattered wavelengths are the same.) In contrast, when the particle size is comparable to the wavelength, Mie elastic scattering occurs and the scattered intensity fluctuates critically with angle between the limits of 0° and 180°. The process is a complicated one and theoretically has been modelled for only spheres and other simple shapes. Even so, it provides the basis of many particle sizing instruments (Chapter 5).

Inelastic scattering also can occur in which the incident and scattered wavelengths are different, e.g. Raman scattering with the extent of the shift being characteristic of the scatterer. This process is used for monitoring aerosols as described in Chapter 4. Reflection and refraction are involved in scattering but are more familiarly invoked in conventional light microscopy for direct particle observation.

1.7 PARTICLE REFRACTIVE INDEX AND PARTICLE DENSITY

Particle refractive index affects light scattering intensity through its real and imaginary components. The real component is the ratio of speed of light in air to speed of light in the substance while the imaginary component is related to the degree that the substance absorbs the light falling on it. Both values change independently with wavelength in a manner characteristic of the substance. For hygroscopic materials the values change also with RH. These dependencies are important in the context of particle sizing discussed in Chapter 5. Particle density (mass per unit volume) is of concern mainly through aerodynamic behaviour. For example, two spherical particles of the same size but having different densities will sediment at different velocities. If sufficiently small (less than about 1 μm diameter) the particle with lower density will demonstrate greater Brownian motion than the other. Usually aerosol particle density is less than the value for the bulk material because of the greater porosity of the former. In turn, particle porosity, and therefore particle density, is a function of the method of particle formation and requires special methods for its determination.

1.8 THERMAL GRADIENTS

When a particle is warmer on one side than the other, as indicated above, the escape velocity of molecules on the former will exceed that on the latter. Due to this velocity differential the particle moves. Transparent particles move towards the source as they act as lenses thereby focusing energy on the distal side. Opaque particles on the other hand move away from the source and down the temperature gradient. In general, particles move away from a hot surface towards a colder surface. This thermophoretic velocity depends on material properties of the aerosol particles as well as the ratio of particle radius (r) to mean free path of the air molecules, when $r < 1$,

$$v = -\frac{D}{3} \cdot \frac{1}{T} \cdot \frac{dT}{dx} \tag{1.21}$$

while for $r \gg 1$,

$$v = -\frac{4D}{27} \cdot \frac{1}{T} \cdot \frac{dT}{dx} \tag{1.22}$$

where, D = diffusion coefficient of air (about 1.85×10^{-5} m/s at STP)

T = °K

$\frac{dT}{dx}$ = temperature gradient

Except in thermal precipitators (q.v.) with temperature gradients of about 100 °C over a few mm or less, thermophoresis for particles greater than about 3 μm is not marked. Even with precipitators the upper size limit is about 5 μm diameter; on the other hand the technique represents a gentle sampling method.

1.9 HYGROSCOPICITY AND HUMIDITY

Most microorganisms being built up from carbohydrates, proteins, lipids and nucleic acid are hygroscopic (i.e. attract water molecules). Likewise materials with which they are associated in the airborne state (e.g. mucus, culture media) similarly are hygroscopic. One consequence is that amounts of water associated with airborne particles bearing microbes depend upon relative humidity (RH). The two are related through water sorption isotherms, an example of which is that in Figure 1.4 for *Serratia marcescens*. Classically, the isotherm is interpreted as follows. Starting at equilibrium at low RH, as RH is increased and equilibrium re-established, a monolayer of water molecules builds up on hydrophilic (i.e. water 'loving') structures and becomes complete where the curve levels out between 20% and 60% RH. At higher RH values more water molecules are bound such that multilayers form, corresponding to the steeply rising portion of the curve at RH > 70%.

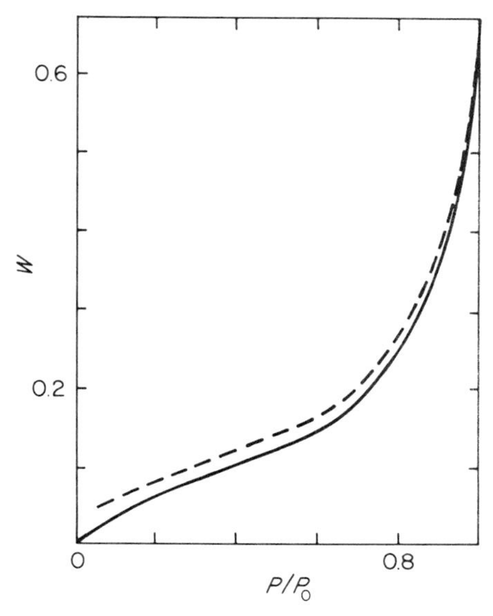

Figure 1.4 Water sorption isotherm of *Serratia marcescens* as a function of water activity (i.e. % RH as a fraction). Weight of water (w) per gram of dry *Serratia marcescens* as a function of water activity (P/P_0). ——— absorption — — — — desorption. (Data of Bateman *et al.* 1962, *J. gen. Microbiol.*, **29**, 207.)

Starting at equilibrium at high RH, reversing the steps corresponds to dehydration (or desiccation) while the first case corresponds to vapour phase rehydration. The two processes while (usually) being reversible are not entirely equivalent. This is expressed by a displacement between the dehydration and rehydration isotherms; the difference is termed hysteresis. It usually reflects a difference in the energetics of the adsorption/absorption and desorption processes. Alternatively, these may not be completely reversible when, for example, desorption (i.e. dehydration) leads to a structural collapse which is irreversible. Then masked adsorption sites are unavailable during rehydration with discontinuities in sorption isotherms being indicative. The formation of a monolayer can be modelled by the Langmuir adsorption isotherm or by more complex expressions (BET). In the case of microorganisms only very limited measurements have been made of their isotherms, so at this time it is probably inappropriate to consider a case more complex than that due to Langmuir. Effectively, the process may be represented as,

$$M + x.H_2O \rightleftharpoons M.x.H_2O \tag{1.23}$$

analogous to hydrate formation of inorganic salts. But unlike the cases of salts which demonstrate marked discontinuities in sorption isotherms, those for macromolecules are smooth because the discontinuities appear small and indistinct. This result arises through the combined action of many functional groups (of the biopolymer) which differ slightly in water affinity (so any discontinuities become smoothed). From equation 1.23, it can be shown that the fraction of water sorption sites filled is given by,

$$\theta = \frac{a_w}{K + a_w} \tag{1.24}$$

where, θ = fraction of sites filled
a_w = water vapour activity
K = equilibrium constant

and water vapour activity is related to the percent relative humidity by the equation,

$$a_w = \frac{\%\mathrm{RH}}{100} \tag{1.25}$$

According to the equation 1.24 when the value of a_w is much less than the value of the equilibrium constant, K, then a_w may be neglected in comparison with K, so that,

$$\theta = \frac{a_w}{K} \tag{1.26}$$

and the fraction of sites filled increases linearly with a_w. As a_w approaches and then exceeds the value of the equilibrium constant, K may be neglected in comparison with the value of a_w, and equation (1.24) becomes,

$$\theta = \frac{a_w}{a_w} = 1$$

i.e. all the sites are filled and any further increase in a_w is without effect. Thus the Langmuir sorption isotherm provides a relatively simple representation of the formation of a monolayer of water molecules. Representation of the build-up of multilayers though is complex and will not be dealt with here.

Hygroscopicity has several ramifications for microbial aerosols both for their physical as well as their biological behaviour. The latter is the subject for the ninth and ensuing chapters while the former concerns particle shape, size and density. Consider an irregular hygroscopic particle produced through freeze-drying, for example. On exposure to high RH it sorbs water with structural collapse and solution. It changes from an irregular shape to that of a sphere of solution. In the process its density changes and approaches unity, while its mass and size increase. As may be anticipated changes such as these alter many physical properties including settling velocity, site of deposition in the lung, etc., light scattering, reflectivity coefficient, aerosol sampling, etc.

The settling velocity of hygroscopic particles, therefore, increases with RH, as can particle reflectivity and scattered light intensity. A consequence is that particle size measured by light scattering can apparently increase markedly with increasing RH, not only because of greater particle diameter but a disproportionately larger scattered intensity as well. The change in particle diameter with RH makes it difficult also to size them microscopically unless RH for test and measurement is maintained the same. Such size changes occur too when hygroscopic aerosol particles enter the respiratory system with the RH close to 100%. As discussed later the landing site depends on particle size and therefore too on particle hygroscopicity.

Hygroscopicity and RH are important for biological effects as well as the physical ones mentioned above; unfortunately determining the RH of a given atmosphere is not particularly easy. Dimmick and Marton (1969) share this view, but like many others ask whether relative or absolute humidity govern aerosol survival and infectivity. Chapters 9–15 try to provide an answer.

Absolute humidity is the actual mass of water (as vapour) per unit volume of air. Relative humidity is the actual vapour pressure of water vapour in the air relative to the maximum water vapour pressure in the air at the same temperature and pressure. The latter is the saturated vapour pressure over a flat surface of pure water at the given temperature and pressure, e.g. for an atmospheric pressure of 760 mmHg the saturated vapour pressure of water is 760 mmHg when the bulk water temperature is at the boiling point (100°C).

A common misconception is that air takes up moisture to generate humidity and at 100% RH the air is saturated with water. Nothing could be further from the truth in that humidity is related only to water vapour pressure which is independent of whether air is or is not present (assuming that both act as perfect gases, which is nearly so). Consequently, a precise definition of relative humidity is that it equals the thermodynamic water activity a_w multiplied by 100. The thermodynamic activity of a substance, for example when in solution, is equal to its concentration multiplied by its activity coefficient. For dilute solutions activity coefficients are very nearly equal to 1 but as the solute concentration increases interactions between solute molecules also increase thereby usually reducing their effectiveness and their activity coefficient. Concentrated solutions therefore rarely behave strictly as might be expected, rather they behave as less concentrated ones would be expected to do. Osmotic pressure of solutions is a good example and an experimental value from which activity coefficients may be determined.

A pure substance such as liquid water has an activity of 1 and when at equilibrium the vapour above it also will have an activity of 1, i.e. saturated vapour pressure is the same in activity as pure liquid water. For a concentrated solution in which the activity of water is reduced to say 0.8, at equilibrium, the water vapour above it also would have an activity of 0.8, otherwise equilibrium would not exist. Such would be achieved with a 29% (w/w) solution of NaCl. This is why saturated salt solutions may be used for controlling RH in comparatively small sealed containers (e.g. Figure 1.5).

A test sample of microbes with a small quantity of water when placed in this container will lose water by distillation to the saturated solution. In turn a small

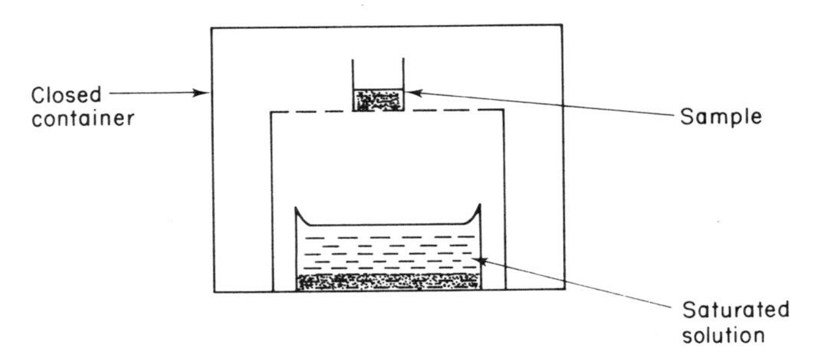

Figure 1.5 Relative humidity control by saturated salt solution.

quantity of the solute dissolves to maintain the equilibrium and then the sample and saturated solution will have the same water activity. Likewise, a dry sample put into the container will take up water through distillation from the saturated solution in which a small quantity of solute comes out of solution to maintain equilibrium. Again the water activity, at equilibrium, in the sample and saturated solution will be the same.

The main advantage of this approach is that the controlled RH is accurately known but has the disadvantage that a constant temperature must be maintained for the long period of equilibration. Replacing saturated solutions with a concentrated solution of sulphuric acid enables a range of water activities to be achieved, their values being derived from tables giving water activity as a function of sulphuric acid specific gravity for example.

Water activity of a solution whether saturated or not may be determined (Figure 1.5) by holding it over a sulphuric acid solution. On attaining equilibrium both solutions have the same water activity which can be determined from the specific gravity of the sulphuric acid solution, for example.

In the current context such systems offer calibration standards as well as the opportunity to study the survival of microorganisms held at precisely defined relative humidities. RH values for several saturated salt solutions will be found in the CRC *Handbook of Chemistry and Physics* and in the monograph by Penman (1958).

The measurement of RH of air flows or of aerosol holding facilities requires methods more rapid than equilibration with sulphuric acid solutions, etc. One alternative, rapid, yet absolute method is to measure dew point by automatic dew point hygrometer. They utilize a clean and highly reflecting surface which is thermo-electrically cooled until water condenses on it, which is detected by changed reflectance. This temperature is the dew point and a direct measure of RH. In automatic instruments the temperature of a mirror cycles and formation and disappearance of dew is monitored. Their main disadvantages are cost and errors from particulate contamination of the mirror. They are suitable for laboratory calibration standards.

Other automatic RH sensors rely on measuring electrical resistance changes

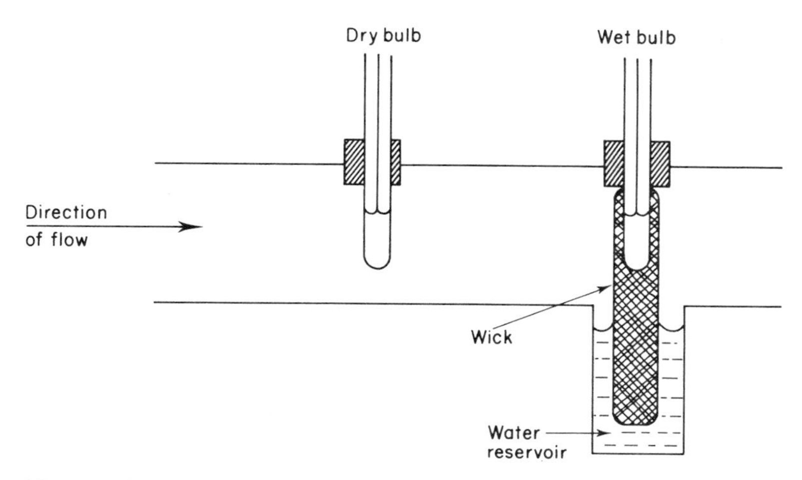

Figure 1.6 Wet and dry bulb system for determining relative humidity.

of films of materials, e.g. salts, paper or the dielectric constant of a capacitor exposed to the air. While such instruments are relatively cheap, in the authors experience none are satisfactory due to their ease of contamination by test aerosols and poor stability of calibration. Perhaps in clean air environments they operate well. The same is true of instruments which measure RH by attenuation of infrared over a short path. But for determining the RH of a contained concentrated aerosol, it is likely to interfere with that measurement. For crudely estimating RH the protein membrane and hair hygrometers are appropriate (Penman, 1958). Being relatively inexpensive they may be discarded if adversely affected by test aerosols. Measurement is in terms of changed length or volume caused by sorption and desorption of water vapour. The main disadvantages are their slow response and limited accuracy.

The most satisfactory method for measuring RH of flowing air likely to contain test aerosols is the wet and dry bulb method, or psychrometry. The principle is that when exposed to an airstream water will cool until the vapour pressure at its surface matches that of the surrounding air. By recording this temperature and comparing its value with that of the ambient air, the value of the RH may be deduced. A convenient method is to compare two matched thin-bulb mercury-in-glass thermometers, one bulb being covered by a wet wick fed from a water reservoir (Figure 1.6). To induce cooling the thermometer bulbs must be ventilated and shielded from extraneous (radiation) heat sources. A velocity of 1.5 — 2 m/s over the thermometer bulbs is adequate but needs to be maintained as the temperature of the wet bulb depends on RH and ventilation rate. The direction of air flow is from dry to wet bulb, as the alternative may result in a lowered dry bulb temperature caused by evaporative cooling of the wet bulb.

Given that care is taken to ensure adequate ventilation and that the wick is clean and wetted by good quality distilled water, that the dry bulb is suspended in the air stream and that the wet bulb does not contact the water in the reservoir, then the temperature recorded by the dry bulb, together with the difference from the wet bulb — the wet bulb depression — provides an accurate measure of RH. Conversion of these temperatures to RH is facilitated by look-up tables (e.g. CRC *Handbook of Chemistry and Physics*). Alternatively they may be calibrated against a dew point hygrometer, in turn standardized against saturated salt solutions, for example. A third possibility is to calculate one's own look-up tables.

Penman (1958) provides the following theory for computation of the psychrometric constant required to convert dry bulb temperature and wet bulb depression to RH. The basic assumption is that all the heat needed to vapourize water of the wick of the wet bulb is drawn from the air. A volume of ambient air reaches the wet bulb and replaces an equal volume of air having the wet bulb temperature and humidity, which is wetter. Let the incident ambient air have a temperature T_a and a water content of x g/g dry air; let the displaced volume of air have temperature T_w and a water content of x_w g/g dry air. Then, the increase in sensible content (measured by the fall in temperature) per gram of incident air is,

$$-\Delta H_s = C_p(T_a - T_w) \tag{1.27}$$

where, C_p = specific heat of air at constant pressure.

The increase in latent heat content per gram of dry air is,

$$\Delta H_l = L(x_w - x) \tag{1.28}$$

where, L = latent heat of vapourization at T_a.

At equilibrium, $-\Delta H_s = \Delta H_l$, and,

$$C_p(T_a - T_w) = L(x_w - x) \tag{1.29}$$

or,

$$x = x_w - \frac{C_p}{L}(T_a - T_w) \tag{1.30}$$

i.e. the August-Apjohn hygrometer equation.

But, $$x = \frac{\phi e}{p}$$

where, ϕ = ratio of densities of water vapour and dry air at the same temperature and pressure,
e = vapour pressure,
p = ambient pressure.

Whence, $$e = e_w - pC_p(T_a - T_w)/\phi L \tag{1.31}$$

or, $$e = e_w - 0.4989\,(T_a - T_w) \tag{1.32}$$

for pressure (mm of mercury) and temperature (°C).

The two thermometers give values of T_a and T_w directly, while the saturation vapour pressure, e_w, can be found for the value T_w from tables (e.g. CRC

Handbook of Chemistry and Physics) and the value of e computed. Then,

$$\text{RH }(\%) = \frac{100e}{e_a} \tag{1.33}$$

where, e_a = the saturation vapour pressure at T_a.

One failing of this approach is that no direct account is taken of ventilation rate of the wet bulb or conduction of heat along the thermometer stem. On the other hand, the derivation is based on a volume of ambient air completely replacing an equal volume of air around the wet bulb. This will occur when there is a noticeable air velocity, e.g. 1.5–2 m/s. Another apparent disadvantage for calculation by, for example, computer is that an extensive table of data of the values of the saturation vapour pressure as a function of temperature would seem to be required. A simple alternative evaluates the constants A, B in the following equation and then calculates the saturation vapour pressure for any given temperature. The equation due to Van't Hoff is,

$$\log_e e_T = A - \frac{B}{T} \tag{1.34}$$

where, $\log_e e_T$ = the natural logarithm of the saturated vapour pressure at temperature $T°$ (Kelvin) (i.e. °C + 273)

A, B = constants

and the constants A, B, evaluated by a least squares fit of this equation to published data (*loc. cit.*).

Validity of the approach is suggested by the work of Cox (1968) in which experimental and calculated values of the psychrometric constant are compared. Values are reproduced in Table 1.2 for atmospheres of air, nitrogen, argon and helium, with the calculated value for air being virtually identical with that used in the tables of Marvin (1941) and of *The Handbook of Chemistry and Physics*. While agreement between experimental and calculated values for air and nitrogen are excellent and for argon are reasonable, those for helium are poor. This suggests that the simple formula of August and Apjohn does not always apply in practice.

Table 1.2. Experimental and calculated values of the psychrometric constants. Measurements were made at 26.5° over a relative humidity range of 50–16%

Gas	A (calc.)	A (expt.)	No. of determinations	S.D.
Air	0.499	0.506	13	0.020
Nitrogen	0.496	0.500	8	0.013
Argon	0.343	0.358	16	0.010
Helium	0.354	0.532	16	0.035

The method of using wet and dry bulb thermometers can provide an accuracy of measurement of 1%, or perhaps slightly better, when extreme care is taken. On the other hand, lack of care can lead to extreme error. As intimated above, matched accurately calibrated thin-bulb mercury-in-glass thermometers with stem correction are needed which preferably can be easily read to 0.1 °C. The thermometer system should not be exposed to extraneous heat sources as the basic tenet is that only the flowing air provides heat to vapourize water of the wet bulb. Likewise, thermometer bulbs should be clear of tubing walls, water reservoirs, etc. Even then a serious source of error arises when the wick and/or water reservoir become contaminated (e.g. by deposited aerosol particles). The wick should be replaced long before it is stiffened through contamination. While corrections for conduction of heat to the wet bulb through the thermometer stem itself are possible, they become really necessary only when at low RH, e.g. about 20% RH or less.

In some applications it may be advantageous, but more costly, to replace mercury-in-glass thermometers with high stability thermistors. These are semiconductor devices characterized by a large negative temperature coefficient of resistance. Having low thermal capacity they respond rapidly to small temperature differences and require only low ventilation rates. This may be disadvantageous though when there are rapid RH fluctuations to which associated circuitry will not respond or when some average value is required. In these circumstances either thermal capacity can be increased or the circuitry time constant raised. In any case, the thermistors will require accurate calibration and maintained stability of their response.

Whichever approach is adopted, when applied to measuring the RH of aerosols, concomitant contamination of the wick, etc. will occur as mentioned above. Therefore, the question arises as to why the air to be measured should not be filtered first to remove potential contaminants. The reasons are that the filter material itself may sorb water and that it might introduce pressure changes, both of which cause serious errors. For similar reasons the air to be measured should not be drawn through narrow pipework.

1.10 CONCLUSIONS

Behaviour of microbiological aerosols is governed by their physical attributes as well as biological ones. The physical factors mainly control where, how and in what quantities the particles reach a particular landing site. Molecular motion, gravitational, thermal and electrostatic fields play important roles as does humidity. Inertial forces and fluid dynamics are more concerned in the landing process while interactions with electromagnetic radiation are invoked largely for particle sizing, observation and analysis.

REFERENCES

Cox, C. S. (1968). *J. Gen. Microbiol.*, **50**, 139–147.

Cox, C. S., Derr, J. S., Fleurie, E. G. and Roderick, R. C. (1970). *Appl. Microbiol.*, **20**, 927–934.

Dimmick, R. L. (1969). In *An Introduction to Experimental Aerobiology*, R. L. Dimmick and Ann B. Akers (Eds), Wiley Interscience, New York, London, pp. 3–21.
Dimmick, R. L. and Marton, G. F. (1969). In *An Introduction to Experimental Aerobiology*, R. L. Dimmick and Ann B. Akers (Eds) Wiley Interscience, New York, London, pp. 46–58.
Dimmick, R. L. and Wang, L. (1969). In *An Introduction to Experimental Aerobiology*, R. L. Dimmick and Ann B. Akers (Eds), Wiley Interscience, New York, London, pp. 164–176.
Goldberg, L. J., Watkins, H. M. S., Boerke, E. E. and Chatigny, M. A. (1958). *Amer. J. Hyg.*, **68**, 85–93.
Marvin, C. F. (1941). *Psychrometric Tables*, W.B. No. 235, United States Government Printing Office, Washington.
Orr, C., Jr. (1966). *Particulate Technology*, Macmillan Co., New York.
Phalen, R. F. (1984). *Inhalation Studies: Foundations and Techniques*, C.R.C. Press, Florida.
Penman, H. L. (1958). *Humidity*, The Institute of Physics: Monographs for Students, Chapman and Hale, London.
Weast, R. C. (1980). *C.R.C. Handbook of Chemistry and Physics*, C.R.C. Press, Florida.

Chapter 2

Experimental techniques

2.1 INTRODUCTION

In the airborne state the survival and infectivity of microorganisms is affected by a large number of environmental factors. Under natural conditions many of them operate simultaneously; in the laboratory the separate control of these factors allows an evaluation of individual effects. Even so, living microorganisms mutate and rapidly adapt to changed growth environments. Consequently, it is often difficult to be sure that a given strain of microorganism will respond consistently to an applied stress when repeatedly tested over a long period of time. Yet, to derive an understanding of the cause of loss of viability and infectivity such consistency of response is required.

In theory, aerosol survival (or infectivity) experiments usually involve, apart from the environmental factors, aerosol generation, aerosol storage and aerosol collection (either using an artificial sampler or a host). Each process represents a potential stress or cause of loss of viability (or infectivity) of the test microorganisms. In order to understand why a particular experimental result is achieved, the effects of each of these processes needs to be established. However, because of the nature of the required experiments, the three processes cannot be independently evaluated. For example, to discover if aerosol collection *per se* were damaging, an aerosol first must be generated. Consequently, any observed result could have been due to the generation process, effects occurring while airborne (even though this may be for a very short time) or to the collection process. Furthermore, aerosol generation could cause sublethal damage which only becomes lethal when another (normally) sublethal stress also is applied.

In practice, the situation is more complex because aerosol generation, either naturally or artificially, requires considerable force which can damage or even fracture fragile structures. In turn the fragility of microorganisms depends upon species and physiological condition while the magnitudes of shear stresses depend upon the methods of generation. Once airborne, the microbes either become desiccated or hydrated depending upon whether they were originally wetter or drier than the storage atmosphere. Under natural conditions microbial aerosols are exposed to radiation, oxygen, ozone and various pollutants (all of which potentially are lethal) and during airborne travel may become entrapped to be later re-aerosolized. Following inhalation by a host aerosol particles hydrate and become retained while artificial recovery requires collection into liquids or onto surfaces which may subject the microorganisms to further shear forces and also to osmotic shock. An important feature of assessment is that some microorganisms, injured as a result of various stresses, may recover fully when given a suitable environment. This property of 'reversible injury' or 'repair' in microorganisms is widespread and the implications of it are important in medical and veterinary science, and in food and pharmaceutical microbiology. On assay, the proportion of the population replicating or initiating disease is a measure, respectively, of viability and infectivity.

Following collection from aerosols, bacteria can differ biochemically from unaerosolized controls in many ways through, for example, release of enzymes, antigens, metal ions and nucleotide components. With so many differences one may ask whether it is possible that loss of viability still is due to only a *single* physical or biochemical lesion rather than a *general* collapse of metabolism. The answer would seem to be in the affirmative for the following reasons.

1. Release of enzymes, metal ions, surface antigens and RNA nucleotide components occurs equally for viable and non-viable bacteria.
2. Loss of ability to
(a) synthesize nucleic acid or
(b) derive energy or
(c) undergo the process of cell division or
(d) invade tissue cells,
must result in failure to multiply or to initiate disease which by definition are, respectively, loss of viability and infectivity.

Perhaps, then, each lesion (2(a)–(d)) plus that of physical disruption represents a fundamental death mechanism. As will be shown in later chapters, given the necessary circumstances each of these lesions does occur as a result of microorganisms being airborne.

2.2 STRAIN OF MICROORGANISM

Microorganisms have a wide variety of form and composition and therefore responses to aerosol stresses. General statements concerning responses, nonetheless, can sometimes be made for a defined stress. For example, when dehydration–rehydration stress is applied (in the absence of radiation, toxic gases, etc.) various strains of *Escherichia coli* survive better at low relative humidity (RH) than at high RH (Cox, 1966a, 1968b), and like *Serratia marcescens* become sensitive to oxygen-induced loss of viability below about 70% RH (Hess, 1965; Cox *et al.*, 1973, 1974). On the other hand, bacteria such as *Francisella tularensis* and *Escherichia coli* survive differently when disseminated as aerosols from the wet and dry states (Cox, 1970, 1971; Cox and Goldberg, 1972).

While gram-negative bacteria may be sensitive to oxygen at low RH bacterial spores, phages and viruses (so far tested) are unaffected in the airborne state by the presence of oxygen. Likewise, bacterial spores are little affected by pollutants while vegetative bacteria, phages and viruses are affected. In contrast, vegetative bacteria, bacterial spores, phages, viruses, etc., in the airborne state all tend to be inactiviated (to varying degrees) by radiation. Even so, the general 'toughness' of spores is apparent also in their ability to withstand desiccation while ether-resistant viruses generally are most stable at high RH, unlike ether-sensitive viruses which tend to be most stable at low RH (Benbough, 1971; Spendlove and Fannin, 1982). However, their virulence is a function also of virus passage (Hearn *et al.*, 1965; Bradish *et al.*, 1971; Schaffer *et al.*, 1976).

A more complete account for particular strains of microorganisms can be found from the tables given by Anderson and Cox (1967) and by Spendlove and Fannin (1982). But, when comparisons are made between the behaviour of different microorganisms, the exact experimental protocol must be taken into account. The sections that follow discuss why this should be.

2.3 GROWTH OF MICROORGANISMS

Even though most properties of microorganisms are affected by their chemical composition (Herbert, 1961), only Brown (1953), Strasters and Winkler (1966) and Dark and Callow (1973) have made direct comparisons of how different growth media affect subsequent aerosol survival. Effects of growth phase, however, have received more attention (Brown, 1953; Goodlow and Leonard, 1961; Cox, 1966a; Cox *et al.*, 1971; Dark and Callow, 1973). At least for *Escherichia coli* strains, it is more important than growth media with resting phase bacteria being more aerostable than log phase ones. For continuously growing *E. coli* growth rate similarly is more important than growth media (Dark and Callow, 1973). Such findings suggest that the age of each bacterium in a population is important for survival capability. In the case of viruses, even though biochemical composition of viruses is more a function of the particular cell line used than the medium supporting those cells (Kates *et al.*, 1962), ability of aerosolized virus to survive is most affected by the nature of the spray fluid (Harper, 1961, 1963, 1965; Benbough, 1969, 1971; Barlow, 1972).

One of the most irritating problems faced by investigators of aerosol survival and infectivity is the apparent variation from experiment to experiment in microbial response to given aerosol stresses. This is especially so for bacteria and one cause is the batch to batch variation in their growth. To alleviate the problem some workers grow a large quantity of the test microorganism and store portions of it in the frozen state. Unfortunately, such a procedure is not altogether satisfactory as a large storage space is required, and the freeze–thaw cycle is in itself a stress. A useful method for bacteria is to rapidly freeze small drops (1/40 ml) of a culture in liquid nitrogen. Following freezing, the drops are harvested and stored at −70 °C (Cox, 1968c). For each experiment a frozen drop is used to inoculate the growth medium (freshly prepared from a large stock of dehydrated medium) which is incubated and harvested under conditions that are as constant as possible. Such a procedure enabled reproducible results to be achieved over a period of more than 10 years (Cox, 1976). Even so, this method does not completely prevent mutation from occurring in the frozen inocula. As a result, the proportion of bacteria in derived populations having, for example, specific repair and growth requirements increases with storage time of the frozen inocula. For medium-term studies continuously growing bacteria may be satisfactory but selection of variants still can occur. In the author's experience the best method for bacteria is that of storing frozen inocula (at −70 °C) prepared from a single culture to which cryoprotectants, such as glycerol or dimethylsulphoxide, were added about 30 min before freezing in liquid nitrogen (see above).

A further factor at least for *Escherichia coli* is that some strains produce variants more readily than others with the variants demonstrating subtle survival differences compared to the parent strains (Cox, 1966b). Whether other microorganisms behave in this way is not certain but other species of bacteria derived from freeze-dried samples have been found to demonstrate analogous behaviour. For viruses, serial passage produces changed responses to aerosolization (Bradish *et al.*, 1971; Schaffer *et al.*, 1976) while according to Heinrich (1979) and Hope-Simpson (1979) the occurrence of microbial variants is common in nature.

2.4 AEROSOL GENERATION

Aerosols may be generated from liquid suspensions (wet dissemination) or from dried materials (dry dissemination). In nature, coughing or sneezing are examples of the former while disturbing dust is an example of the latter. In the laboratory wet dissemination is accomplished by two-fluid (air, liquid) atomizers, spinning top or disc, vibrating needle or orifice, or ultrasound (May, 1949, 1966, 1973; Wolf, 1961; DeOme *et al.*, 1944; Green and Lane, 1964; Dimmick, 1969). Dry dissemination using explosions has been described by Beebe (1959) or air jets by Dimmick (1959, 1969); Cox (1970) and Cox *et al.* (1970, Figure 2.1) while Crider *et al.* (1968) report a continuous generator. The dry aerosols invariably are polydisperse and highly charged electrostatically (Cox *et al.*, 1970; Stein *et al.*, 1973) unless ionizing radiation is used to dispel charges (Cox *et al.*, 1970); deagglomeration and fracture problems are common (Derr, 1965).

Two-fluid atomizers invariably lead to polydisperse aerosols and impose high shear forces. They provide relatively large output and when baffled (Figure 2.2) the degree of polydispersity is minimized. In contrast, spinning top (Figure 2.3), vibrating needle (Figure 2.4) and Berglund-Liu vibrating orifice generators provide monodisperse aerosols while imposing low shear forces. But their output is relatively low.

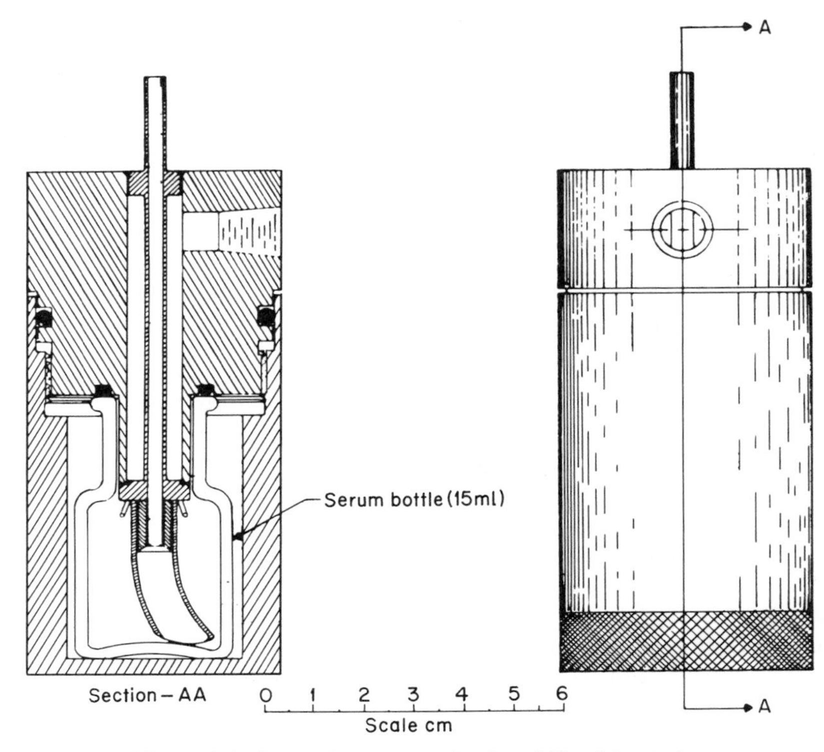

Figure 2.1 Aerosol generator for lyophilized bacteria.

In the past, two-fluid atomizers (e.g. Collison (1935)) were first choice dispersers, but nowadays vibrating orifice monodisperse generators (e.g. Berglund-Liu) have become highly developed and are reasonably reliable. When their relatively low output would not be a disadvantage, they are to be preferred as imposed shear stress is less than that in two-fluid atomizers and their output is monodisperse. However, problems still can arise owing to the orifice becoming blocked by clumped microorganisms, etc.

2.5 AEROSOL PARTICLE SIZE

When aerosols are generated by disseminating fluids droplets are formed which comprise mainly water. Water being highly volatile evaporates, its degree and rate of evaporation depending upon ambient relative humidity (RH) as well as the nature and concentration of solutes present in the spray fluid (Cox, 1965). Contrary to statements in many papers, aqueous droplets can equilibriate comparatively slowly and depending upon conditions, true equilibration may take several minutes (Cox, unpublished data) or several seconds (Cox, 1966a). With dry dissemination, biological aerosol particles due to their hygroscopicity adsorb and absorb water from the atmosphere. The rate

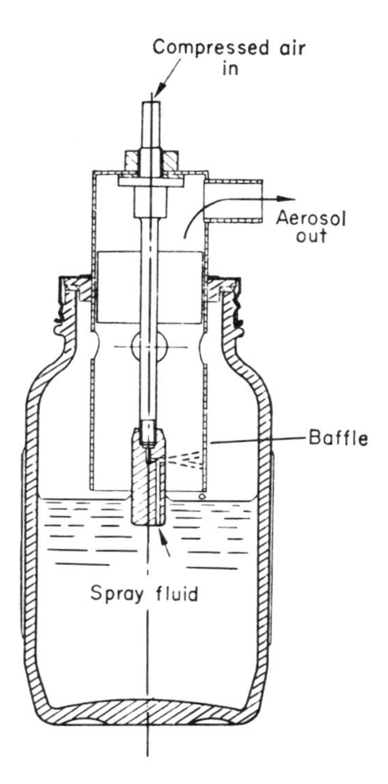

Figure 2.2 Collison atomizer.

and extent of water gain depends upon similar factors as for water loss (Cox, 1965) but in general the rate of rehydration is slower than the corresponding rate of evaporation.

Determination of the true particle size distribution can be difficult and is dealt with in Chapter 5. For infectivity studies in hosts an additional problem occurs because during inhalation particles rehumidify and this results in much larger particle size. This subject is considered in Chapters 5 and 9.

For a given fluid disseminator the equilibrated particle size distribution mainly depends upon the nature and mass of solutes present in the spray fluid. Microorganisms suspended in distilled water produce the finest aerosols while fluids containing much solute produce the coarsest aerosols. Adjustment of solute concentration, therefore, gives some control of particle size distribution. For dry disseminators particle size distribution is largely controlled by the fineness of the dry powder, the nature of any solutes in it, agglomeration processes (Derr, 1965; Cox *et al.*, 1970) and the ambient RH.

Sizes of aerosol particles can influence survival and infectivity of microorganisms contained within them although such effects have not been widely studied. Dunklin and Puck found the death rate of pneumococci in 1.6 μm diameter particles to be lower than that in 3.2 μm diameter particles. By contrast, in solar radiation microorganisms survive better in large than in small

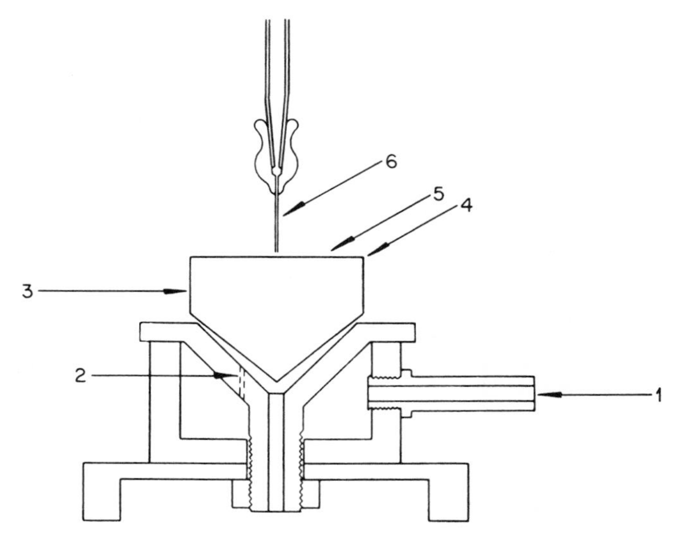

Figure 2.3 Spinning top atomizer (schematic).
1: compressed air inlet; 2: compressed air jet; 3: spinning top; 4: 90° edge; 5: plane turned surface; 6: liquid feed pipe. (From Green and Lane (1964). *Particulate Clouds: Dusts, Smokes and Mists*, E. and F. N. Spon Ltd.) Reproduced by permission of E. & F. N. Spon Ltd.

particles (Goodlow and Leonard, 1961) as they do when exposed to OAF (see Chapter 12; Druett, 1970; Benbough and Hood, 1971).

Effects of particle size upon infectivity are more involved and are of great importance. They are discussed in Chapter 15 which deals specifically with infectivity of microbial aerosols.

2.6 SPRAY FLUIDS

Under natural conditions microorganisms in aerosol particles are surrounded by dried natural fluids such as saliva. Due to the chemical complexity of such materials it is often difficult to resolve how each component may affect microbial survival and infectivity. Probably without exception the nature of the matrix within which aerosolized microorganisms reside will influence subsequent survival and infectivity, either directly or through particle size effects, or both.

In the laboratory simple fluids including distilled water are employed to determine the mode of action of suspending or spray fluids and for elucidating death mechanisms. The response of microbes sprayed from a suspension in distilled water (or dry disseminated from freeze-dried preparations derived from such suspensions) usually is taken as the simplest situation or even the microorganism's fundamental response to aerosolization. However, osmotically labile microorganisms may die on suspension in hypotonic fluids in which

Figure 2.4 Vibrating needle monodisperse droplet generator.
A: sintered glass taper; B: vibrating needle; C: electromagnet; D: micromanipulator; F: constant feed syringe drive. (From Zentner, *Bact. Rev.* (1961), **25**, 188.) Reproduced by permission of American Society for Microbiology.

case resolving the contribution of the suspending fluid to the observed response can be difficult (e.g. Cox, 1971). An alternative approach, which in these cases has proved to be a powerful one, is that of selectively removing components from complex fluids (Harper, 1961, 1963, 1965; Benbough, 1969, 1971; Barlow, 1972).

Many compounds added to disseminating fluids or powders have been examined for their effects upon subsequent survival and infectivity of aerosolized microorganisms. They include spent culture fluids, simple sugars, polyhydric alcohols, surface active agents, amino acids, metal chelating agents, vitamins, free radical scavengers and simple salts. Some were found to be extremely toxic while others conferred various levels of protection against aerosol stresses. Those found to afford best protection over the widest range of conditions include spent culture media, di- and tri-saccharides and the polyhydric alcohols sorbitol and inositol. Possible reasons for their stabilizing action will be found in Chapters 10 and 11.

2.7 AEROSOL STORAGE

Particles in the size range 1–20 μm diameter settle under gravity at an appreciable rate (Chamberlain, 1967; Table 2.1) and various techniques have

Table 2.1 Terminal velocity and relaxation time of spheres of unit density

Diameter (μm)	Terminal velocity (cm/s)	Relaxation time (s)
1	3.5×10^{-3}	3.6×10^{-6}
2	1.3×10^{-2}	1.3×10^{-5}
3	2.9×10^{-2}	3.0×10^{-5}
5	7.8×10^{-2}	8.0×10^{-5}
10	3.0×10^{-1}	3.1×10^{-4}
20	1.2	1.2×10^{-3}
30	2.7	2.7×10^{-3}
50	7.1	7.2×10^{-3}

been used to reduce it. For storage times of up to a few minutes dynamic aerosols (i.e. aerosols which are continuously generated into a flowing airstream) can be used (Henderson, 1952; Hatch and Dimmick, 1965, 1966; Figure 2.5). For longer storage periods, vertical wind tunnels (Druett and May, 1952; Figure 2.6), rotating drums (Goldberg *et al.*, 1958; Goldberg, 1970, 1971; Figure 2.7) or large stirred holding vessels (e.g. Hood, 1971, 1973, 1974) are suitable. However, in all of these, the size distribution of polydisperse aerosols changes with time (i.e. the median diameter decreases)

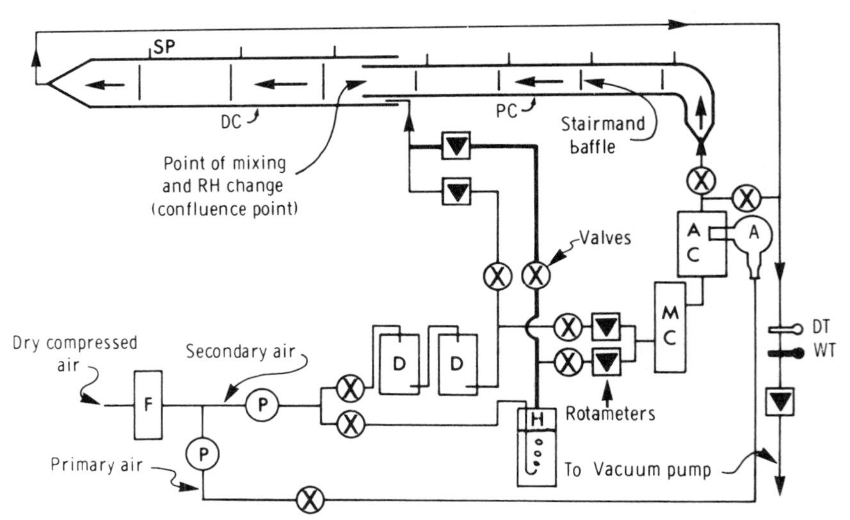

Figure 2.5 Dual aerosol transport apparatus (DATA).
PC: primary aerosol duct; DC: diluted aerosol duct; A: atomizer; AC: aerosol mixing chamber; MC: air mixing chamber; D: dryer; P: pressure regulator; F: filter; SP: sampling ports; H: humidifying chamber; DT, WT: wet-dry bulb psychrometer. (From *An Introduction to Experimental Aerobiology*, R. L. Dimmick and Ann B. Akers (Eds), Wiley Interscience, 1969.) Reprinted by permission of John Wiley & Sons Inc.

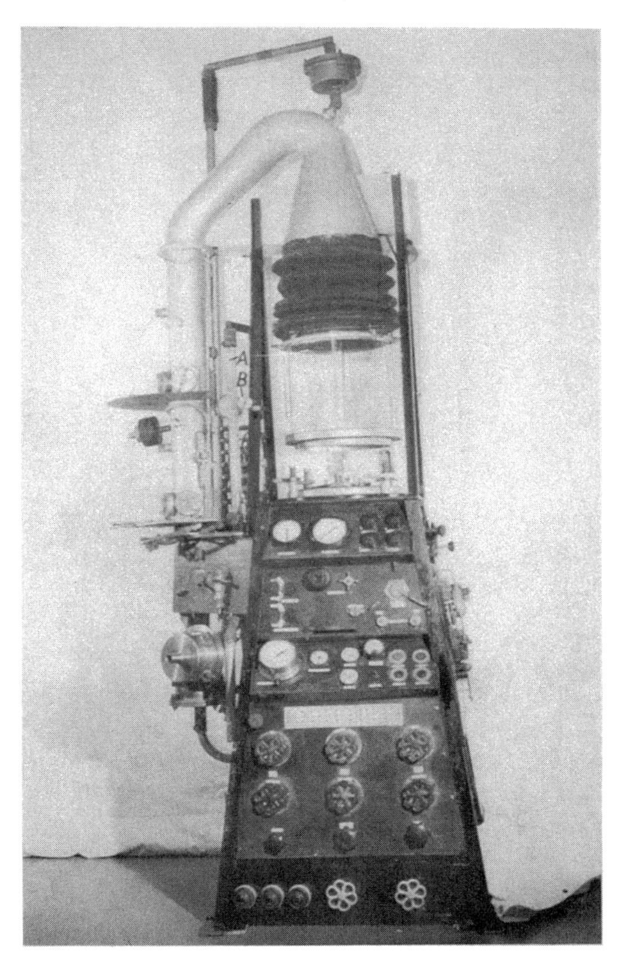

Figure 2.6 Vertical wind tunnel.

due to the preferential loss of large particles. Therefore, any study involving effects of particle size upon aerosol survival or infectivity preferably should be performed with monodisperse aerosols. Alternatively, microthreads consisting of ultrafine spider threads wound onto stainless steel frames (May and Druett, 1968; Figure 2.8) can be used to support aerosol droplets. In this technique the frames are loaded into a 'sow' through which aerosols are passed and particles become attached to the spider threads producing a 'captive aerosol'. A further advantage of this technique is that the 'captive aerosols' can be easily exposed to different environments (e.g. rooms, open air, etc.) for extended periods of time without physical loss of aerosol particles. The sticking to the stainless steel frames of some of the aerosol particles, however, can produce errors in viability estimations, while the survival of *E. coli* (Hood, 1971) and of Semliki

Figure 2.7 Henderson apparatus coupled to a 75 l rotating drum.

Forest virus (Benbough and Hood, 1971) held on microthreads is lower than that in the airborne state. Also the technique is not entirely suitable for animal infectivity studies. Even so, it is an extremely useful tool for aerobiologists and its development was largely responsible for the discovery and identification of the Open Air Factor (see Chapter 12).

Very large droplets (about 150 μm diameter) may be supported on fine glass fibres (Cox, 1965; Silver, 1965). This method has the major advantages of virtually shear-free droplet formation and collection. It has been extremely useful in determining causes of death (Cox, 1965; Silver, 1965), as discussed in Chapter 10.

2.8 AEROSOL COLLECTION

Aerosols need to be collected so that the physical, chemical and biological states of airborne microbes can be determined but many difficulties are associated with collecting aerosols. For example, if the test aerosol is polydisperse, even using isokinetic sampling (i.e. when the aerosol being sampled is moving with the same velocity as the air flowing into the sampler so that all sizes of aerosol particles are sampled with equal probability) samplers are not equally efficient for all particle sizes. Particles measuring 20 μm diameter are more difficult to sample than 2 μm diameter particles while submicron particles require specialized samplers. Consequently, results can be biased due to these effects alone. In practice, it is usually necessary to restrict

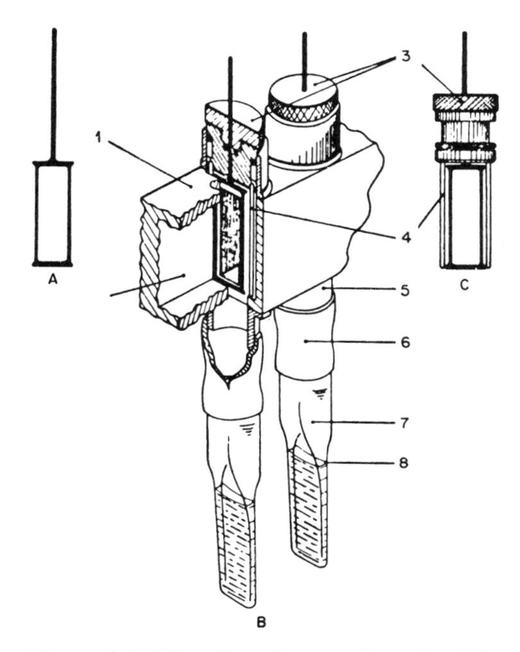

Figure 2.8 Microthread apparatus. A: stainless steel frame, B: isometric cut-away sketch of 'sow' tube showing position of frames and sampling cells, C: frame mounted in removable O-ring sealed cap; 1, 2: rectangular sectioned 'sow', 3: removable O-ring sealed caps holding frames, 6: connecting rubber tube, 7: sampling cell (glass), 8: collecting fluid.

the range of particle sizes investigated in a given experiment either through control of aerosol generation or collection.

The impinger (Figure 2.9) is a widely used sampler and different types were compared by Shipe *et al.* (1959) and by Tyler and Shipe (1959). Their results show marked differences in the numbers of viable bacteria recovered by the different impingers from a given aerosol. Nonetheless, Brachman *et al.* (1964) recommended the AGI-30 as a standard. Other collection devices involve impaction onto surfaces, filtration, sedimentation, centrifugation and electrostatic and thermal precipitation (Errington and Powell, 1969; Green and Lane, 1964; Wolf *et al.*, 1959; Bachelor, 1960; Anderson and Cox, 1967; May, 1964, 1967, 1972; Noble, 1967; Akers and Won, 1969; Davies, 1971). Peto and Powell (1970) provide tables for correction of Andersen sampler data while two improvements to collection techniques have been reported. One is a method for reducing the rate of water evaporation from the nutrient agar surfaces used in some types of slit samplers (May, 1969; Thomas, 1970a, b, c). The other is a method for rehydrating aerosol particles by passing them

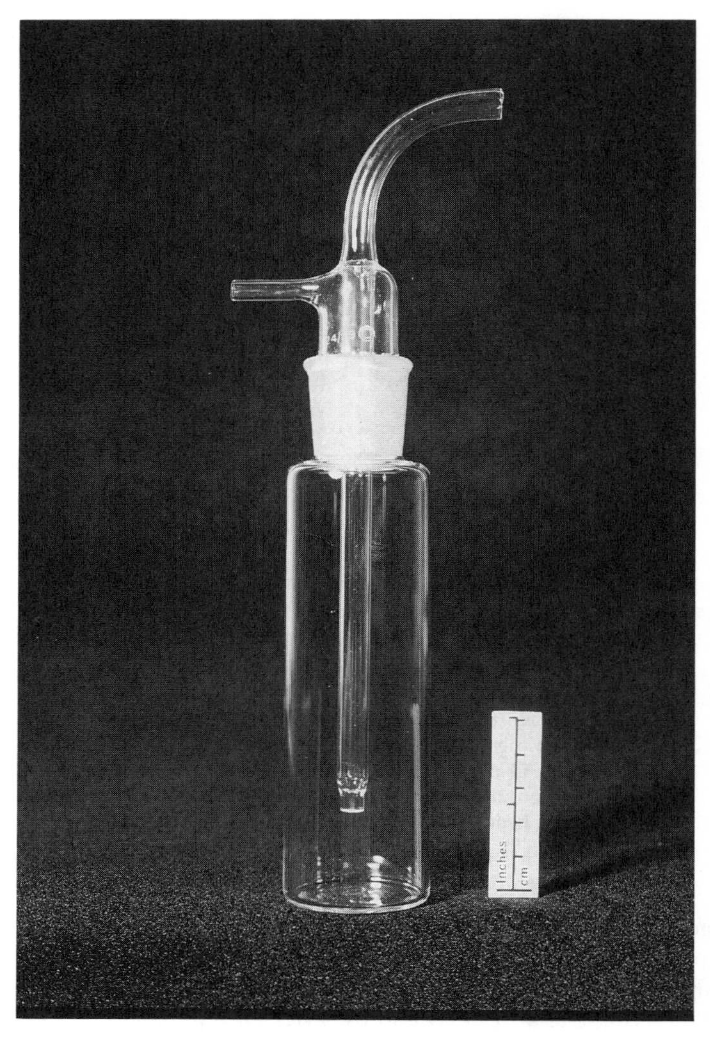

Figure 2.9 AGI-30 impinger.

through a humid atmosphere before collection by an impinger or other sampler (Cox, 1966b, 1967, 1968b; Hatch and Warren, 1969; Maltman and Webb, 1971; Goldberg and Ford, 1973; Figure 2.10). This technique also would be an alternative to that of May and that of Thomas described above for reducing water evaporation.

Accounts of various aerosol samplers will be found in review articles by Wolf *et al.* (1959), Anderson and Cox (1967), Akers and Won (1969), Strange and Cox (1976), May (1967, 1972) and Spendlove and Fannin (1982), as well as in Chapter 3.

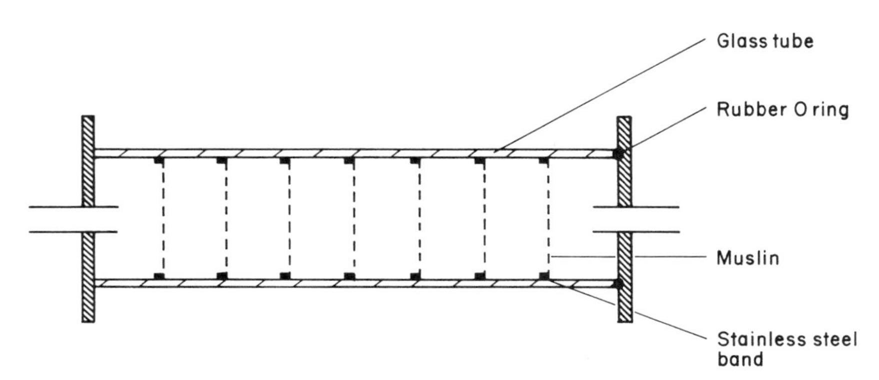

Figure 2.10 Chamber of 2 l capacity used for making humidity changes prior to sampling by impinger.

2.9 COLLECTING FLUIDS

As indicated in Section 2.6 the nature of the spray fluid can greatly influence subsequent aerosol survival and infectivity; similarly, the nature of the collecting fluid can be important.

Liquid collecting fluids often are based upon simple salt solutions, with additives such as proteins, antifoams, etc., or are complex media, plus antifoams. For samplers relying upon impaction, nutrient agar usually is the collecting surface of choice. When sampling aerosols at low RH, or at low temperatures, evaporation inhibitors (e.g. Thomas, 1970a, b, c; May, 1969) or antifreezes (e.g. Lee and Garbett, 1966; Vlodavets *et al.*, 1958; Won and Ross, 1966, 1968) may be required. Even at room temperature under certain conditions collecting fluids incorporating high concentrations (e.g. 1 M/l) of sucrose can give much higher viable recoveries when compared to the same fluid without sucrose (I. H. Silver, personal communication; Cox, 1965, 1966a, b, 1967) as can collecting fluids containing similar concentrations of glycerol (Cox, 1967; Maltman and Webb, 1971). However, such effects depend upon the spray fluid. For example, the viability of *Escherichia coli* strains wet disseminated from supensions in distilled water into nitrogen, argon or helium atmospheres (i.e. under conditions when effects due to toxic gaseous components are absent) and stored in radiation shielded containers are not greatly influenced by the addition of sucrose to phosphate collecting fluids (Cox, 1966a, 1968a, 1976). On the other hand, when sprayed from suspension in 0.3 M raffinose (a trisaccharide) instead of 0.13 M raffinose, viability can be markedly enhanced through such an addition (Cox, 1966a). The effect also is dependent upon RH as well as strain of *E. coli* (Cox, 1966a). In contrast, for *Francisella tularensis* and Semliki Forest virus the presence of high concentrations of sucrose in collecting fluids is not beneficial and may even be detrimental (Cox, 1971, 1976), whereas hypertonic fluids enhance the viability of *Klebsiella pneumoniae* (Maltman and Webb, 1971).

One reason for this apparently complex situation was very elegantly shown by Record *et al.* (1962). Using freeze-dried *E. coli*, Record *et al.* found that rapid rehydration in dilute solutions (low osmotic pressure) resulted in low viability, cell lysis and liberation of spheroplasts, whereas slow rehydration gave much higher viability and a concomitant reduction in the number of spheroplasts. The effect also depended upon the composition of the suspending medium in which the cells were freeze-dried. A partial explanation is as follows. During slow dehydration solutes outside the cell wall concentrate more rapidly than those within the cell, with the result that mass transfer of small solute molecules to the cell interior may occur. Larger solutes penetrate the cell wall but not the cytoplasmic membrane while very large molecules remain external to the cell wall. On rapid rehydration water being a comparatively small molecule quickly enters the cell interior. The osmotic pressure differential between cell interior and exterior is a function of this rate, as well as the rate of egress of solute from the cytoplasm, and from the space between the cell wall and cytoplasmic membrane. In addition, the rate of diffusion of molecules away from the immediate vicinity of the cell wall will affect this differential osmotic pressure. Therefore, depending upon the actual solutes present and their distribution, rapid rehydration or collection into hypotonic fluids may induce severe osmotic stress. This shock may be sufficient to cause osmotic rupture of cell walls and spheroplast formation, or even lysis of cytoplasmic membranes. In general slow rates of rehydration are those most likely to cause least osmotic shock while hypertonic collecting fluids similarly are likely to reduce it; however, if their hypertonicity is high, unless these are diluted gradually osmotic lysis may still occur.

The above is discussed more fully by Record *et al.* (1962) for freeze-dried bacteria and for large droplets (100 μm diameter) by Cox (1965). For the smaller droplets found in aerosols the situation apparently is not the same. The beneficial effects of slow rehydration and of hypertonic collecting fluids on the basis of the above explanation might be expected to progressively increase as the degree of desiccation increases (i.e. as the storage RH decreases). This, at least for *E. coli* strains and *Klebsiella pneumoniae*, is not the case. When *E. coli* are sprayed from supension in 0.3 M raffinose (under conditions when the raffinose is added immediately before spraying, and only dehydration–rehydration stresses are likely to occur) the beneficial effects of hypertonic collecting fluids are found only in the RH range of about 60–100% with maximum benefit occurring at about 80% RH (see Figure 2.11). For the Jepp strain addition of sucrose to the collecting fluid, while beneficial in the above range, becomes detrimental at RH values below about 60% RH (Figure 2.12). This response occurs also when sodium glutamate replaces raffinose. Such results apparently are not consistent with spheroplast formation as discussed previously for freeze-dried *E. coli*. Apart from differences in dehydration rate, another reason for possible discrepancy is that an additional factor is involved in bacterial susceptibility to hypertonic collection phenomena. The data given in Table 2.2 indicate that susceptibility to hypotonic collection stress is a function also of the time of contact before spraying of bacteria, and spray fluid

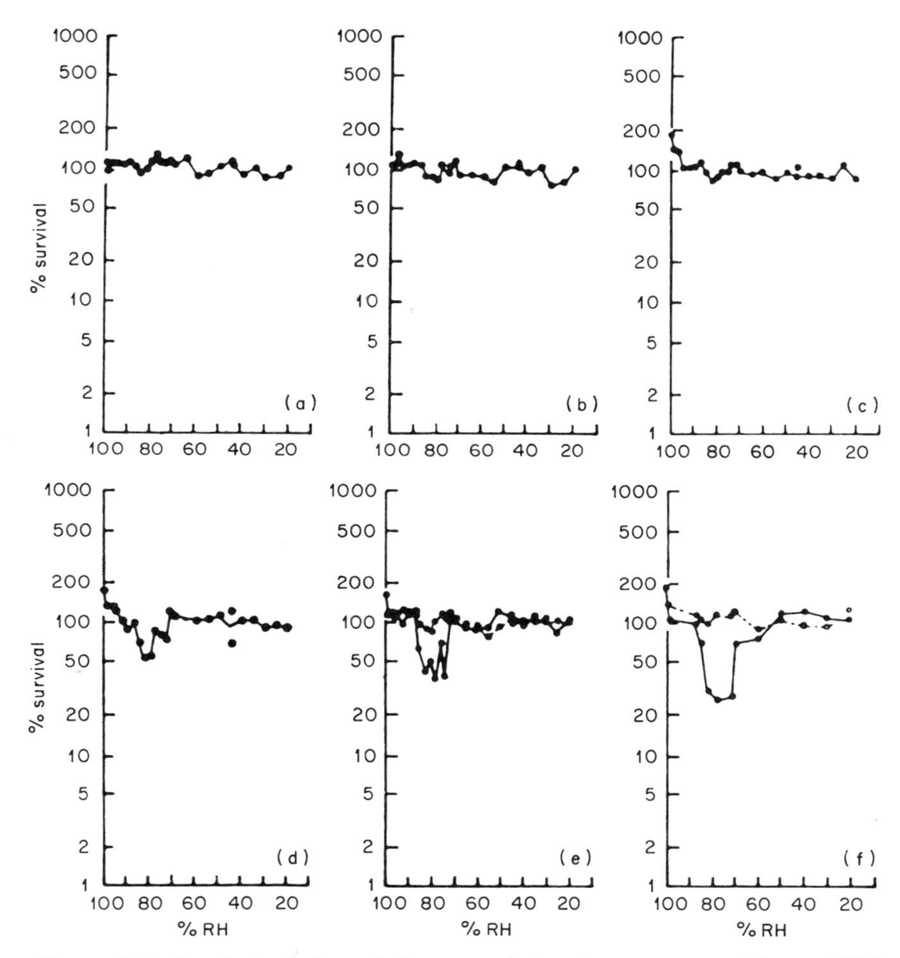

Figure 2.11 Survival of *E. coli* (commune) in nitrogen sprayed from 0.3 M raffinose at aerosol ages of (a) 0.3 s, (b) 3 s, (c) 2 min, (d) 15 min, (e) 30 min, (f) 3 h. ●: Collection into phosphate buffer, ○: collection into 1 M sucrose in phosphate buffer.

additive. At short equilibration times (0–30 min) the viability of *E. coli* Jepp stored at 82% RH is higher when sampled into the sucrose collecting fluid whereas for an equilibration time of 60 minutes viability is virtually 100% and independent of the presence of sucrose in the collecting fluid (Table 2.2). A similar effect of contact time has been found for freeze-dried *Serratia marcescens* (Heckly *et al.*, 1967).

Behaviour similar to that for *E. coli* has been observed with *Klebsiella pneumoniae* (Maltman and Webb, 1971). For their experiments these workers incubated cells and additives for 10 minutes at 37 °C before spraying. On storage at 55% RH collection into hypertonic collecting fluids increased the

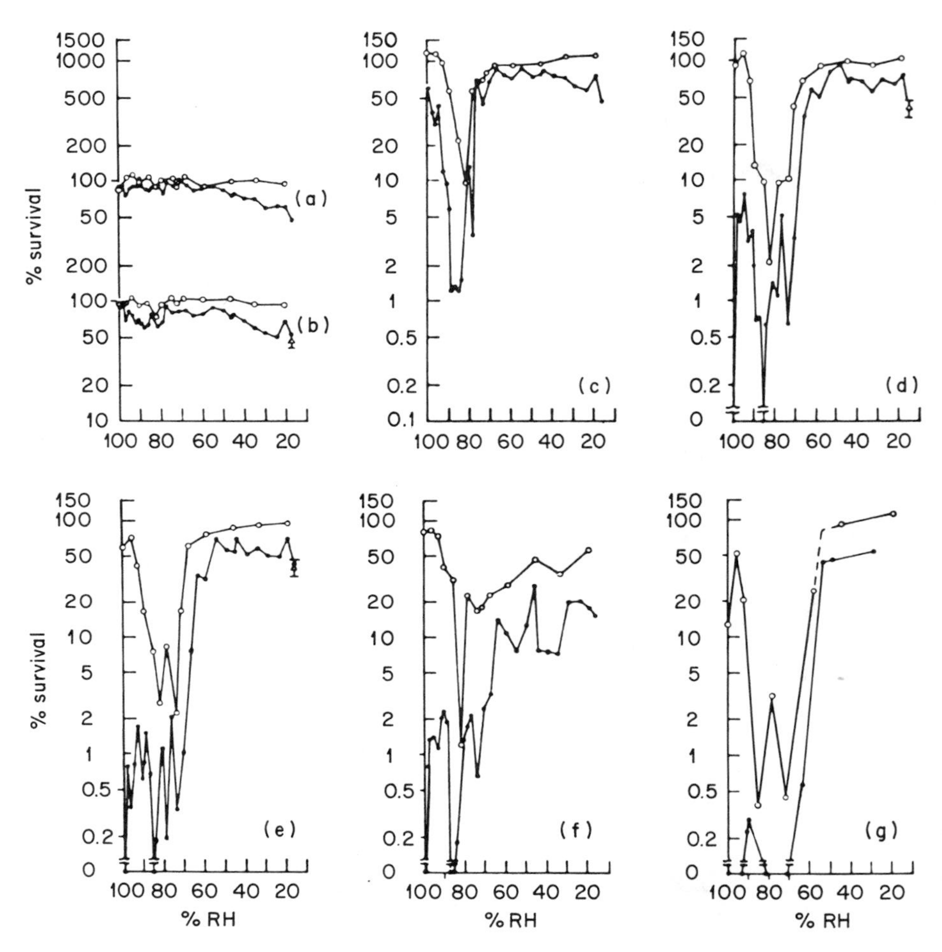

Figure 2.12 Survival of *E. coli* (Jepp) in nitrogen at aerosol ages of (a) 0.3 s, (b) 3 s, (c) 2 min, (d) 15 min, (e) 30 min, (f) 31.5 min, (g) 3 h. (a)–(e) and (g) collected in phosphate buffer, (f) collected in 1 M sucrose in phosphate buffer. ●: Sprayed from distilled water, ○: sprayed from 0.13 M raffinose in distilled water, ⫱: Arithmetic mean and standard deviation for 14 determinations when sprayed from distilled water.

number of viable *K. pneumoniae*. Unfortunately, Maltman and Webb (1971) do not report results for the influence of time of incubation of cells and additives prior to spraying.

This phenomenon of changed viability when using hypertonic collecting fluids has not been widely studied with viruses. A report by Cox (1976) for Semliki Forest virus suggests little effect and therefore osmotic collection stress may be confined to bacteria. Its complex nature for bacteria is suggested by the dependence upon the spray fluid, time of contact between bacteria and additive prior to spraying and the aerosol storage RH. For the reasons given at

Table 2.2 The influence of equilibration time with 0.13 M-sodium glutamate on the aerosol survival of *Escherichia coli* JEPP stored in nitrogen at 82% RH and 26.5°: aerosol age 25 min

Equilibration time (min.)	Collecting fluid	Survival (%)		
		Storage RH	100% RH*	30% RH*
0	PB†	0.82	8.0	1.7
	PBS†	25	6.8	0.15
30	PB	7.3	34	35
	PBS	57	45	28
60	PB	91	90	60
	PBS	83	95	100

*Aerosol stored at 82% RH and shifted to 100 or 30% RH prior to collection.
†PB = phosphate buffer; PBS = phosphate buffer + sucrose (M).

the end of the next section (2.10), any further discussion of the role of collecting fluids will be found in later chapters.

2.10 REHUMIDIFICATION

The term rehumidification is used to mean the rehydration process which occurs when dry aerosol particles are exposed to an atmosphere of high RH before sampling into liquid, i.e. a vapour phase rehydration rather than a liquid phase rehydration. This distinction is required because the two processes are not equivalent. The former is a comparatively slow one involving adsorption and absorption of water vapour, whereas the latter is rapid addition of the aerosol particles to bulk liquid water. Studies of the consequences of these two methods of rehydration are required not only to try to identify causes of loss of viability but also because rehumidification takes place when aerosol particles enter host respiratory systems.

J. R. Maltman (personal communication) suggested that rehumidification could be achieved by passing microbial aerosols through static cylindrical chambers (Figure 2.10) while M. T. Hatch (personal communication) found flash evaporators (Figure 2.13) suitable also. Both techniques have shown that rehumidification can produce both beneficial and detrimental effects upon viability, depending upon the test microorganism and the spray fluid. For example, the viability of T3 coliphage and of *Francisella pestis* bacteriophage when stored at low RH is increased by a factor of about 10^4 if the aerosol is rehumidified before sampling with an impinger (Hatch and Warren, 1969). Similar but less dramatic effects have been observed for several different bacteria but in some instances rehumidification can decrease viability (Strange and Cox, 1976). This has also been found for certain viruses (Benbough, 1971).

As for effects of collecting fluids the same factors also have a role in whether or not rehumidification influences observed viability. In general, rehumidifica-

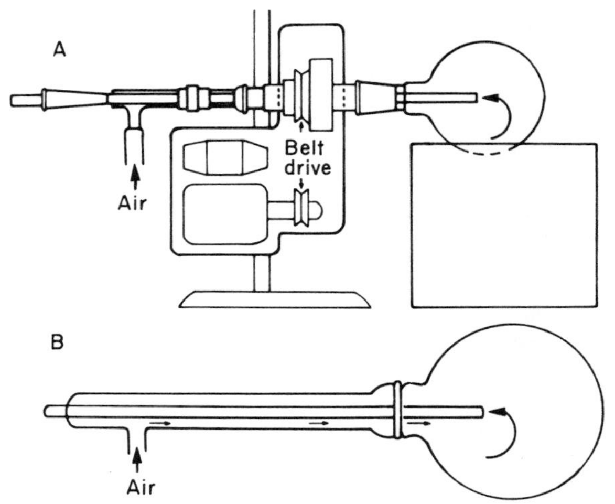

Figure 2.13 Flash evaporator rehumidification system. (From Hatch and Warren (1969), *Appl. Microbiol.*, **17**, 685.) Reproduced by permission of American Society for Microbiology.

tion only makes a marked difference when spent culture media (or other protecting additives) form the matrix containing the aerosolized microorganisms, although *Klebsiella pneumoniae* sprayed from suspension in distilled water in nitrogen survives better with rehumidification (Goldberg and Ford, 1973). Also, the beneficial action of hypertonic collecting fluids is less marked, or even absent, when combined with the rehumidification technique (Cox, 1965, 1966b, 1967).

From the above discussion it is apparent that the manner in which microbial aerosols are rehydrated can markedly influence their apparent survival and infectivity. Since an inescapable feature of all aerosol experiments (designed to study microbial survival and infectivity) is the need to sample, any detailed discussion of this rehydration process must be related to the particular stresses imposed before sampling. Therefore, more detailed analyses of aerosol collection will be found in those chapters dealing with specific stresses imposed by environmental factors and also in Chapter 14 which considers the crucial role played by microbial repair mechanisms.

2.11 VIABILITY ASSAY AND TRACERS

Experimental aerobiology depends upon accurate and reliable methods for determining total and viable numbers of microorganisms in samples recovered from aerosols. This is because the number of viable cells in an aerosol decreases with storage time through both physical and biological (activity) losses. These must be separately determined if the true viability and infectivity of microbial aerosols is to be obtained.

Physical and biological losses occur from different causes. Physical losses result from deposition onto surfaces through impaction, settling, convective flow, Brownian diffusion, electrostatic attraction, particle aggregation, etc. In addition, the removal of an aerosol sample from a storage container dilutes the remaining aerosol which, if appreciable, also must be taken into account.

Many investigators calibrate a particular piece of equipment for physical losses and assume that from experiment to experiment they are consistent over the calibrated range. This approach, while experimentally convenient, is rarely satisfactory for two major reasons. To make them comparative viable concentrations need to be normalized and with this approach normalization usually is to the number of viable cells per unit volume of stored aerosol sampled after several seconds or even 1 or 2 minutes. However, under some conditions 99.9% of the population may lose viability within 0.1 s and therefore the normalization can be grossly in error, resulting in very distorted impressions of aerosol viability. Furthermore, since the rate of physical loss itself depends upon many factors, including cleanliness of container surfaces, particle size distribution, etc., minor changes in experimental techniques can seriously change physical losses. A significant factor in the many discrepancies between results obtained by different workers may be due to their use of this approach to allow for physical decay.

In practice, physical losses in a given aerosol seldom need to be determined directly, however, and therefore an alternative method to estimate viability (or infectivity) is to compare the total and viable (or infective) numbers of microorganisms in recovered populations. This ratio then is normalized to that ratio in the preparation used to generate the aerosol. If necessary the actual physical losses can be derived from disseminator efficiency, sampler efficiency (as a function of particle size) and total cell counts.

In principle total cell counts in a sample may be obtained by direct observation with a light or electron microscope or an automatic cell counter (e.g. Cytoflurograph, Coulter). However, in practice such methods are difficult since very dilute microbial suspensions usually are involved, lysis of microorganisms may occur following collection and foreign particles may interfere. Even so, with suitable precautions this approach was applied successfully by Cox and Baldwin (1964, 1966) for bacteria sampled directly onto slide culture cells (Postgate *et al.*, 1961).

More general methods for determining total microbial numbers in recovered aerosol samples rely on tracer techniques in which material that does not undergo biological loss is added to the microbial suspension. Determinations of the tracer and microbial contents of aerosol spray and collecting fluids provide total and viable numbers of recovered microorganisms.

Tracer methods involve the use of spores, radioisotopes, dyes or activity of enzymes. However, problems can arise with each of these and an ideal tracer has still to be found. A fundamental feature of the tracer technique is that the physical behaviour of the tracer and test organism is identical. Very often this is sufficiently true for the technique to be applicable, but on occasions it may not be true for a variety of reasons.

Bacterial spores (Harper and Morton, 1952) such as *Bacillus subtilis* var. *niger* are widely used as tracers but under some conditions their use can produce artefacts. One instance is with dry disseminated aerosols when due to electrostatic phenomena the physical loss of spores differs greatly from that of the test organisms *Escherichia coli* and *Francisella tularensis* (Cox *et al.*, 1970). Another is because *B. subtilis* spores sometimes lose viability when wet disseminated. For example, when mixed suspensions of *E. coli* and *B. subtilis* spores in distilled water were disseminated into atmospheres of high relative humidity (RH) up to 50% of the spores died or became heat sensitive at aerosol ages of up to 2 min. After that time the rate of loss of spore viability was extremely low. The extent of this loss of viability varied between batches (Cox, 1966a). Rapid initial death of these spores also was found by Anderson (1966) at high and also mid-range RH. However, these findings may also have been due to differences in these experiments in the physical behaviour of spores and the *E. coli* used as a tracer. Nevertheless, the major advantages of spore tracers are the sensitivity of the technique and ease of assay.

As an alternative P^{32} labelling (Harper and Morton, 1952; Harper *et al.*, 1958), S^{35} labelling (Miller *et al.*, 1961), and C^{14} labelling (Anderson, 1966) of microorganisms have been used. In order to minimize radiation damage the level of radioactivity must be kept low with the result that comparatively large samples of aerosol are needed to obtain statistically significant radioactivity assays. Also, at least for *E. coli* B, labelling with C^{14} proved unsatisfactory due to a lack of consistency of results (Cox, 1968a). Many of these disadvantages may be overcome with the technique of Strange *et al.* (1972) which uses radioactively labelled homologous antibody, or that of Mayhew and Hahon (1970) using immunofluorescence.

Methods involving dyes (e.g. Dunklin and Puck, 1948; Henderson, 1952; Wolfe, 1961) are less sensitive than radioactive methods and necessitate adding the dye to the spray suspension or incorporating it in dry powders. Due to the extremely high solute concentrations which occur in dehydrated aerosol particles many dyes become toxic and also can sensitize microorganisms to inactivation by light. Of the tracer methods, the one that uses dyes probably is the least satisfactory.

From assaying the aerostable enzyme β-galactosidase, Anderson and Crouch (1966) were able to determine total bacterial numbers in aerosol samples and while the technique has many attractions it is limited by its insensitivity.

Having determined or compensated for physical losses with one of the above methods recovered samples are assayed for viability. Viability is a useful but ill-defined concept in that it is a measure of the ability of microorganisms to replicate (i.e. to form colonies or plaques) under certain chosen growth conditions. For example, suppose a sample of a mixed aerosol of subtilis spores and *E. coli* is collected by impinger and 1 ml of the collecting fluid is spread over the surface of a chemically defined nutrient agar contained in a petri dish (or '*plate*'). The plate after overnight incubation shows 100 orange colonies of *B. subtilis* and 10 white colonies of *E. coli* then if an equal number of spores and coli organisms were in the spray fluid the viability normalized to that ratio (equal to 1) would be 10%. However, suppose that on a complex medium agar 1 ml of that collecting fluid resulted in the growth of 100 orange colonies of subtilis and 50 white colonies of *E. coli* then the viability would now appear to be 50%, i.e. the two assay methods applied to similar samples give 10 and 50% for the viability of the recovered *E. coli*. At first sight, perhaps, this type of result may seem absurd but actually is fairly common. It can arise, for example, from (a) the production of mutants (by aerosolization) having requirements for complex growth factors present only in the complex medium or (b) repair mechanisms which require complex molecules for their operation.

Such effects sometimes make comparisons of results by different workers extremely difficult and can lead to apparent contradiction in data. However, in trying to establish causes of death these uncertainties in viability assays must be considered. Even attempting to eliminate such possibilities through the use of complex media is not always successful due to the fastidiousness of repair mechanisms and their often highly specific requirements (Morichi, 1969; Morichi and Irie, 1973; Morichi *et al.*, 1973; Hambleton, 1970, 1971).

2.12 INFECTIVITY ASSAY

Aerosol infectivity is assessed by exposing groups of test animals to aerosols for known periods of time. The respiratory dose is calculated from the animal's breathing rate coupled with an aerosol retention factor (Harper and Morton, 1953), and the viable aerosol concentration determined from an artifically collected sample taken during animal exposure. But many difficulties are associated with collecting aerosols artificially (Sections 2.8 to 2.10). As discussed in other chapters, some causes of loss of viability are directly attributable to artificial sampling processes in contrast to the more gentle collection processes operating on inhalation of microbial aerosols by host animals. A further difficulty is, therefore, correlation of the number of viable microorganisms deposited in respiratory systems of animals with that recovered from the same aerosol by artificial samplers. Determination of the LD_{50} (i.e. the average number of viable microorganisms received per animal which causes lethality in half of the exposed animals, or the 50% lethal dose) of a pathogen inherently assumes that natural and artificial sampling processes are equivalent. This situation may be closely approached by combining the rehumidification technique with an efficient sampler which imposes low shear stresses. As far as is known, however, no infectivity studies have been performed where such an arrangement was used for establishing challenge levels (or dose received). Yet, depending upon experimental conditions, rehumidification markedly affects the viability of microorganisms artificially recovered from aerosols (for example, see Chapter 10).

2.13 CONCLUSIONS

The Aerobiological Pathway exposes microorganisms to a large number of environmental stresses, which operate simultaneously. They include the take-off and landing processes as well as desiccation/hydration and exposure to radiation, oxygen, ozone, pollutants, etc. While some microorganisms succumb to these stresses, others demonstrate reversible injury or repair (q.v.). In the laboratory, the separate control of these factors leads to an evaluation of their individual effects, of the metabolic state of microorganisms and of aerosol particle size. Numerous sophisticated techniques need to be employed to unravel the complexities of events which occur on the generation of microbial aerosols.

REFERENCES

Akers, A. B. and Won, W. D. (1969). In *An Introduction to Experimental Aerobiology*, R. L. Dimmick and Ann B. Akers (Eds), Wiley Interscience, New York, London, pp. 59–99.
Anderson, J. D. (1966). *J. Gen. Microbiol.*, **45**, 303–313.
Anderson, J. D. and Cox, C. S. (1967). *Symp. Soc. Gen. Microbiol.*, **17**, 203–226.
Anderson, J. D. and Crouch, G. T. (1966). *J. Gen. Microbiol.*, **47**, 49–52.
Bachelor, H. W. (1960). *Adv. Appl. Microbiol.*, **2**, 31–64.
Barlow, D. F. (1972). *J. Gen. Virol.*, **15**, 17–24.
Beebe, J. M. (1959). *J. Bact.*, **78**, 18–24.
Benbough, J. E. (1969). *J. Gen. Virol.*, **4**, 473–477.
Benbough, J. E. (1971). *J. Gen. Virol.*, **10**, 209–220.
Benbough, J. E. and Hood, A. M. (1971). *J. Hyg. (Camb.).*, **69**, 619–626.
Brachman, P. S., Ehrlich, R., Eichenwald, H. F., Gabelli, V. J., Kethley, T. W., Madin, S. H., Maltman, J. R., Middlebrook, G., Morton, J. D., Silver, I. H. and Wolfe, E. K. (1964). *Science*, **144**, 1295.
Bradish, C. J., Allner, K. and Maber, H. B. (1971). *J. Gen. Virol.*, **12**, 141–160.
Brown, A. D. (1953). *Austral. J. Biol. Sci.*, **6**, 470–485.
Chamberlain, A. C. (1967). *Symp. Soc. Gen. Microbiol.*, **17**, 138–164.
Collison, W. E. (1935). *Inhalation Therapy Technique*, Heinemann, London.
Cox, C. S. (1965). In *First International Symposium on Aerobiology*, R. L. Dimmick (Ed), Naval Biological Laboratory, Naval Supply Center, Oakland, California, pp. 345–368.
Cox, C. S. (1966a). *J. Gen. Microbiol.*, **43**, 383–399.
Cox, C. S. (1966b). *J. Gen. Microbiol.*, **45**, 283–288.
Cox, C. S. (1967). *J. Gen. Microbiol.*, **49**, 109–114.
Cox, C. S. (1968a). *J. Gen. Microbiol.*, **50**, 139–147.
Cox, C. S. (1968b). *J. Gen. Microbiol.*, **54**, 169–175.
Cox, C. S. (1968c). *Nature (Lond.).*, **220**, 1139.
Cox, C. S. (1970). *Appl. Microbiol.*, **19**, 604–607.
Cox, C. S. (1971). *Appl. Microbiol.*, **21**, 482–486.
Cox, C. S. (1976). *Appl. Environ. Microbiol.*, **31**, 836–846.
Cox, C. S. and Baldwin, F. (1964). *Nature (Lond.).*, **202**, 1135.
Cox, C. S. and Baldwin, F. (1966). *J. Gen. Microbiol.*, **44**, 15–22.
Cox, C. S. and Goldberg, L. J. (1972). *Appl. Microbiol.*, **23**, 1–3.
Cox, C. S., Baxter, J. and Maidment, B. J. (1973). *J. Gen. Microbiol.*, **75**, 179–185.
Cox, C. S., Bondurant, M. C. and Hatch, M. T. (1971). *J. Hyg. (Camb.).*, **69**, 661–672.

Cox, C. S., Gagen, S. J. and Baxter, J. (1974). *Can. J. Microbiol.*, **20**, 1529–1534.
Cox, C. S., Derr, J. S., Fleurie, E. G. and Roderick, R. C. (1970). *Appl. Microbiol.*, **20**, 927–934.
Crider, W. L., Berkley, N. P. and Strong, A. A. (1968). *Rev. Sci. Instrum.*, **39**, 152–155.
Dark, F. A. and Callow, D. S. (1973). In *Fourth International Symposium on Aerobiology*, J. F. Ph Hers and K. C. Winkler (Eds), Oosthoek, Utrecht, Netherlands, pp. 97–99.
Davies, R. R. (1971). In *Methods in Microbiology*, vol. 4, J. R. Norris and D. W. Ribbons (Eds), Academic Press, London, New York. pp. 367–404.
DeOme, K. B. *et al.* (1944). *Amer. J. Hyg.*, **40**, 239–250.
Derr, J. S. (1965). In *First International Symposium on Aerobiology*, R. L. Dimmick (Ed), Naval Biological Laboratory, Naval Supply Center, Oakland, California, pp. 227–261.
Dimmick, R. L. (1959). *Arch. Ind. Hlth.*, **20**, 8–14.
Dimmick, R. L. (1969). In *An Introduction to Experimental Aerobiology*, R. L. Dimmick and Ann B. Akers (Eds), Wiley Interscience, New York, London, pp. 22–45.
Druett, H. A. (1970). In *Third International Symposium on Aerobiology*, I. H. Silver (Ed.), Academic Press, London, New York, p. 212.
Druett, H. A. and May, K. R. (1952). *J. Hyg. (Camb.).*, **50**, 69–81.
Dunklin, E. W. and Puck, T. T. (1948). *J. Exptal. Med.*, **87**, 87–101.
Errington, F. P. and Powell, E. O. (1969). *J. Hyg. (Camb.).*, **67**, 387–399.
Goldberg, L. J. (1970). In *Third International Symposium on Aerobiology*, I. H. Silver (Ed.), Academic Press, London, New York, p. 268.
Goldberg, L. J. (1971). *Appl. Microbiol.*, **21**, 244–252.
Goldberg, L. J. and Ford, I. (1973). In *Fourth International Symposium on Aerobiology*, J. F. Ph. Hers and K. C. Winkler (Eds), Oosthoek, Utrecht, Netherlands, pp. 86–89.
Goldberg, L. J., Watkins, H. M. S., Boerke, E. E. and Chatigny, M. A. (1958). *Amer. J. Hyg.*, **68**, 85–93.
Goodlow, R. G. and Leonard, F. A. (1961). *Bact. Rev.*, **25**, 182–187.
Green, H. L. and Lane, W. R. (1964). *Particulate Clouds: Dusts, Smokes and Mists*, E. and F. N. Spon, London.
Hambleton, P. (1970). *J. Gen. Microbiol.*, **61**, 197–204.
Hambleton, P. (1971). *J. Gen. Microbiol.*, **69**, 81–88.
Harper, G. J. (1961). *J. Hyg. (Camb.).*, **59**, 479–486.
Harper, G. J. (1963). *Arch. Ges. Virusforsch.*, **13**, 64–71.
Harper, G. J. (1965). In *First International Symposium on Aerobiology*, R. L. Dimmick (Ed), Naval Biological Laboratory, Naval Supply Center, Oakland, California, pp. 335–343.
Harper, G. J. and Morton, J. D. (1952). *J. Gen. Microbiol.*, **7**, 98–106.
Harper, G. J. and Morton, J. D. (1953). *J. Hyg. (Camb.).*, **51**, 372–385.
Harper, G. J. and Morton, J. D. (1962). *J. Hyg. (Camb.).*, **60**, 249–257.
Harper, G. J., Hood, A. M. and Morton, J. D. (1958). *J. Hyg. (Camb.).*, **56**, 364–370.
Hatch, M. T. and Dimmick, R. L. (1965). In *First International Symposium on Aerobiology*, R. L. Dimmick (Ed), Naval Biological Laboratory, Naval Supply Center, Oakland, California, pp. 265–268.
Hatch, M. T. and Dimmick, R. L. (1966). *Bact. Rev.*, **30**, 597–603.
Hatch, M. T. and Warren, J. C. (1969). *Appl. Microbiol.*, **17**, 685–689.
Hearn, H. J. J., Soper, W. T. and Miller, W. S. (1965). *Proc. Soc. Exp. Biol. Med.*, **119**, 319–322.
Heckly, R. J., Dimmick, R. L. and Guard, N. (1967). *Appl. Microbiol.*, **15**, 1235–1239.
Heinrich, M. R. (1979). In *Biometeorological Survey*, vol. 1, 1973–1978. Part A, *Human Biometeorology*, S. W. Tromp and Janneke J. Bouma (Eds), Heyden, London, Philadelphia, pp. 162–169.
Henderson, D. W. (1952). *J. Hyg. (Camb.).*, **50**, 53–68.
Herbert, D. (1961). *Symp. Soc. Gen. Microbiol.*, **11**, 391–416.

Hess, G. E. (1965). *Appl. Microbiol.*, **13**, 781–787.
Hood, A. M. (1971). *J. Hyg. (Camb.).*, **69**, 607–617.
Hood, A. M. (1973). In *Fourth International Symposium on Aerobiology*, J. F. Ph Hers and K. C. Winkler (Eds), Oosthoek, Utrecht, Netherlands, pp. 149–151.
Hood, A. M. (1974). *J. Hyg. (Camb.).*, **72**, 53–60.
Hope-Simpson, R. E. (1979). In *Biometeorological Survey*, vol. 1, 1973–1978. Part A, *Human Biometeorology*, S. W. Tromp and Janneke J. Bouma (Eds), Heyden, London, Philadelphia, pp. 170–185.
Kates, M., Allison, A. C., Tyrrell, D. A. J. and James, A. T. (1962). *Cold Spring Harbor Symp. Quant. Biol.*, **27**, 293–301.
Lee, R. E. and Garbett, M. (1966). *Appl. Microbiol.*, **14**, 133–134.
Maltman, J. R. and Webb, S. J. (1971). *Can. J. Microbiol.*, **17**, 1443–1450.
May, K. R. (1949). *J. Appl. Phys.*, **20**, 932–938.
May, K. R. (1964). *Appl. Microbiol.*, **12**, 37–43.
May, K. R. (1966). *J. Sci. Instru.*, **43**, 841–842.
May, K. R. (1967). *Symp. Soc. Gen. Microbiol.*, **17**, 60–80.
May, K. R. (1969). *Appl. Microbiol.*, **18**, 513–514.
May, K. R. (1972). In *Assessment of Airborne Particles. Fundamentals and Implications to Inhalation Therapy*, T. T. Mercer, P. E. Morrow and W. Stober (Eds), Thomas, Springfield, Illinois, pp. 420–484.
May, K. R. (1973). *Aerosol Sci.*, **4**, 235–243.
May, K. R. and Druett, H. A. (1968). *J. Gen. Microbiol.*, **51**, 353–366.
Mayhew, C. J. and Hahon, N. (1970). *Appl. Microbiol.*, **20**, 313–316.
Miller, W. S., Scherff, R. A., Piepoli, C. R. and Idoine, L. S. (1961). *Appl. Microbiol.*, **9**, 248–252.
Morichi, T. (1969). In *Freezing and Drying of Microorganisms*, T. Nei (Ed), University of Tokyo Press, Tokyo, pp. 53–68.
Morichi, T. and Irie, R. (1973). *Cryobiol.*, **10**, 393–399.
Morichi, T., Okamoto, T. and Irie, R. (1973). In *Freeze-drying of Biological Materials*. pp. 47–59. Paris: Institut International du Froid.
Noble, W. C. (1967). *Symp. Soc. Gen. Microbiol.*, **17**, 81–108.
Peto, S. and Powell, E. O. (1970). *J. Appl. Bact.*, **33**, 582–598.
Postgate, J. R., Crumpton, J. E. and Hunter, J. R. (1961). *J. Gen. Microbiol.*, **24**, 15–24.
Record, B. R., Taylor, R. and Miller, D. S. (1962). *J. Gen. Microbiol.*, **28**, 585–598.
Schaffer, F. L., Soergel, M. E. and Straube, D. C. (1976). *Arch. Virol.*, **51**, 263–273.
Shipe, E. L., Tyler, M. E. and Champman, D. N. (1959). *Appl. Microbiol.*, **7**, 349–354.
Silver, I. H. (1965). In *First International Symposium on Aerobiology*, R. L. Dimmick (Ed), Naval Biological Laboratory, Naval Supply Center, Oakland, California, pp. 319–333.
Spendlove, J. C. and Fannin, K. F. (1982). In *Methods in Environmental Virology*, C. P. Gerba and S. M. Goyal (Eds), Marcel Dekker, New York, Basel, pp. 261–329.
Stein, R. L., Rybeck, W. H. and Sparks, A. W. (1973). *J. Coll. Int. Sci.*, **42**, 441–447.
Strange, R. E. and Cox, C. S. (1976). *Symp. Soc. Gen. Microbiol.*, **26**, 111–154.
Strange, R. E., Benbough, J. E., Hambleton, P. and Martin, K. L. (1972). *J. Gen. Microbiol.*, **72**, 117–125.
Strasters, K. C. and Winkler, K. C. (1966). *Bact. Rev.*, **30**, 674–677.
Thomas, G. (1970a). In *Third International Symposium on Aerobiology*, I. H. Silver (Ed.), Academic Press, London, New York, p. 266.
Thomas, G. (1970b). *J. Hyg. (Camb.).*, **68**, 273–282.
Thomas, G. (1970c). *J. Hyg. (Camb.).*, **68**, 511–517.
Tyler, M. E. and Shipe, E. L. (1959). *Appl. Microbiol.*, **7**, 337–348.
Vlodavets, V. V., Zuikova, E. I. and Motova, M. A. (1958). *Microbiologiya*, **27**, 632; *Biol. Abstr.*, **34**, 13,726 (1959).

Wolf, W. R. (1961). *Rev. Sci. Instrum.*, **32**, 1124–1129.
Wolf, H. W., Skaliy, P., Hall, L. B., Harris, M. M., Decker, H. M., Buchanan, L. M. and Dahlgren, C. M. (1959). *Sampling Microbiological Aerosols. Public Health Service Publications*, Washington, No. 60.
Wolfe, E. K. (1961). *Bact. Rev.*, **25**, 194–202.
Won, W. D. and Ross, H. (1966). *Cryobiol.*, **3**, 88–93.
Won, W. D. and Ross, H. (1968). *Cryobiol.*, **4**, 337–340.

Chapter 3

Aerosol samplers

3.1 INTRODUCTION

Artificial aerosol sampling is fundamental to experimental aerobiology. To determine the outcome of aerosol generation and the ensuing airborne state samples of that aerosol invariably need to be collected. As indicated in Chapter 1 physical attributes of aerosols play dominant roles in their efficient collection and in ensuring that samples truly represent the whole (e.g. both have the same particle size distributions). Additionally, sampling processes can modify profoundly observed biological behaviour (e.g. survival, infectivity) of collected samples as indicated in Chapter 2, and emphasized in later chapters. The importance, subtleness, and difficulties of aerosol sampling cannot be overstressed. This chapter covers most of the appropriate aerosol samplers.

3.2 EFFICIENCY OF SAMPLING

A given sampler must be judged at least in terms of capability to physically collect microbial aerosols while minimizing sampling stresses so that biological activity is not impaired. Quite often these two requirements are not compatible. One example is the Porton impinger (Section 3.5). This all-glass sampler has a short length of capillary tubing held above a liquid surface and the air jet produced by applying suction strikes the base of the container. The particles become trapped by impingement into the violently agitated collecting fluid and impaction onto the frequently washed glass base. In its original version the clearance between the bottom of the capillary and the glass base was 5 mm. While for sampling aerosols the Porton impinger was highly efficient, in the process it caused many bacteria to die. By increasing the gap to 30 mm viability was preserved much better but at the expense of sampling efficiency which on a physical basis was reduced slightly. For this device high efficiency of sampling and freedom of trauma for sampled microbes usually are incompatible.

Damage as a result of collection by impinger is demonstrated by comparing results obtained for various impingers, e.g. Shipe, Midget, All-glass, (i.e. raised Porton, AGI-30) and capillary impingers. Data shown in Table 3.1

Table 3.1 Recoveries in four samplers of small or heterogeneous aerosol droplets of *Serratia marcescens* and *Bacillus subtilis* spores. (Shipe, Tyler and Chapman; *Appl. Microbiol.*, (1959) **7**, 349. Reproduced with permission of American Society for Microbiology)

Aerosol	Type of sampler	Plate count per L aerosol at indicated period (min)			
		2–3	4–5	8–9	16–17
S. marcescens in mixed aerosol (small droplets*)	Shipe sampler	4700†	4550	3500	1820
	Midget impinger	4500	3430	2950	1980
	All-glass impinger	3780	3250	2060	1710
	Capillary impinger	1560	1480	1210	700
B. subtilis in mixed aerosol (small droplets*)	Shipe sampler	5340	6930	4230	4800
	Midget impinger	4100	3700	4000	3600
	All-glass impinger	5440	4780	4230	4280
	Capillary impinger	5900	4700	5000	4860
S. marcescens aerosol (heterogeneous droplets‡)	Shipe sampler	20 400	16 600	10 900	5900
	Midget impinger	12 800	12 300	12 900	4000
	All-glass impinger	6800	5400	4000	2800
	Capillary impinger	6800	6700	5000	3300

*Less than 3.0 μm MMD (mass median diameter), Devilbiss No. 40 nebulizer.
†Each value is mean of 5 samples. Suspension counts: mixed, *S. marcescens*, 11.7×10^8 per ml; *B. subtilis*, 7.1×10^8 per ml; *S. marcesens* only, 24.6×10^8 per ml.
‡Greater than 3.0 μm MMD, explosive disseminator in test chamber.

indicate that viable recoveries (by these samplers) of hardy spores of *Bacillus subtilis* usually agreed to within a factor of 1.3. For vegetative *Serratia marcescens*, a relatively sensitive bacterium, viable recoveries were spread over a much wider range with the capillary impinger giving a poor performance and the Shipe impinger providing highest recoveries. It is pertinent that the Shipe impinger differs fundamentally from the others in that aerosols are directed tangentially (rather than at right angles) to the collecting fluid. Such a configuration reduces impingement at near sonic velocity of sampled aerosols onto the glass base.

The examples above are indicative of a more general trend when sampling microbial aerosols, namely that observed results most likely will depend on the precise sampling method employed. It is necessary, therefore, to thoroughly investigate what effects any given sampler has for each situation where it is applied. The subleties are such that apparently minor changes in experimental protocol can cause a given sampler (such as an impinger where shear forces are high) to change from preserving viability to reducing it for sampled microorganisms. In general, aerosol collection devices which have lowest shear forces provide samples of microorganisms having highest viability. On the other hand, these samplers usually have lowest physical efficiencies in terms of numbers of airborne particles collected.

3.3 ISOKINETIC SAMPLING

Operating most samplers and sampling probes (Fuchs, 1975) raises the immediate difficulty of ensuring that a representative sample of the aerosol of interest is collected. Apart from possible problems due to adequacy of mixing, etc., there is that of making certain that particles of all sizes have an equal probability of entering the sampler. It is achieved when the velocity of air in the sampler inlet equals that of air being sampled, i.e. isokinetic. Then, the streamlines display no directional changes, as shown in Figure 3.1. If the velocity of the sampling air is greater than that of the moving airstream small particles will predominate because they more easily cross the streamlines. If the velocity is slower, larger particles predominate because unlike smaller particles they do not follow the curvature of the streamlines around the sampling nozzle. Fuchs (1975) provides a good review of these problems. Anisokinetic sampling and the magnitude of the sampling error has been examined and reviewed by a number of investigators including Davies (1964, 1968a, b), Dennis (1976), Fuchs (1964), Hinds (1982), May (1967), Peterson (1978), Suggs (1978), Vitols (1966) and Watson (1954). Nozzle design is crucial for isokinetic sampling with most commercial designs being sharp edged with a chamfer angle of 30° or slightly less (Figure 3.2). Davies (1968a, b) provides the following equation for estimating error due to anisokinetic sampling:

$$\frac{C_s}{C_a} = \frac{(v_a/v_s) - 1}{4\psi + 1} \tag{3.1}$$

where, C_s = particulate concentration in sampler air

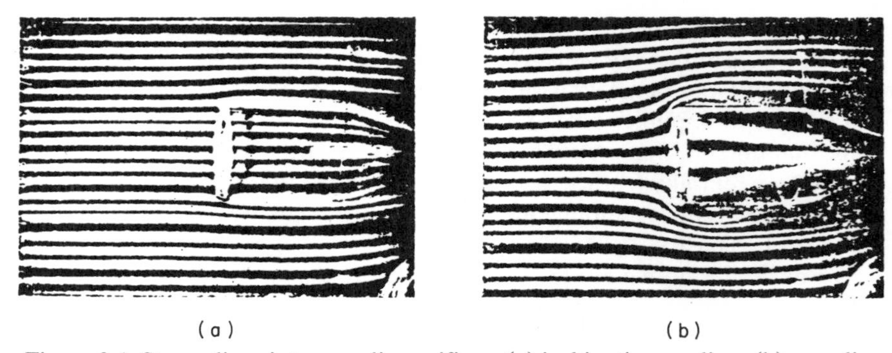

Figure 3.1 Streamlines into sampling orifices: (a) isokinetic sampling; (b) sampling at half air speed. (From May (1945), *J. Sci. Instrum.*, **22**, 187). Reproduced by permission of the Institute of Physics.

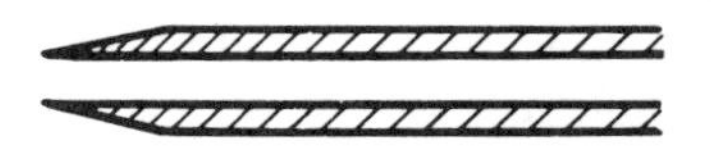

Figure 3.2 Sharp edged sampling probe (chamfer angle 30°, or less).

C_a = that in ambient airstream

v_s = velocity of sampler air

v_a = velocity of ambient airstream

ψ = inertial impaction parameter (Stokes number)

$$\psi = \frac{d_p^2 C(\rho_p - \rho_a)\, v_a}{18\eta D} \qquad (3.2)$$

(According to Fuchs, 1978, the coefficient should be 9 rather than 18.)

where, d_p = particle diameter

C = Cunningham slip correction factor

ρ_p = particle density

ρ_a = air density

η = air viscosity

D = diameter of nozzle opening

The equation predicts an error of 20% when the value of v_a/v_s is from 0.5 to 2 and when $\psi < 0.1$; i.e. for particles within the range 1–5 μm diameter the errors are tolerable, but increase as this range is widened and can attain 300% (Suggs, 1978).

Laminar flow conditions, assumed for the above, may apply in ducts and pipes if properly configured but rarely do so out-of-doors. Here, wind speed variations of ±40% of the mean can rapidly occur while directional changes of ±20° horizontally and ±15° vertically are common. Under these conditions isokinetic sampling is difficult as exemplified by the data of May (1967) for a nozzle about 1–2 cm diameter carrying a flow of 5 m/s and sampling an aerosol comprising unit density spheres (Figure 3.3 and 3.4).

Their shape is explained as for streamlines and laminar flow (see above) while the effect of yaw is analogous. That the intake efficiency is less than unity reflects the probability of particle impaction onto the leading edge of nozzles. Since this probability increases with decreasing nozzle diameter, narrow nozzles are to be avoided. Given that the nozzle (by means of an associated windvane) always faces into the wind and that its flow rate is adjusted to be isokinetic with the mean wind speed, then sampling errors tend to cancel because the wind speed fluctuates above and below that mean. May (1960) describes a device suitable for rapidly following changes in wind direction. However, in order to have low mass it incorporates a filter to collect aerosol particles and therefore it is unsuitable for sampling vegetative microbes unless they are hardy (see Section 3.11).

Another problem with anisokinetic sampling is that particle trajectories seldom are parallel to the axis of the intake nozzle and pipe connecting it to the actual sampler. The result is collisions between particles, especially the larger

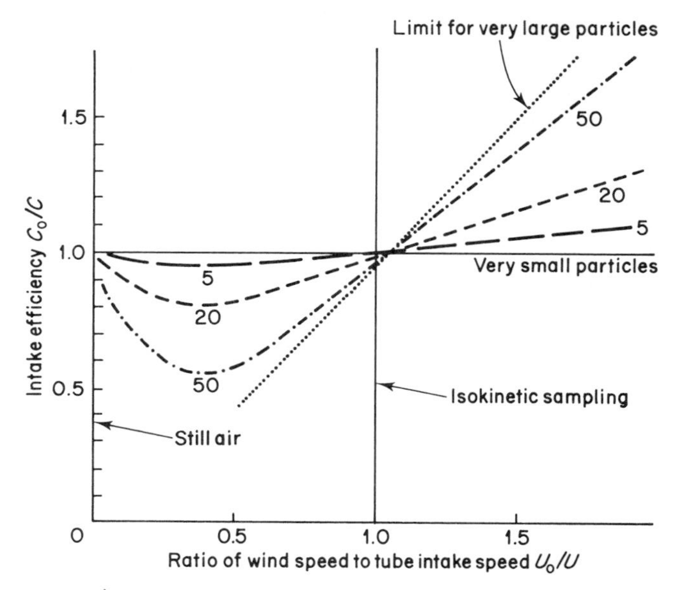

Figure 3.3 Effect of changing wind speed past a horizontal sampling nozzle for particles of different diameter (μm). Curves apply to a nozzle of 1–2 cm diameter sampling at about 5 m/s. (From May (1967), *17th Symp. Soc. Gen. Microbiol.*) Reproduced by permission of HMSO.

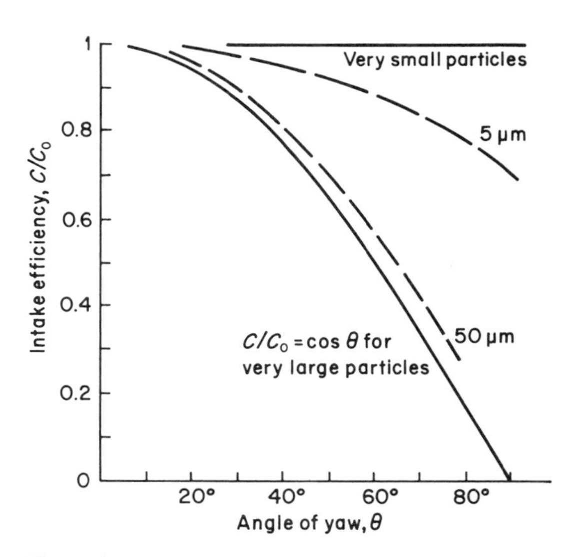

Figure 3.4 Effect of yaw on a nozzle similar to that in Figure 3.3; isokinetic sampling. (From May (1967), *17th Symp. Soc. Gen. Microbiol.*) Reproduced by permission of HMSO.

ones, and nozzle or pipe walls. These wall losses when coupled with those losses associated with intake nozzles operated anisokinetically often elicit unacceptable sampler performance. Ameliorative action of a kind is to separately wash out the intake nozzle and pipe walls should it be imperative to operate samplers anisokinetically. Results from assay of these washings then can be included with results from the sampler proper. An alternative and preferable approach is that of stagnation point sampling discussed in the following section.

3.4 STAGNATION POINT SAMPLING

Reference to Figure 3.3 indicates that particles of respirable size are sampled efficiently from still air (i.e. wind speed of zero) by a horizontal nozzle. Such an outcome is very different to that for anisokinetic sampling of unsteady airstreams and suggests a remedy. That is, by bringing moving aerosols to rest in the vicinity of the sampler intake nozzle efficient sampling should be achieved for a range of wind speeds and directions.

May (1967) indicates that when a hemicylindrical baffle is placed immediately behind the sampler intake, approaching air is brought nearly to rest just in front of the nozzle through a cushion effect inside the baffle. If possible, the baffle should be large so that arresting the airstream is effective. Fuchs (1975) also addresses this method of sampling which has promise.

3.5 IMPINGERS

The emphasis placed in the past on the impinger (Figure 3.5) as first choice sampler for microbial aerosols is epitomized by the fact that it provided the motif for the *First International Symposium on Aerobiology, 1963*. Furthermore, Brachman *et al.* (1964) recommended that data obtained with other samplers should be compared with those determined with a standard reference sampler, viz. the AGI-30, all-glass impinger (Wolf *et al.*, 1959). The reasons were largely historical, together with cheapness, availability and simple design of the AGI-30.

Greenburg and Smith (1922) developed the first impinger and since then many slightly different versions have been described, while May and Druett (1953) constructed a pre-impinger for dividing the total sample into two size fractions (>5 μm and 1–5 μm diameter). Impingers trap particles by the principle of high-speed impaction originally with the jet submerged in a collecting fluid. The jet is operated as a critical orifice so that particles impinge at or near sonic velocity. Silverman and Franklin (1942) showed that erroneous results often arose due to particle disruption and fragmentation, especially for aggregates. For sampling microorganisms the Porton raised impinger was developed at the Microbiological Research Establishment, Porton Down, England (May and Harper, 1957). The main difference was increasing the gap between the sampling jet and the base of the impinger to 30 mm (hence AGI-30) (Figure 3.5) to reduce physical trauma.

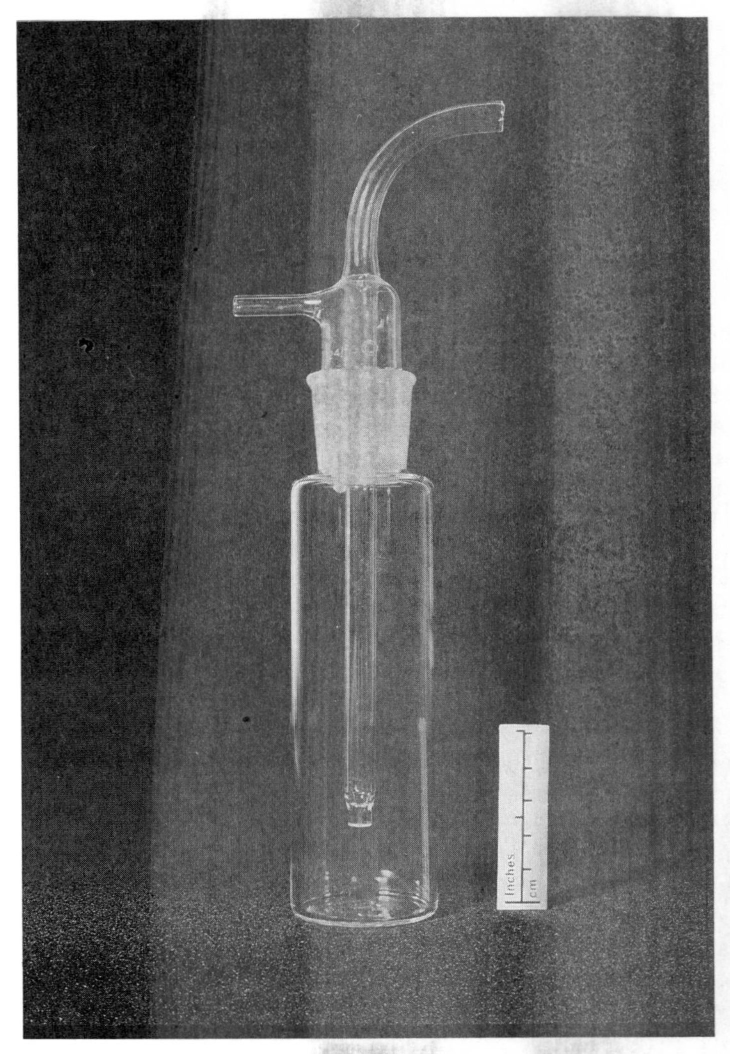

Figure 3.5 AGI-30 impinger.

Most impingers (e.g. Porton raised impinger, AGI-30) are designed to operate by drawing aerosols through an inlet tube curved to simulate the nasal passage and thence through a jet. The jet held 30 mm above the impinger base consists of a short piece of capillary tube. When the pressure drop across this capillary attains a minimum of half an atmosphere the flow through it becomes sonic and therefore rate limiting (i.e. it behaves as a critical orifice). Once calibrated the impinger does not need to be constantly monitored by flowmeter provided the pressure drop across the jet is maintained at half an atmosphere, or greater. The usual sampling rate for the AGI-30 is 12.5 l/min when it is efficient for microbial particles in the respirable size range (i.e. 0.8 to 15 μm), especially when used with a pre-impinger. But, as larger particles are collected on the curved inlet they are recovered by washing, e.g. by pipetting a known volume of collecting fluid into the impinger inlet which then flows slowly through the jet and into the impinger base.

The usual volume of collecting fluid is 20 ml but depending on application may be reduced to as little as 2 ml. A frequent reason for doing so is to increase the concentration (per ml) of collected microorganisms. An alternative approach here is to extend sampling times but then other effects arise. One such is that collecting fluids concentrate due to evaporation which also causes cooling. At low RH and low temperatures collecting fluids actually can freeze. Whether ensuing cold shock and possible osmotic shock are likely to alter the survival/infectivity of collected microorganisms will depend on actual circumstances. But, as indicated in later chapters, the most likely outcome is that survival/infectivity will be decreased.

A frequent if not constant concern is whether the action of collecting microbial aerosols with impingers or other samplers causes trauma leading to loss of biological activity. Very often, as discussed in later chapters, such concerns are well founded and led Shipe *et al.* (1959) through studies of the viability of bacteria sampled by different impingers (Section 3.2) to the development of the Shipe impinger. Shipe reasoned that some impingement of vegetative bacteria onto the bottom of an AGI-30 might kill or damage them. Also that collected bacteria may be lost through splashing of the collecting fluid while others (>3 μm particles) would be lost in the neck. The Shipe impinger was designed, therefore, to minimize losses due to intake and splashing and its efficiency in this regard is supported by the data presented in Table 3.1. Observed decrease in viable recoveries with aerosol age has been ascribed to fall-out of larger particles (Tyler and Shipe, 1959; Tyler *et al.*, 1959). But, as described in detail in later chapters, loss of viability while airborne must have contributed perhaps significantly.

A more recent development aimed at improved microbial sampling is the three-stage glass impinger (May, 1966) (Figure 3.6). The three stages are intended to correspond with the three principal deposition sites of the human respiratory system, viz. stage 1 corresponds to the upper respiratory tract, stage 2 to the bronchioles and stage 3 to the alveoli. The impingement process for each stage is made gentle to minimize impact trauma and viability loss. One disadvantage in some applications is its complex construction and ensuing cost. However, it is offset by the sampler being more portable than the AGI-30 + pre-impinger and by being less subject to loss of collected sample through spilling of the pre-impinger contents, an action that is difficult to avoid in practice. Other advantages accrue including lower evaporation rate and likelihood of freezing of collecting fluids. The sampler is available in three sizes with corresponding flow rates of 55, 20 and 10 l/min.

In operation air drawn into the intake tube flows over a sintered glass impaction (or impingement) disc washed continuously by the agitated collecting fluid. Larger particles impact to become dispersed in the collecting fluid. The air

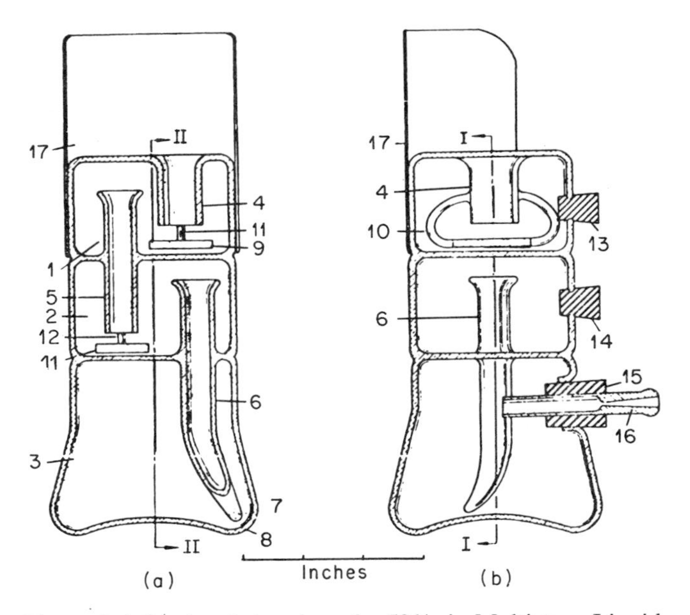

Figure 3.6 Sectional drawing of a 50 l/min Multistage Liquid Impinger. A and B are sectional side elevations at right angles to each other in the directions I-II and II-II, respectively. The air inlet tube 4 is smoothly curved to promote laminar flow and has a flat ground lower end. The straight tube 5, also with a smoothly curved bell-mouth, a flat ground lower end and a bore of 10 mm, is sealed into the flat floor of stage 1. Tube 6 with a smooth bell-mouth is sealed into the floor of stage 2. At its lower end it bends and tapers smoothly and continuously to the nozzle 7. Two circular discs, 9 and 11, of coarse sintered glass 3 mm thick are held 1 mm above the floor of their respective chambers. The discs 9 and 11 are twice the diameter of the bores of respective tubes 4 and 5 and are separated from the flat ends of these tubes by a distance equal to three-eighths of the bore. Access holes to each chamber are sealed by rubber bungs 13, 14 and 15. The lowest bung 15 is fitted with a tube 16 for connection to a suitable pump. (From Rajhans (1978), *Air Sampling Instruments*, 5th edn, American Conference of Governmental Industrial Hygienists.) Reproduced by permission of American Conference of Governmental Industrial Hygienists.

then flows in like manner into the second stage intake tube which is narrower than that for the first stage. As a consequence of higher air speed and particle velocity the larger of the particles penetrating stage 1 are impacted onto the sintered glass disc of stage 2. As for the first stage collected particles are dispersed in the stage 2 collecting fluid. Particles too small to impact at stage 2 pass into the gently tapered jet of stage 3 set tangentially to the collecting fluid surface of stage 3. This configuration causes the collecting fluid to circulate or swirl round the liquid chamber of this stage. As the jet velocity is the minimum possible for sampling 1 μm particles with high efficiency, particles are sampled as gently as is practicable, commensurate with high collection efficiency.

Splashing and frothing of collecting fluids is minimized by the design provided the sampler is mounted vertically and level and the chambers are filled accurately with the correct volumes. As the flow-rate control is external to this sampler (unlike the AGI-30 for example) the pressure drop across it is minimal (about 5 cm water gauge) which helps to prevent evaporation of collecting fluids. The design represents a marked advance over more conventional impingers but unfortunately its complex construction in glass and ensuing cost represent disadvantages over classical impingers. Even so, these days it is probably the first choice sampler of aerosols of fragile microbes.

3.6 IMPACTORS

Inertial forces are responsible for impactor action as discussed in detail by Ranz and Wong (1952), May (1945) and Fuchs (1978), while Davies and Aylward (1951), Davies *et al.* (1951), Marple and Liu (1974) and Marple *et al.* (1974) have determined fluid flow and particle trajectories in impactor jets. For sampling submicron particles operation of an impactor jet at reduced pressure (when the Cunningham slip correction factor is enhanced) is possible (Stern *et al.*, 1962; Parker and Buchholz, 1968). Under these conditions, particles may experience condensation of water vapour on their surface. Resulting increases may occur in particle size but for reliable use the pressure drop across such jets has to be carefully monitored and maintained constant.

Impaction has been employed for sampling aerosols since before the turn of the century (Rajhans, 1978). Of the early instruments, the Bausch and Lomb dust counter of 1938 (Gurney *et al.*, 1938) deserves particular mention as an integral part of it was a 200 x darkfield microscope. In addition, moist blotting paper was used to humidify the air before passing through the impaction slit. On exit the resulting gaseous expansion caused moisture condensation on the particles with the result that impaction slides did not require adhesive. The approach is still pertinent today.

One potential disadvantage of single stage impactors is that particles of all sizes are sampled together. Particle size information then requires microscopic examination which can be tedious when large numbers of samples demand analysis. One remedial option was that of May (1945) who introduced the cascade impactor, i.e. a series of four progressively finer jet impactors connected in series (see below). Since 1945 numerous versions have been developed, including a five-stage sampler by Brink (1958), a six-stage sampler by Mitchell and Pilcher (1958) and others by Lippmann (1961), Carson and Paulus (1974), Ranz and Wong (1952), Laskin (1949), Lundgren (1967), Mercer *et al.* (1962) and Wilcox (1953). That of Andersen (1958) has, depending on model, up to 10 stages with each having as many as 400 circular jets. Other circular jet impactors are those of Mercer and Stafford (1969), Lippmann (1959) and Sierra Inc. In more recent time, May (1975) has introduced his 'ultimate' cascade impactor.

Originally, the major advantage of multiple stage cascade impactors was that they can provide particle size data from mass rather than microscopic analyses as required by single stage impactors. However, since then automated image analysis techniques have become highly developed (Chapter 5) and their application removes that advantage. Whether to use a single or multiple stage impactor, nowadays, depends more on the particular circumstances applying for the given experiment. Impactors such as those mentioned above basically consist of stages with single jets or stages with multiple jets. As an example of the former the May cascade impactor will be described while the Andersen stacked sieve sampler will provide an example of the latter. Descriptions of many of the impactors mentioned above will be found in articles by Rajhans (1978), Green and Lane (1964), Fuchs (1975, 1978), Technical Reports Series No. 179 (1978), May (1967, 1972), Mercer (1973), Hesketh (1977), Dennis (1976), Frielander (1977), Lundgren *et al.* (1979), Liu (1976) and Hinds (1982).

As mentioned in Section 1.5, an impactor basically is a jet, usually tapered, either circular or rectangular, below which is an impaction plate or surface (Figure 1.3). Particle-laden air sucked through the jet is directed at the impaction plate so that particles impact onto it. If designed properly, most particles larger than a given size impact while those smaller remain airborne and follow the streamlines. In a cascade impactor like that of May the aerosol passes through four impaction stages connected in series, with successive stages having increased air velocity through the corresponding jet. Consequently, largest particles impact on the first stage whereas smaller ones impact on the last impaction stage. Smallest particles (submicron) may be too small to be impacted and some samplers (e.g. Ultimate cascade, Andersen) have a filter as the final stage for trapping them. To properly understand the workings of impaction jets, detailed fluid dynamic analyses of them have been made, most recently by Marple and Liu (1974) and Marple *et al.* (1974). Additionally, Marple modelled their streamlines and velocity profiles by replacing air with water and appropriately scaling physical dimensions of the jet. Comparisons were made with jets operating in air and with data generated theoretically by solving the Navier-Stokes equations by a finite difference method. Marple's approach is to express the equations in terms of vorticity and stream functions. Derived differential equations are transformed to a finite difference form and solved by a relaxation method over a grid of node points for the field of interest. Summarizing their conclusions, Marple and Liu (1974) and Marple *et al.* (1974) found the length of the jet throat to have little effect for Reynolds numbers greater than about 500. Below this value, the effect was marked. Jet-to-plate distance above 1.5 jet widths (slit) or 1.0 jet diameters (round) did not markedly affect impactor performance (e.g. cut-off diameter). Cut-off diameter is that diameter at which 50% of particles are sampled. They found also that calculated and experimental streamlines usually agreed well. Examples are given in Figure 3.7 and it may be seen that as Re increased the streamlines move closer to the impaction plate. Corresponding to these

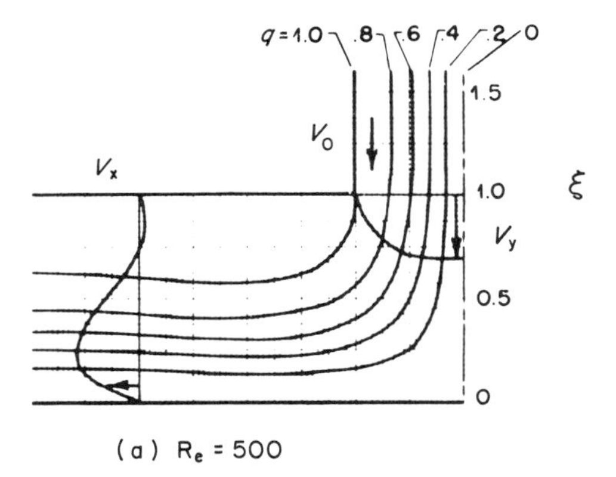

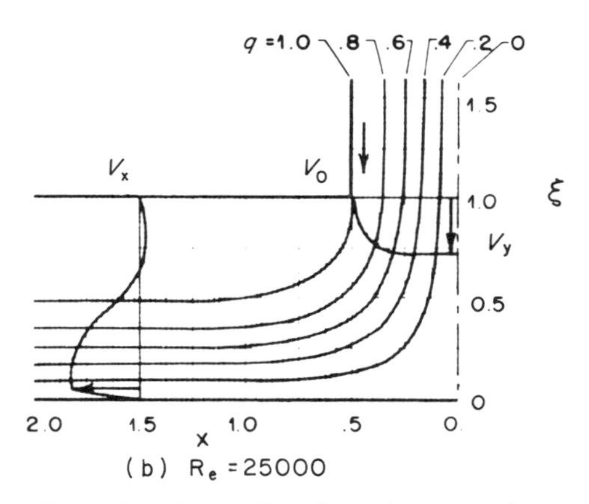

Figure 3.7 Streamlines in an impactor jet as a function of Reynolds number. (From Marple (1970), Ph.D. Thesis, *A fundamental Study of Inertial Impactors*, University of Minnesota.) Reproduced by permission of the author.

streamlines, aerosol particle trajectories can be calculated (Marple *et al.*, 1974). So that comparison with experimental data was possible, these were obtained with a special impactor differing from conventional ones in that aerosols were introduced at a point. By means of a micrometer attachment the entry point could be set at a specific location between the impactor axis and the wall (Figure 3.8). Measurement of the impaction point on the plate gave the end point of a particle trajectory as a function of entry point and other variables (e.g. S/W, F, etc.). To ensure that particles remained at their impaction points the plate was coated with an adhesive layer.

Particles for these studies were 6–14 μm diameter styrene-divinylbenzene spheres aerosolized with a spinning disc device and brought to the Boltzmann

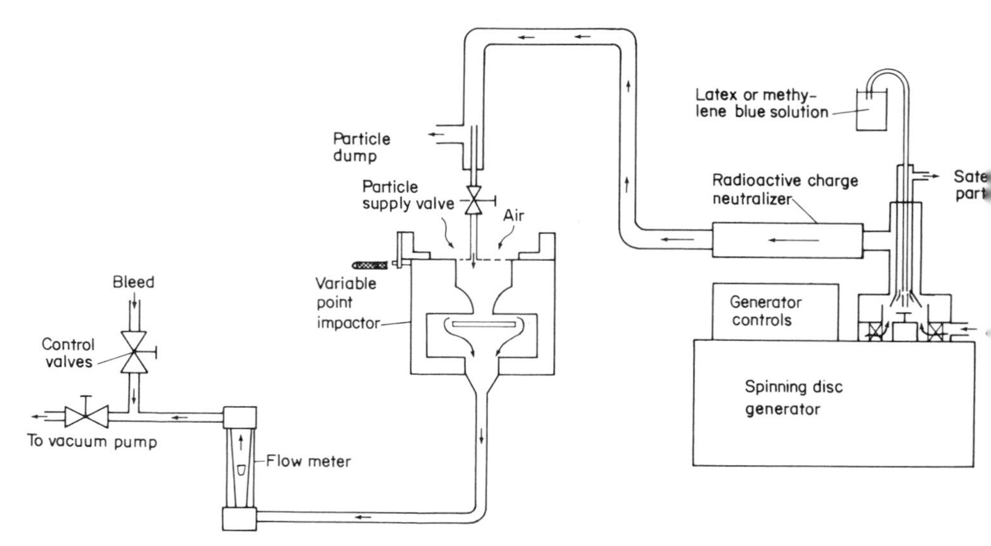

Figure 3.8 Variable point of entry of a particle into an impactor jet. (From Marple (1970), Ph.D. Thesis, *A Fundamental Study of Inertial Impactors*, University of Minnesota.) Reproduced by permission of the author.

charge distribution by passing them through a Kr^{85} charge neutralizer. That is, the radioactive source causes the air to become ionized with approximately equal numbers of positive and negative air ions. Their collisions with aerosol particles form approximately equal numbers of positively and negatively charged aerosol particles. Then, the distribution of charges is symmetrical with the most likely charge per particle being zero, while an average number of charges (positive or negative) $\bar{q}$ per particle of diameter, d, is,

$$\bar{q} = 2.37\ d^{\frac{1}{2}} \quad (3.3)$$

(electron charges) (μm)

and the overall net charge is approximately zero at equilibrium. In this manner effects due to particle charge are minimized but not necessarily eliminated.

By means of these techniques Marple *et al.* obtained good agreement between experimental and calculated deposition patterns. In general, when sampling polydisperse aerosols, larger particles tend to be found at the deposit mid-line and smallest particles at the deposit edge. Gradation of particle size tends to occur between these limits. One cause for the breakdown of this generalization is that particles do not always remain at their impaction points thereby causing the appearance of haloes, for example, in deposits (May, 1975). Such non-uniformity can lead to problems when sizing deposited particles microscopically.

The reason why particles may leave their impaction points basically is due to low particle adhesion to the impaction surface. As a result particle bounce occurs (Dahneke, 1973; Rao and Whitby, 1977, 1978; Esmen *et al.*, 1978; Fuchs, 1978) or particles are blown from their impaction points (Rao and Whitby, 1977, 1978). These effects can arise even when an adhesive is applied to the impaction plate, although the effects are greatest with no adhesive. The best adhesives seem to be high viscosity oils but these can be pushed from the impaction area because of the action of an impactor air jet. One alternative is to use glass fibre filters as the impaction surface but according to Rao and Whitby (1977, 1978), while this practice significantly reduces particle bounce, it also modifies the impactor collection characteristics, e.g. d_{50} value.

Because particle bounce and blow-off reduce collection efficiency and in cascade impactors result in particles being deposited on the 'wrong' stage, they can produce serious artefacts in experimental data. Quite often it may not be immediately obvious that such defects are operating unless samples collected by cascade impactor are examined microscopically. Then, appearance of haloes or large particles in deposits of stages with much smaller d_{50} value is indicative. When it occurs the remedy is either to find a suitable adhesive or other impaction surface or to use prehumidification etc.

According to May (1975) stage d_{50} values may be estimated from the relatively simple empirical equations,

$$d_{50} = 45(\mathrm{W}^3/\mathrm{F})^{\frac{1}{2}} \qquad \text{for circular jets} \tag{3.4}$$

$$d_{50} = 75\mathrm{W}(\mathrm{L}/\mathrm{F})^{\frac{1}{2}} \qquad \text{for slit jets} \tag{3.5}$$

where, W = width (or diameter) (cm)

L = length (cm)

F = flow rate through jet (l/min).

The values of d_{50} are unit density aerodynamic particle cut-off sizes at close to normal atmospheric temperature and pressure with S/W between 0.5 and 3 (S = slit width (cm)). For particles less than about 1 μm diameter, F should be multiplied by the Cunningham slip correction factor, C. While the equations are empirical they are related to solutions for the Stoke's number corresponding to a collection efficiency of 50%, namely,

$$\psi_{0.5} = \frac{8F}{\pi \mathrm{W}^3} \cdot (mB)_{0.5} \qquad \text{for circular jets} \tag{3.6}$$

and,

$$\psi_{0.5} = \frac{2F}{\pi \mathrm{L}\mathrm{W}^2} \cdot (mB)_{0.5} \qquad \text{for slit jets} \tag{3.7}$$

where, $(mB)_{0.5}$ = particle mass × particle mobility, i.e. relaxation time of particles for which collection efficiency is 50%

and, $\psi^{\frac{1}{2}}$ is proportional to particle size.

Analysis methods for data obtained with multistage impactors can appear relatively simple. For example, when using effective cut-off aerodynamic diameter (i.e. d_{50}) for interpretation it is assumed that all particles collected by a given stage, regardless of particle shape or density, have aerodynamic diameters greater than the stage d_{50} value. Having measured the maximum particle size in the deposit (e.g. microscopically) together with the mass of aerosol deposited on each stage, the cumulative percent mass is plotted on log-probability paper as a function of d_{50} value for the four stages. The data points usually lie on a reasonable straight line and a mass median aerodynamic diameter may be read from the graph, i.e. the d_{50} value corresponding to 50% mass. The standard deviation also can be calculated.

Unfortunately, although quite often used, this method tends to give larger than true size because of the assumption mentioned above. In practice, a given stage actually samples some particles having aerodynamic diameters less than the stage d_{50} value because of the S-shaped collection efficiency-particle size curve (Figure 3.9). This means that in a cascade impactor, for example, with its four stages some of the particles depositing on the first stage are of a size more allied with the d_{50} value of the second stage. Therefore, part of the mass deposited on stage 1 ideally should have deposited on stage 2, and so on, for all stages. The error in analysis depends on several factors including the true particle size distribution and the actual shape of the collection efficiency-size curves for each stage as well as the number of these stages.

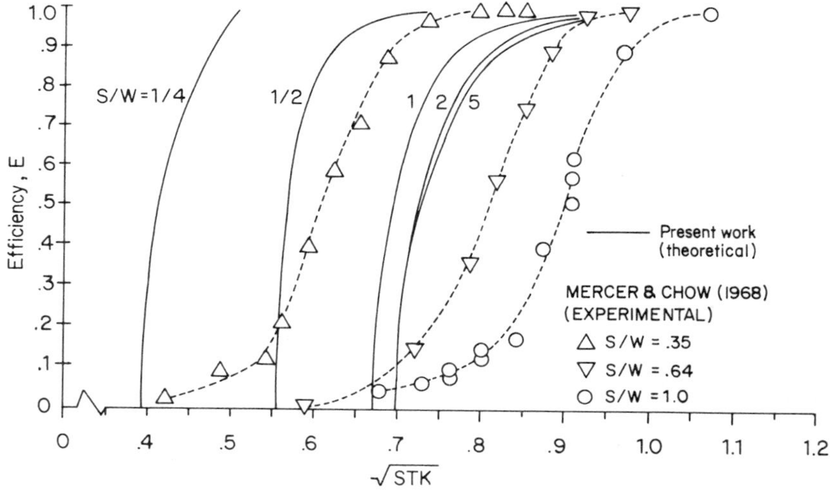

Figure 3.9 Comparison of theoretical and experimental efficiency curves for the rectangular impactor. (From Marple (1970), Ph.D. Thesis, *A Fundamental Study of Inertial Impactors*, University of Minnesota.) Reproduced by permission of the author.

For cascade impactors when sampling aerosols having log-normal size distributions, correction factors are available (Mercer, 1964; Soole, 1971), while Picknett (1972) provides an alternative method of analysis in which a mixture of monodisperse aerosols is used to represent the size distribution of the test aerosol. Fuchs (1978) indicates that the method of Picknett like that of others is not entirely satisfactory and that measuring the sizes of particles microscopically is the only method that can give the true size distribution of polydisperse aerosols sampled with multistage samplers.

3.7 STACKED SIEVE SAMPLERS

In similar vein to the recommendation adopted at the First International Symposium on Aerobiology that the AGI-30 should be the standard liquid impinger, the Andersen (1958) stacked sieve viable sampler was chosen as the standard device for collecting microbial aerosols onto solid surfaces (Brachman *et al.*, 1964). This sampler consists of a series of six stages, each composed of a plate having a large number of similar holes (i.e. circular jets). Each plate is held above a petri dish containing nutrient agar with successive plates having smaller holes. At constant flow (1 ft^3/min or 28.3 l/min) through the sampler the air velocity increases correspondingly with successive stages. Largest particles are deposited on the top stage and smallest on the bottom stage. After sampler operation petri dishes are removed and placed in an incubator and bacterial colonies counted. For viruses, etc., these first need to be recovered by washing the agar surface and then assayed.

While this sampler is appropriate for bacteria results can suffer through a colony arising from more than a single particle, i.e. when two or more bacterial particles are impacted by the same circular jet. For these cases observed colony counts can be corrected to allow for the effect provided there are less colonies than circular jets per stage (i.e. when the sampler is not overloaded). Peto and Powell (1970) provide tables for true versus observed colony counts. Even so, in practice, it is not always easy to arrange for the aerosol concentration and sampling time combination to produce neither too many nor too few colonies.

Other difficulties can arise if plastic petri dishes are used with this sampler as according to Andersen (1958) these can cause electrostatic particle deposition on to their exterior surface and onto the sampler walls. On the other hand, glass and aluminium dishes are satisfactory provided dishes are filled with the correct volume (27 ml) of agar to give the required gap between plate and impactor surface. In the case of the Andersen and Andersen (1962) aerosol monitor a neoprene ring is employed to contain the nutrient agar. According to Leif and Hebert (quoted by Akers and Won, 1969) the presence of neoprene inhibits the growth of *F. tularensis*, *S. lutea* and *Br. suis* but not *B. subtilis*. Findings such as these emphasize again the often subtle effects that can arise and the need to test materials for potential toxicity.

Efficient collection of larger particles on stages 1 and 2 does not occur with the original Andersen sampler. May (1964) added a new stage (stage zero) as well as changing the hole pattern and hole sizes of stages 1 and 2 to alleviate difficulties with the original design. Unless sampling times are short or the RH high, the areas of nutrient agar directly under each hole of a stage can rapidly dry. Fastidious microorganisms consequently may fail to grow on the upper stages but grow on the lower stages because the RH of the air as it passes through the sampler increases. This result occurs through water evaporation from the nutrient agar. Covering the agar with a water evaporation retardant OED can reduce the problem (May, 1973).

The main use of the Andersen sampler is to determine particle size distributions of microbial aerosols. For bacteria, colony counts per stage may be treated analytically as described in the previous section. For viruses, etc., when assayed from petri dish washings, the number of viable/infectious units per stage must be divided by the cube of the d_{50} value for that stage. This procedure converts number of viruses, etc., to relative number of particles per stage as required to compute aerosol particle mas median diameter, for example. Provided the test microorganism is aerostable or losing viability/infectivity slowly no other corrections to the data are required before their analysis. However, under other conditions allowance may need to be made for viable/infectivity decay should it be a function of particle size. For example, if the decay rate were lower in large particles than in small particles, without correction for this effect the derived particle size distribution would be biased to larger particle size. In this regard the use of *B. subtilis* var. *niger* spores as tracer organism circumvents many problems associated with using Andersen biological samplers. But even then the petri dishes need to be level and filled precisely with 27 ml agar to give the correct plate to agar surface distance.

While it may seem from the above account that using an Andersen sampler requires inordinate attention to detail, this is the rule rather than the exception for *all* aerosol sampling and automatic particle sizing devices. To reiterate once more, artificially sampling aerosols without introducing artefacts is probably the most difficult task faced by an aerosol scientist.

3.8 CENTRIFUGAL SAMPLERS

Centrifugal samplers (e.g. Cyclone, Conifuge, Conicycle) impose a circular path on aerosol particles thereby increasing their effective mass compared to that under gravity, e.g.

$$SF = \frac{F_c}{F_g} = \frac{V_e^2}{R} \tag{3.8}$$

where, SF = separation factor
F_c = centrifugal force
F_g = gravitational force
V_e = tangential velocity of the particle
R = radius of curvature of particle trajectory.

In practice, SF can achieve values of 300–5000 and therefore centrifugal samplers are suitable for sampling respirable size aerosols. Cyclone samplers are one of the simplest of aerosol collectors, having no moving parts. Depending on their size, they operate over a range 1 to 400 l/min, usually with a lower size limit of about 2 μm (Rajhams, 1978), although when specially modified will collect submicron particles (Chatigny, 1978). There are three

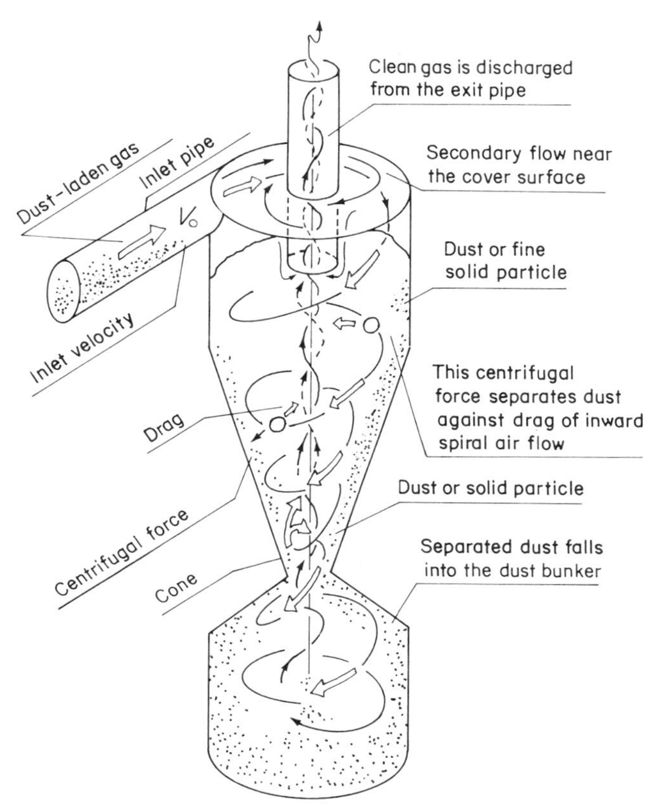

Figure 3.10 Flow pattern in the returned flow tangential inlet cyclone. (From Ogawa (1984), *Separation of Particles from Air and Gases*, C.R.C. Press, Florida.) Reproduced by permission of CRC Press Inc.

main types of cyclone, axial flow, returned flow tangential inlet and returned flow axial flow (Ogawa, 1984). The most common type used for sampling microbial aerosols is the returned flow type with tangential inlet (Figure 3.10). It operates by applying suction at the exhaust pipe when a controlled flow of air enters the inlet, strikes the body of the cyclone and acquires a tangential velocity component. Particles entrained in this air behave similarly and are carried to the cyclone walls. Largest particles tend to be deposited first and at the top of the cyclone.

When sampling microbial aerosols (Errington and Powell, 1969) it is usual to inject into the air inlet a metered flow of suitable fluid to form a thin liquid layer on the inside wall of the cyclone. (A wetting agent usually has to be incorporated). Deposited particles then are carried by the liquid flow to the bottom of the cyclone, while the collected liquid may be recirculated to increase recovered concentrations or pumped to a bottle or fraction collector.

When sampling dusts and other solid particles, the liquid injection may be omitted when the dry powder collects at the bottom of the cyclone. While the sampling efficiency of returned flow tangential inlet cyclones can reach 75% (Ogawa, 1984), losses of particles through the exhaust sometimes need to be avoided. For this reason the axial flow cyclones were designed which basically are uniflow cylinders with radial vanes for imparting the tangential velocity component. Returned flow axial flow cyclones are similar except exhaust air escapes through a central pipe as for a returned flow tangential inlet cyclone.

In order to properly design cyclone samplers account has to be taken of air velocity distributions which in turn depend on physical dimensions and flow rate of the cyclone sampler. Ogawa (1984) provides a thorough account of theories for deriving them together with theories predicting efficiency-particle size relationships.

A comparatively recent development of the cyclone is the rotary flow (or tornado) cyclone (Pieper, 1977) with a sampling efficiency of very nearly 100% for particles as small as 0.7 μm. It basically consists of a vertical cylinder at the bottom of which is an axial air inlet with guide vanes to impart helical air flow. A secondary rotational flow is directed downward to intensify the primary vortex flow, and the two flows exhaust at the top. By the enhanced vorticity, the tangential velocity component of the aerosol particles carried by the primary air flow is enhanced as is particle deposition. Ogawa (1984) provides additional details of this design which gives a collection efficiency of 99.7% for particle sizes down to below 1 μm, a performance superior to that for conventional cyclones.

During their calibration a possible point to look for is that the particle burden can increase air flow by 30% in a manner dependent on particle concentration, size and composition (Littlejohn and Smith, 1978). Even so, Lippmann and Kydoniens (1970) successfully operated six 10 mm nylon cyclones simultaneously in parallel in a single housing. Each cyclone operated at a different flow rate (0.9 to 5 l/min) and with its own backing filter, thereby providing particle size distributions over the particle size range 1 to 10 μm (analogous to a multistage impactor sampler).

An aerosol centrifuge of the Goetz aerosol spectrometer type consists of a vertical aluminium cone, grooved with a well-fitting conical cover. The unit spins about a vertical axis at up to 24 000 rev/min analogously to a bench conical centrifuge. Air is sucked in through an inlet at the top thence through the channels and expelled through a jet orifice at the bottom. This orifice controls the air flow rate.

As particles flow through the aerosol centrifuge their increasing acceleration causes them to be deposited on the channel wall formed by a thin removable foil. When the collecting foil is laid out particles of a given size are found over the length l_d — the cut-off value of an impactor stage.

The conifuge differs in that it consists of two metal cones, one inside the other, and fixed together but with a gap between them. When rotated about a vertical axis air is drawn in at the top, flows through the annular gap between the two cones and exhausts at the base through jet orifices. This relatively clean air is fed back to the inlet except for a relatively small constant flow which is pumped away and replaced by an equal flow of sample air through the inlet at the top. This sample flow spreads as a thin layer over the outer surface of the inner cone. Particles in this air owing to their acceleration move through the thicker layer of clean air towards the inner surface of the outer cone. Particles of a given size deposit as a thin narrow horizontal band, its position being characterized by particle aerodynamic diameter. Typical operating conditions would be 5000 rev/min, total flow 1 l/min, aerosol flow 100 cc/min, size range 0.05 to 4 μm.

There are available several different variations of the basic aerosol centrifuge and conifuge, the biggest advantage of the conifuge being the accurate sizing of submicron aerosols. Their major disadvantages are complexity of design and associated high cost, inlet losses leading to uncertainties about the relevance of observed to actual size distribution, and in providing aerosol samples of a relatively inconvenient form.

3.9 ELECTROSTATIC SAMPLERS

These devices differ from all other samplers in that electrical rather than mechanical forces are purposely applied to separate particles from airstreams. Because they operate with little pressure drop, electrostatic samplers are much more efficient electrically than impingers, impactors, cyclones, etc. Those samplers, due to their large pressure drops, require high-power vacuum pumps whereas electrostatic precipitators operate with low-power air movers (Furtado and Rusch, 1978).

Their principle of operation is that particles to be collected are given one or more charges and then accelerated towards an electrode of opposite polarity. Particle charging can arise, for example, through friction or ionizing radiation. However, the best and most widely applied method is corona discharge from a wire at very high potential. A high electrical field around this wire causes ionization of air molecules and a corona glow or discharge. For highest sampling efficiency unipolar corona charging is necessary and a usual configuration is a fine wire or point mounted coaxially in a larger radius cylinder.

Depending on whether charges on the wire are negative or positive, so negative or positive coronas are generated. Negative coronas appear as a series of localized glowing areas fairly uniformly displaced along the wire and increase in number with increasing voltage and current. Positive coronas in contrast, are a smooth uniform glow but require higher voltages than for negative corona discharges.

As the negative voltage on the central wire is increased from zero no current flows until a minimum value is attained, corresponding to the ionization voltage of air:

$$M \rightarrow M^{\oplus} + 1 \text{ electron}$$

Further increases in voltage cause an electron avalanche together with increased current and corona discharge along the wire. This corona is confined to the immediate vicinity of the wire because it is only there that electrical fields are sufficiently high to cause air ionization. Electrons so formed pass through the corona towards the positive electrode, colliding with neutral molecules on the way,

$$\in + \mathrm{M} \rightarrow \mathrm{M}^{\ominus}$$

On the other hand, the positive ions created originally move to collide with the negative wire thereby generating more electrons by the energy of those collisions. Such secondary electron emissions maintain the negative corona. The radius of the corona, according to Ogawa (1984), can be estimated from the equation:

$$R_c = R_w + (1.15 \times 10^{-5} \times (V - V_c)) \tag{3.9}$$

where, R_c = corona radius
R_w = wire radius
V_c = corona starting voltage
V = applied voltage.

For example, if $R_w = 0.05$ cm, $V_c = 5$ Kv and $V = 90$ Kv, then corona radius = 1.32 cm. For $R_w = 0.6$ cm and $V_c = 25$ Kv then $R = 1.58$ cm. This author also provides equations for calculating corona starting voltages. The way in which particles become charged through corona discharge is by at least two mechanisms: field (or bombardment) charging and diffusion charging. The former depends on collision of air ions moving through the electric field with particles. According to Lippmann (1978a) the maximum number of electron charges (N_M) that a particle can acquire is given by the equation:

$$N_M = \left(1 + 2\,\frac{\varepsilon - 1}{\varepsilon - 2} \cdot \frac{E_0 r^2}{e}\right) \tag{3.10}$$

where, ε = particle dielectric constant
E_0 = electric field (V/cm)
r = particle radius (cm)
e = charge per electron,

while the actual number (N) acquired is:

$$N = N_M\left(\frac{\pi N_0 e K t}{\pi N_0 e K t + 1}\right) \tag{3.11}$$

where, N_0 = ion density in charging region (ions/cm^3)
K = ion mobility (cm^2/V s^{-1})
t = time (s).

Diffusion charging of an initially uncharged particle arising by virtue of Brownian motion, is:

Table 3.2 The number of elementary charges absorbed by a particle during the time t(s) Reproduced by permission of American Conference of Government Industrial Hygienists

Particle diameter (μm)	By ion bombardment t(s)				By ion diffusion t(s)			
	0.01	0.1	1.0	10.0	0.01	0.1	1.0	10.0
0.2	0.7	2.0	2.4	2.5	3.0	7.0	11.0	15.0
2.0	72.0	200.0	244.0	250.0	70.0	110.0	150.0	190.0
20.0	72 000.0	20 000.0	24 000.0	25 000.0	1100.0	1500.0	1900.0	2300.0

$$N = \frac{rkT}{e} \ln\left(1 + \frac{\pi rCN_0e^2t}{kT}\right) \tag{3.12}$$

where, k = Boltzman's constant (erg/°K)
T = temperature (°K)
C = root mean square velocity of ions (cm/s).

As may be seen from the data of Table 3.2 taken from Lippmann (1978a), diffusion charging predominates for particles less than 1 μm diameter while ion bombardment dominates for particles larger than 1 μm diameter. But, for particles 0.1 μm diameter and less, the predicted number of charges per particle approaches unity, whereas in practice this number is much higher. The discrepancy is related to inaccurate assumptions concerning the initial particle charge acquisition mechanism but is not of concern here. However, ozone production in these samplers is of great interest. It is well known that corona discharge produces oxides of nitrogen and ozone at levels of several parts per million and as discussed in Chapters 12 and 13, ozone and its reaction products can be highly toxic for airborne microorganisms. Therefore, when used for collecting samples for viability/infectivity assays, electrostatic samplers may induce losses due to formation of oxidizers. Morris *et al.* (1961) ran into this particular difficulty with their electrostatic precipitator. Even so, several of these precipitators are manufactured specifically for sampling aerosols of microorganisms. They are discussed in detail by Lippmann (1978a), while Ogawa (1984) describes industrial devices.

3.10 THERMAL PRECIPITATORS

Thermal precipitators collect aerosol particles by passing them through a narrow channel possessing a marked temperature gradient perpendicular to the direction of flow. Particles travel down this temperature gradient because of thermophoresis (Section 1.8). Thermophoretic velocity for a spherical particle of larger radius (r) with respect to the mean free path (λ) of air molecules (i.e. Knudsen number = $\lambda/r < 0.05$) according to Epstein (1929) is:

$$\upsilon = \frac{2K_g}{5P(2K_g + K_p)} \cdot \frac{dT}{dx} \tag{3.13}$$

where, K_g = thermal conductivity of the gas

K_p = thermal conductivity of the particle

P = gas pressure

$\frac{dT}{dx}$ = temperature gradient

In practice, prediction is good for low conductivity particles but more rigorous mathematical treatments are required for other cases. Given sufficient temperature gradient (e.g. 200 °C/mm), thermal precipitators collect virtually *all* particles between 0.01 μm and about 5 μm. Conveniently, the sampler may be as simple as a fine nichrome wire between two glass coverslips, with a 0.5 mm gap and an airflow of 6 cm^3/min. But for sampling microbial aerosols, Kethley *et al.* (1952) used a thermal precipitator in which the temperature gradient is maintained by two circular plates. Convection current effects are minimized in their design by heating the upper plate and maintenance of viability/infectivity aided by cooling the lower collection plate. Flow rates of 300 cm^3/min and a gap of 0.38 mm were employed together with filter paper soaked in nutrient as the collection surface. According to Orr *et al.* (1956) *S. marcescens* and *B. subtilis* spores were sampled in a manner comparable to that of an AGI-30 impinger.

Advantages of this sampler include most importantly a very gentle impaction together with a very high collection efficiency for even submicron particles. These when coupled with that of flexibility of the nature of the collection surface (which may be changed according to the type of sample analysis) makes thermal precipitation an attractive sampling procedure. However, in some applications it may be disadvantageous due to relatively low sampling rate (6–1000 cm^3/min) and possible thermal degradation of the aerosol being sampled. Reported applications of this sampler to the collection of microbial aerosols seem to be very limited, especially when it is used in conjunction with a rehumidification chamber (q.v.) which would prevent drying out of collection surfaces during extended sampling times.

3.11 FILTERS

In the general area of aerosol sampling filtration is the most widely applied technique. Reasons include low cost and simplicity as all that is required is a filter mounted in an appropriate holder, flow rate controller (e.g. critical orifice) and a source of suction (e.g. pump, vacuum bottle). Choice of filter and pump provides control of particle sizes collected together with the density of the deposited sample. Many different types of filters are available and a good summary of those suitable for collecting aerosols will be found in the article by

Lippmann (1978b). Basic types are cellulose fibre, glass fibre, mixed fibre and plastic fibre filters and membrane filters. Fibre filters are characterized by having relatively large spaces between individual fibres within the filter bed whereas membrane filters consist of membranes having relatively uniform small pores. The former have low air resistance while for the latter it is high. Another difference is that fibre filters trap particles throughout the filter bed whereas they are trapped on the incident face of membrane filters. For some applications (e.g. microscopic particle size analyses) this attribute of membrane filters can be advantageous.

Very closely related to membrane filters, which are available with numerous chemical compositions, are nucleopore filters. These, rather than being formed from gels like membrane filters, are made by irradiating thin polycarbonate film in a nuclear reactor. Radiation passing through the polycarbonate thereby forms minute holes which are enlarged by controlled chemical etching. In this way nucleopore filters possess very uniformly sized and simply structured pores.

In contrast to air monitoring etc. studies where filtration has proved satisfactory for sampling, those concerned with microbial survival/infectivity have indicated that it is a most unsatisfactory technique. As mentioned by Anderson and Cox (1967) and Strange and Cox (1976) and discussed by Akers and Won (1969), difficulties arise due to non-consistant recovery of microorganisms from the filters, effects of air-flow on viability and release of growth inhibitors from filter materials. In addition, at least for some bacteria, observed viability on recovery depends also on the number of cells on the filter (Webb, 1965). (Although not known at the time of these experiments, this effect is likely to have been due to switching on or off of repair mechanisms—see Chapter 14).

Except in circumstances where very hardy microbes are to be collected, sampling by filtration should be avoided otherwise experimental artefacts are likely to be introduced into studies concerned with the survival or infectivity of airborne microorganisms.

3.12 SEQUENTIAL AND TAPE SAMPLERS

The Casella 'Airborne Bacterial Sampler' Mk II and the New Brunswick 'Microbiological Air Sampler' are sequential samplers. They operate, basically, by having a nutrient agar filled petri dish mounted on a turntable which turns slowly under an impaction slit. Air at a known rate is drawn through this slit. Exposed dishes following incubation provide colony counts which may be correlated with sampling rate and time. The 'Moving Slide Impactor' of Meteorology Research Inc. is somewhat analogous but has two identical slits sampling onto a moving slide. The two simultaneous samples allow for two different sample treatments. Depending on their respective controllers, these three samplers may be operated continuously for up to 5 min,

60 min and 8.5 h, respectively, or as programmed, e.g. to take a 1 min sample every 30 min.

For continuously sampling into liquids, cyclone samplers are discussed in Section 3.8 while the continuous-flow liquid impinger of Goldberg and Watkins (1965) is an alternative. Sequential samples are obtained by connecting their liquid output to an automatic fraction collector, for example.

Morrow Brown (1973) designed and constructed a sequential sampler with a culture plate 2 ft (0.6 m) in diameter and a spiral path (for impacted samples) of nearly 7 ft (2.1 m) in length. In this manner, 20 l of air were sampled every 30 min for 24 h. An alternative is to have a number of samplers coupled to a controller which automatically switches them on and off sequentially according to a programme. When the viability of microorganisms collected by these techniques is of concern, a protocol needs to be developed for handling the samples. For example, when the Casella sampler is stepped such that samples are taken periodically over a period of hours, how can one ensure that microorganisms sampled in the first hour or so are still viable several hours later when the petri dish is removed and incubated?

When preservation of viability of microorganisms in sequential samples is not of primary concern, then the four stage Lundgren Impactor with its slowly rotating collector drums may be appropriate. It allows continuous sampling up to 24 h. Also applicable are continuous samplers in which tape from a feed spool passes under a jet and thence to a motor-driven take-up spool. For convenience and to protect collected samples, cassettes of adhesed or filter tapes are available. Perry (1978) provides a useful summary of commercially available tape samplers. Such devices, if required, can be used in conjunction with some of the aerosol monitoring techniques described in Chapter 4.

3.13 HIGH VOLUME SAMPLERS

As found by Winkler (1968) and by Thomas (1970a, b), and discussed by May (1973), successful collection of natural microbial aerosols, because of their low concentrations, requires collection from large volumes of air. Thomas's approach was to operate a laboratory sampler for extended periods whereas others have used high volume samplers. May (1973) describes several of these including his 'Pagoda' sampler while the high volume cyclone sampler is described in Section 3.8. Fontagnes and Isoard (1973) describe one cyclone set-up in which 20 000 litres of air are treated per minute.

High volume samplers as a class, in the main, are of three types, electrostatic, cyclone (or scrubbers) and multiple jet. The Litton large-volume sampler (LVS) is an example of the first type. It operates at flow rates up to 10 000 l of air/min. Particulate matter in a corona discharge is deposited onto a grounded disc rotated at 200 or 300 rev/min and covered by a thin flowing film of metered collecting fluid. Even though used successfully by Winkler (1968), Gochenour (1966) in his critical appraisal of the LVS indicates that it is only a qualitative sampler. The Andersen Hi-volume sampler classifies particles into four size fractions in an analogous manner to the Andersen stacked sieve sampler, but operating at about 500 l/min. In contrast, the BGI 30 High Volume Cascade Impactor has four stages each with a large rectangular slot while Environmental Research Corporation supply a multiple slit liquid impaction high volume sampler. Sierra Instruments Inc. provide a multiple slit multistage cascade impactor operating at up to about 2000 l/min. As may be seen from these examples, high volume samplers rely on the same collection principles as their low volume laboratory counterparts.

Whether it is more appropriate to use a high volume device operating for a relatively short period or a low volume sampler for a relatively long period, will depend on application. In either case, when sampling directly onto agar loss of water needs to be prevented. This may be achieved with rehumidification prior to sampling or by an evaporation retardant such as OED applied to the agar surface (May, 1969, 1973).

3.14 CALIBRATION METHODS

Aerosol samplers need to be calibrated in terms of flow rate and of collection efficiency as a function of particle size and shape. Flow rate conveniently can be fixed with a critical orifice, i.e. an orifice through which the air speed is sonic and therefore invariant. Even then, though, critical orifices should be calibrated for flow rate. Flow rate often is measured with a flowmeter, either flow rate or flow velocity. An example of the former is a rotameter and of the latter a thermoanemometer. They are provided with a manufacturer's calibration but should still be checked against a spirometer or gasometer, for example. This meter comprises a cylinder closed at one end with the open end under liquid. Its weight is counterbalanced so that there is minimum resistance to movement as air passes in or out of it. Volume changes are calculated from cylinder height differences multiplied by cylinder cross-sectional area. Spirometers are available in a range of sizes and are primary volume standards. One primary flow standard is the nearly frictionless piston formed with a soap bubble in a burette, for example. The volume of air displaced per unit of time (i.e. the flow rate) is measured in terms of the time taken for a bubble to sweep a known volume.

Perhaps more convenient is the Wet Test Meter in which a drum with partitions is half-submerged in a liquid (often water) with openings at the centre and periphery of each radial chamber. Air entering at the centre flows into a submerged chamber causing it to rise and rotate the drum. Extent of rotation is recorded on a dial. For a fixed liquid level, the meter when horizontal is accurate to about 1% or better. An alternative is a domestic gas meter.

A rotameter is a vertically mounted slightly tapered tube in which sits a float. The tube is wider at the top than the bottom so air entering at the bottom lifts the float until the pressure drop across the annulus between it and the tube wall supports the float. The higher the flow the higher the float rises with a corresponding increase in annular area. The tube usually is glass etched with a scale which for most accurate work should be calibrated against a primary standard. Rotameters are available to cover flows from a few cm^3/min to a few thousand litres of air/min.

For measuring air velocity the Pitot tube often is used as a reference, as it does not necessarily require calibration. It comprises a narrow open tube which is inserted into the axis of flow. This impact tube is surrounded by a concentric static pressure tube having eight holes equally spaced in a plane eight diameters from the impact tube opening. The difference in the two pressures is the velocity pressure (P) which is related to the linear velocity, V, by the equation due to Bernouilli,

$$V = \sqrt{2gP} \tag{3.14}$$

where, g = gravitational constant

or,

$$V = 1097 \sqrt{\frac{h}{\rho}} \tag{3.15}$$

when, h = velocity pressure

ρ = density of air

Hot wire anemometers also are employed to measure air velocity as can laser velocimeters relying on measuring the Doppler shift of laser light scattered by either naturally occurring or purposely generated aerosol particles.

Having determined its flow rate by means of one of the above approaches, the sampler collection efficiency needs to be established. One approach is to compare the test sampler against one of known performance. However, the best method probably is to generate monodisperse aerosols (of known size, or size determined microscopically, for example) into a chamber. These aerosols then are collected isokinetically by the test sampler backed by an absolute sampler, such as a filter. Sampler efficiency as a function of particle size is calculated from the ratio of the particles retained by the sampler to this value plus that for the absolute sampler. Washings from the walls of the test sampler provide values for wall losses, while washings from the walls of tubing connecting the two samplers together allow for particles otherwise lost.

While such procedures may seem fairly straightforward, in practice, great care is required to achieve meaningful data. Ideally, the whereabouts of all particles involved in the test should be ascertained. A frequent difficulty is that of providing an absolute sampler possessing a flow rate the same as that for the test sampler. For example, membrane filters, unless of very large surface area, are limited to flow rates of a few litres per minute. A detailed account of procedures, including aerosol generation, will be found in the excellent article by Drew and Lippmann (1978).

3.15 CONCLUSIONS

Aerosol sampling is fundamental to experimental aerobiology and yet to perform it correctly is most difficult. The large number of different samplers and the profusion of papers on the subject bears testament. Under laboratory conditions isokinetic collection is essential if representative samples of aerosols are to be collected. In the field, isokinetic or stagnation point collection likewise is highly desirable but difficult to achieve owing to variable wind speed and direction. For microbial aerosols, difficulties associated with expedient collection are perplexed through the need to preserve biological as well as physical integrity. As will be evident in later chapters, in certain instances artificial sampling rather than the airborne state is a major cause of death of airborne microorganisms.

REFERENCES

Akers, A. B. and Won, W. D. (1969). In *An Introduction to Experimental Aerobiology*, R. L. Dimmick and Ann. B. Akers (Eds), Wiley-Interscience, New York, London, pp. 59–99.

Andersen, A. A. (1958). *J. Bact.*, **76**, 471–484.

Andersen, A. A. and Andersen, M. R. (1962). *Appl. Microbiol.*, **10**, 181–184.

Anderson, J. D. and Cox, C. S. (1967). *Symp. Soc. Gen. Microbiol.*, **17**, 203–226.

Brachman, P. S. and 10 others (1964). *Science*, **144**, 1295.

Brink, J. A. (1958). *Ind. Eng. Chem.*, **50**, 645.

Carson, G. A. and Paulus, H. J. (1974). *Amer. Ind. Hyg. Assoc.* **35**(5), 262–268.

Chatigny, M. A. (1978). In *Air Sampling Instruments* 5th edn, Amer. Conference of Gov. Ind. Hygienists, Cincinnati, Ohio, pp. E-1 to E-10.

Dahneke, B. (1973). *J. Coll. and Interface Sci.*, **45**, 584–590.

Davies, C. N. (1964). *Proc. R. Soc.*, **279**A, 413–419.

Davies, C. N. (1968a). *Brit. J. Appl. Phys.*, **21**, 921–932.

Davies, C. N. (1968b). *Recent Advances in Aerosol Research*, Macmillan, New York.

Davies, C. N. and Aylward, M. (1951). *Proc. Phys. Soc. (Lond.) B*, **64**, 889–911.

Davies, C. N., Aylward, M. and Leacey, D. (1951). *Arch. Ind. Hyg. Occ. Med.* **4**, 354.

Dennis, R. (1976). *Handbook on Aerosols*, U.S. Energy Research and Development Administration, Oak Ridge, Tennessee.

Drew, R. T. and Lippmann, M. (1978). In *Air Sampling Instruments*, 5th edn., American Conference of Gov. Ind. Hygienists, Cincinnati, Ohio, pp. I-1 to I-38.

Epstein, P. (1929). *Z. Phys.*, **54**, 537.

Errington, F. and Powell, E. O. (1969). *J. Hyg.*, **67**, 387–399.

Esmen, N. A., Ziegler, P. and Whitfield, R. (1978). *J. Aerosol. Sci.*, **9**, 547–556.

Fontanges, R. and Isoard, P. (1973). In *Fourth International Symposium on Aerobiology*, J. F. Ph. Hers and K. C. Winkler (Eds), Oosthoek, Utrecht, Netherlands, pp. 62–66.

Frielander, S. K. (1977). *Smoke, Dust and Haze*. Wiley, New York.

Fuchs, N. A. (1964). *The Mechanics of Aerosols*, Pergamon, Oxford.

Fuchs, N. A. (1975). *Atmos. Environ.*, **9**, 692–707.

Fuchs, N. A. (1978). In *Fundamentals of Aerosol Science*, D. T. Shaw (Ed.), Wiley, New York, Chichester, pp. 1–81.

Furtado, V. C. and Rusch, G. (1978). In *Air Sampling Instruments*, 5th edn. Amer. Conf. of Gov. Ind. Hygienists. pp. K-1 to K-52.

Gochenour, W. S., Jr. (1966). *Bact. Rev.*, **30**, 584–586.

Goldberg, L. J. and Watkins, H. M. S. (1965). In *First International Symposium on Aerobiology*, R. L. Dimmick (Ed.), Naval Biological Laboratory, Naval Supply Center, Oakland, California, pp. 211–216.

Green, H. L. and Lane, W. R. (1964). *Particulate Clouds, Dusts, Smokes and Mists*, 2nd edn, E. and F. Spon, London.

Greenberg, L. and Smith, G. W. (1922). A new instrument for sampling aerial dust. *U.S. Bureau of Mines*, RI 2392.
Gurney, S. W., Williams, C. R. and Meigs, R. R. (1938). *J. Ind. Hyg. Toxicol.*, **20**, 24.
Hesketh, H. E. (1977). *Fine Particles in Gaseous Media*, Ann Arbor Science, Ann Arbor.
Hinds, W. C. (1982). *Aerosol Technology*, Wiley, New York.
International Atomic Energy Agency. (1978). *Particle size analysis in estimating the significance of airborne contamination. Technical Report Series No. 179*, Vienna.
Kethley, T. W., Gordon, M. T. and Orr, C., Jr. (1952). *Science*, **116**, 368–369.
Laskin, S. (1949). In *Pharmacology and Toxicology of Uranium Compounds*, C. Voegtlin and H. C. Hodge (Eds), McGraw-Hill, New York, pp. 463–505.
Lippmann, M. (1959). *Amer. Ind. Hyg. Assoc. J.*, **20**, 406.
Lippmann, M. (1961). *Amer. Ind. Hyg. Assoc. J.*, **22**, 348.
Lippmann, M. (1978a). In *Air Sampling Instruments*, Amer. Conf. of Gov. Ind. Hygienists, Cincinnati, Ohio, pp. P-1 to P-20.
Lippmann, M. (1978b). In *Air Sampling Instruments*, Amer. Conf. of Gov. Ind. Hygienists, Cincinnati, Ohio, pp. N-1 to N-22.
Lippmann, M. and Kydoniens, A. (1970). *Amer. Ind. Hyg. Assoc. J.*, **31**, 730.
Littlefield, J. G. and Shrink, H. H. (1937). Bureau of Mines midget impinger for dust sampling. *U.S. Bureau of Mines*, RI 3360.
Littlejohn, R. F. and Smith, R. (1978). *Proc. Instit. Mech. Engrs.*, **192**, 243–250.
Liu, B. Y. H. (1976). *Fine Particles*, Academic Press, New York.
Lundgren, D. A. (1967). *J. Air. Pollut. Contr. Assoc.*, **17**, 4.
Lundgren, D. A., Harris, F. S., Jr., Marlow, W. H., Lippmann, M., Clarke, W. E. and Durham, M. D. (1979). *Aerosol Measurements*, University Press of Florida, Gainsville.
Marple, V. A. and Liu, B. Y. H. (1974). *Environ. Sci. and Technol.*, **8**, 648–654.
Marple, V. A., Liu, B. H. and Whitby, K. T. (1974). *Trans. A.S.M.E.*, **96**, 394–400.
May, K. R. (1945). *J. Sci. Instru.*, **22**, 187–195.
May, K. R. (1960). *Ann. Occup. Hyg.*, **2**, 93–106.
May, K. R. (1964). *Appl. Microbiol.*, **12**, 37–43.
May, K. R. (1966). *Bact. Rev.*, **30**, 559–570.
May, K. R. (1967). *Symp. Soc. Gen. Microbiol.*, **17**, 60–80.
May, K. R. (1969). *Appl. Microbiol.*, **18**, 513–514.
May, K. R. (1972). In *Assessment of Airborne Particles. Fundamentals, Applications and Implications to Inhalation Therapy*, T. T. Mercer, P. E. Morrow and W. Stöber (Eds), Thomas, Springfield, Illinois, pp. 480–494.
May, K. R. (1973). In *Fourth International Symposium Aerobiology*, J. F. Ph. Hers and K. C. Winkler (Eds), Oosthoek, Utrecht, Netherlands, pp. 27–32.
May, K. R. (1975). *J. Aerosol. Sci.*, **6**, 413–419.
May, K. R. and Druett, H. A. (1953). *Brit. J. Ind. Med.*, **10**, 142–151.
May, K. R. and Harper, G. J. (1957). *Brit. J. Ind. Med.*, **14**, 287–297.
Mercer, T. T. (1964). *Ann. Occup. Hyg.*, **7**, 115.
Mercer, T. T. (1973). *Aerosol Technology in Hazard Evolution*, Academic Press, New York.
Mercer, T. T. and Stafford, R. G. (1969). *Ann. Occup. Hyg.*, **12**, 1.
Mercer, T. T., Tillery, M. L. and Ballew, C. W. (1962). A cascade impactor operating at low volumetric flow rates. *Lovelace Foundation Report*, LF-5, December.
Mitchell, R. I. and Pilcher, J. M. (1958). In *Proceedings of Fifth Atomic Energy Commission Air Cleaning Conference*, TIO-7551, Office of Technical Services, Dept. of Commerce, Washington, pp. 67–84.
Morris, E. J., Darlow, H. M., Peel, J. F. H. and Wright, W. C. (1961). *J. Hyg.* **59**, 487–496.

Morrow Brown, H. (1973). In *Fourth International Symposium on Aerobiology* J. F. Ph. Hers and W. C. Winkler (Eds), Oosthoek, Utrecht, Netherlands, pp. 57–58.
Ogawa, A. (1984). *Separation of Particles from Air and Gases*, vols I and II, CRC Press, Florida.
Orr, C., Jr., Gordon, M. T. and Kordecki, M. (1956). *Appl. Microbiol.*, **4**, 116–118.
Parker, G. W. and Buchholz, H. (1968). Size classification of sub-micron particles by a low pressure cascade impactor. ORNL-4226.
Perry, W. H. (1978). In *Air Sampling Instruments*, 5th edn, Amer. Conf. of Gov. Ind. Hygienists, Cincinnati, Ohio, pp. M1–M32.
Peterson, C. M. (1978). In *Air Sampling Instruments*, 5th edn, Amer. Conf. of Gov. Ind. Hygienists, Cincinnati, Ohio, pp. F-1 to F-11.
Peto, S. and Powell, E. O. (1970). *J. appl. Bact.*, **33**, 582–598.
Picknett, R. G. (1972). *J. Aerosol. Sci.*, **3**, 185–198.
Pieper, R. (1977). The Tornado dust collector. Fourteen years practical experience. VDI-Berichte (VDI-Rep.), **294**, 93.
Rajhans, G. S. (1978). In *Air Sampling Instruments*, 5th edn. Publ. by Amer. Conf. of Gov. Ind. Hygienists, pp. O-1 to O-51.
Ranz, W. E. and Wong, J. B. (1952). *Arch. Ind. Hyg. Occ. Med.*, **5**, 462.
Rao, A. K. and Whitby, K. T. (1977). *Amer. Ind. Hyg. Assoc. J.*, **38**, 174–179.
Rao, A. K. and Whitby, K. T. (1978). *J. Aerosol. Sci.*, **9**, 547–556.
Shipe, E. L., Tyler, M. E. and Chapman, D. N. (1959). *Appl. Microbiol.*, **7**, 349–354.
Silverman, L. and Franklin, W. (1942). *J. Ind. Hyg. and Toxicol.*, **24**, 80.
Soole, B. W. (1971). *J. Aerosol Sci.*, **2**, 1–14.
Stern, S. C., Zeller, H. W. and Schekman, A. I. (1962). *I and EC Fundamentals*, **1**, 273–277.
Strange, R. E. and Cox, C. S. (1976). *Symp. Soc. Gen. Microbiol.*, **26**, 111–154.
Suggs, H. J. (1978). In *Air Sampling Instruments*, 5th edn, Amer. Conf. of Gov. Ind. Hygienists, Cincinnati, Ohio, pp. L-1 to L-21.
Thomas, G. (1970a). In *Third International Symposium on Aerobiology*, I. H. Silver (Ed.), Academic Press, New York, London, p. 266.
Thomas, G. (1970b). *J. Hyg. (Camb.).*, **68**, 511–517.
Tyler, M. E. and Shipe, E. L. (1959). *Appl. Microbiol.*, **7**, 337–348.
Tyler, M. E., Shipe, E. L. and Pamter, R. B. (1959). *Appl. Microbiol.*, **7**, 355–362.
Vitols, V. (1966), *J. Air. Pollut. Control. Assoc.*, **16**, 79.
Watson, H. H. (1954). *Amer. Ind. Hyg. Assoc. Quat.*, **15**, 1.
Webb, S. J. (1965). *Bound Water in Biological Integrity*, Thomas, Springfield, Illinois.
Wilcox, J. D. (1963). *A.M.A. Arch. Ind. Hyg. Occup. Med.*, **7**, 376.
Winkler, W. G. (1968). *Bull. Wildlife Disease Assoc.*, **4**, 37–40.
Wolf, H. W., Skaliy, P., Hall, L. B., Harris, M. M., Decker, H. M., Buchanan, L. M. and Dahlgren, C. M. (1959). *Sampling Microbiological Aerosols. Public Health Monograph No. 60*. (Publ. No. 686), U.S. Govt. Printing Office, Washington.

Chapter 4

Aerosol monitoring methods

4.1 INTRODUCTION

The need to monitor aerosols occurs in virtually all aspects of work concerned with the airborne state whether it is to measure particle size distribution, particle concentration, diffusion rates of aerosols through buildings, efficiency of safety systems, or environmental pollution, etc. Such diversity is matched by a corresponding one for aerosol monitoring methods. This chapter deals with those concerned with quantitative analyses whereas methods applicable to the determination of aerosol particle size are covered in the following chapter.

4.2 PHYSICAL METHODS

Physical methods are applicable in particular to quantitative analyses of aerosols in, for example, testing of efficiency of safety systems (such as those described in Chapter 6) or measuring particle diffusion rates through, into or out of buildings (Chapter 6). As discussed further in Chapter 6, while interest may be in how microbial aerosols behave, it is not always possible to use them for experimental evaluation. Instead, inorganic or organic materials aerodynamically behaving in the same way as their microbial counterparts are substituted. These materials, like those described in Chapter 1, are tracers (Harper, 1980).

The basic approach is to generate a tracer aerosol inside a room, for example, and collect aerosol samples at other points within and outside the building. Knowledge of the amount of aerosol generated and of that collected provides estimates of aerosol hazards within and outside buildings. Likewise, knowing amounts inside and outside a safety cabinet permits its efficiency to be evaluated. Tracer aerosols are collected with samplers or fibre or membrane filters (Chapter 3) and then assayed.

The sampling method employed, depends on application, assay method, its sensitivity and whether natural aerosols are likely to interfere in the assay. As an example, for mass analysis filters are convenient and sampling time may be adjusted commensurate with required sample mass. When an aerosol size distribution is required, stacked sieve, cascade impactor or three-stage glass impinger samplers are suitable. Should aerosol concentrations be low high sample volumes will be required, while if natural aerosol particulates interfere with the assay, sampling into liquid followed by centrifugation or chromatography, for example, may be necessary. But, given the variety of potential tracers and assay methods, a combination little influenced by contaminating natural aerosol particles is likely.

Physical methods appropriate for quantitative chemical analyses of collected samples include X-ray fluorescence, emission and atomic absorption spectroscopy, laser spectroscopy, radioactivity, mass spectroscopy, neutron activation analysis, X-ray diffraction, electron and ion microprobe, Raman microprobe, IR spectroscopy, UV fluorescence, chemiluminescence, thermal analysis, electron microscopy and light microscopy. Discussion of some of these will be found in *Analysis of Airborne Particles by Physical Methods* edited by H. Malissa (1978) and in *Environmental Pollution Analysis* by P. D. Goulden (1978). Here, an outline of each will be given so that the reader may become aware of some of their relative advantages and disadvantages.

X-ray fluorescence is an emission spectroscopic method appropriate for relatively large samples or even single particles when used in conjunction with an electron microscope. The principle of the method is that constituent atoms have their electrons excited out of their ground state by an energy source such as X-rays, electrons, or accelerated particle beams, e.g. protons. Their energy content must be sufficient to promote electrons from an inner orbital (usually K or L shell) to one of higher energy. The fall of electrons from this outer shell back to the ground state is accompanied by emissions of X-rays. Wavelengths or energies of these emission lines are characteristic of each element, while their intensity is proportional to amounts of elements being irradiated. By detecting wavelengths and corresponding intensities of emitted X-rays using wavelength-dispersive or energy-dispersive spectrometers, elemental compositions of samples are determined quantitatively. For samples collected onto filters detection limits range from hundreds of ng/cm^2 for elements such as sodium to a few ng/cm^2 for transition elements.

Advantages include ease of sample acquisition, little or no sample preparation, simplicity and relative speed of analysis. Disadvantages range from confusion between lines from tracer and filter and background materials to particle size and sample thickness effects due to limited penetrations by X-rays. Nonetheless, overall relative accuracy should be within $\pm 10\%$ while use of dedicated computers and time-averaging techniques can improve accuracy.

Laser, emission and atomic absorption spectroscopy (Corney, 1977; Stenholm, 1978) also depend on changes of electron energy levels. In emission methods, substances can be excited by their evaporation from a graphite electrode or by passing aerosols of them into a flame. Excited electrons remain in that state for about 10^{-8} s, then return to the ground state with photon emission. Photon energy or wavelength or frequency is characteristic of each element, while the intensity is related to elemental quantity. In absorption methods, light from sources having very intense lines of small half-width (e.g.

hollow cathode lamps) is passed through the vaporized material and extent of light absorption measured. Each lamp is specific for one element, therefore unlike emission methods only one element at a time can be measured. Except for noble gases, halogens, oxygen, nitrogen and hydrogen, each of the 92 natural elements can be excited by electric arc. But, whichever excitation method is employed, in practice, confusion can arise between overlap of lines from different elements, although use of the strongest and most characteristic of them reduces the problem. For direct analyses of aerosols, μg quantities of them are required so it is not a particularly sensitive method, while its advantages and disadvantages are similar to those for X-ray fluorescence.

Radioactive tracers are neither particularly suitable (because of problems caused by radiation) nor are they particularly sensitive. If used, sampling by filtration is convenient while assay is by Geiger-Müller tube systems, scintillation counters, solid state detection or auto-radiograph techniques.

Mass spectroscopy identifies and quantifies atomic species on the basis of their mass which represents a fundamental property. For elemental analyses the method covers virtually all atoms of the Periodic Table. It provides high sensitivity, a very low detection limit of about 1 to 10 atoms in a total of 10^9 atoms, and an accuracy comparable to that of $\pm 10\%$ as for X-ray fluorescence. In operation, materials for analysis are converted to ionic clouds by sparks, lasers, etc., and accelerated and collimated into suitably shaped ion beams. Such a beam passes through an electrostatic and magnetic analyser at the exit of which the ions of the resolved beam strike an ion-sensitive detector. Associated computer and software turn detected signals into elemental or rather ion masses and amounts. Impactor samples are suitable for analysis by this means which is rapid for single particles but time-consuming for many. Its strength lies in being able to differentiate fine differences between materials and therefore is highly appropriate where problems arise from confusion by background particles. A potential disadvantage in some applications, as for X-ray fluorescence, is that analysed materials are exposed to vacuum.

Neutron activation analysis is a useful, highly accurate and sensitive technique applicable for analysis of ng quantities. The technique quantitatively measures radioactive species produced in the sample by neutron-induced changes; it can be non-destructive and automatic. The γ-emitting radionuclides formed on neutron activation are detected by solid-state Ge(Li) diode sensors. But the activation process can generate many radionuclides in addition to those from the tracer because of background particles, etc., in samples. For quantitative analyses these must be distinguished which may be achieved conveniently through differences in γ-ray energy and utilization of high resolution Ge(Li) spectrometers. Disintegration rate is calculated from the area under each γ-ray peak and elemental concentration found by comparison with activities of known elemental standards.

The actual neutron source can be problematical as suitable ones are nuclear reactors, particle accelerators and isotopes. Each source is characterized by energy and intensity of emitted neutrons and those of highest neutron flux lead to highest sensitivity and shortest assay times. Depending upon the element being estimated, the total procedure can demand several days. Its greatest strength, though, is that ng quantities of tracer sampled by impactor or filter, for example, can be measured reliably and accurately.

Light microscopy offers a relatively cheap and simple assay method having high sensitivity through measurements on single particles. McCrone and Delly (1973) describe the approach in detail and show how sophisticated it can become. The more important properties for identification of particles greater than about 1–2 μm diameter, i.e. a mass of 10^{-11} to 10^{-10} g are transparency, opacity, colour, refractive index, birefringence, size, shape and particle morphology. A tracer particle of particular anisotropy when examined under a polarizing microscope having crossed polars will display a colour characteristic of it and easily recognized by a skilled microscopist. Dispersion staining is another analytical technique and describes colour effects produced when a transparent object is immersed in a liquid having a refractive index a little different to that of the particle and viewed by transmitted white light. The colour pattern at the particle edges is characteristic of many isotropic and anisotropic compounds. For this approach, samples collected onto glass slides of a cascade impactor are highly suitable, while if the microscope is equipped with a heated stage then particle melting point can identify tracer particles. Potentially, a major disadvantage with light microscopy is the difficulty of quantification of identified tracer particles. One fairly simple solution is to utilize monodisperse tracer aerosols when particle counts readily can be converted to particle mass. Otherwise, a specialized microscope such as a Vickers M81 scanning microinterferometer may be necessary. Even then the complete procedure can become laborious when many samples are involved, unless automated image analysis techniques (q.v.) are invoked.

The use of electron microscopes, either transmission or scanning, when coupled with energy-dispersive X-ray spectrometry is mentioned above in connection with X-ray fluorescence. It is a powerful but time-consuming analytical method combining particle size and morphology with elemental analysis. A variation is the X-ray valence band spectrum which originates from electron transitions between the valence band and an inner orbital. These spectra contain information relating to elemental analysis as well as the chemical bonding of that element. X-ray valence band spectra, therefore, can provide unique identification of tracer particles (i.e. compound specific) together with tracer mass.

Electron diffraction patterns from individual particles produced in TEM and STEM electron microscopes can be analysed to provide parameters relating to the crystal lattice. Often these patterns can be used as a 'fingerprint' for a specific substance and aid in the discrimination between tracer and background particles, especially when particles are submicron in size. Electron energy-loss spectra also are appropriate in these cases. They occur because electrons transmitted through particles undergo loss of discrete packets of energy owing to ionizations of constituent elements and to excitation of oscillation of valence band electrons. Such energy spectra of the transmitted electrons demonstrate discrete peaks typical of elements excited. Yet, this method does not seem to be widely applied to chemical analyses of aerosol particles. Analogous to electron diffraction is X-ray diffraction. It has been applied extensively to analysis of crystalline particulates and many spectra are documented. X-ray diffractometers are favoured instruments for quantitative work due to their high precision, convenience and automation, although standards are essential. Sample preparation is minimal following collection by impactor or fibre or membrane filter, but a detection limit of about 5 μg/cm^2 may impose limitations of application.

X-ray photoelectron spectroscopy is the study of the kinetic energy distribution of photoelectrons expelled from material irradiated with monoenergetic X-rays. Determinations of photoelectron kinetic energy provides direct measurement of electron binding energy. In turn, the energy is characteristic of each element while the intensity is related to the concentration of the atoms of that element in the active sample volume. Absolute sensitivity of the technique is in the picogram range for materials located on the particle surface. Sampling onto fibre or membrane filters is suitable for this analysis, but necessary exposure of samples to vacuum causes loss of volatile materials.

Infrared spectroscopy in principle is applicable to analysis of polyatomic molecules or ions whose dipole moment changes during bond vibrations, e.g. stretching. Organic or inorganic tracer molecules can be assayed on the basis of their transmission spectra usually after the collected samples (1 μg to 1 mg) are mixed with KBr, NaCl, etc., ground and compressed to give a pellet 5–10 mm diameter. Samples collected directly on to membrane filters can be assayed by attenuated total reflectance or transmission of, for example, a cellulose nitrate filter. Unfortunately, results depend upon sample preparation because of effects of grinding, pellet density, particle size, uniformity of deposit and extent of surface coverage on membrane filters. In the pellet method there is the additional problem of possible chemical reaction between tracer and halide. This halide is required as a 'solvent', i.e. it reduces the refractive index difference between tracer particle and surrounding medium compared to when it is in air.

Tracer materials which are organic can be extracted from collected samples by organic solvents thereby reducing potential confusion with any contaminating background inorganic particles. Extract evaporation concentrates samples which are suitable for analysis by FTIR spectrophotometry. The method is relatively convenient and when severe interference is caused by background particles chromatographic techniques are applicable for further sample purification. IR spectroscopy has been employed extensively in the analytical praxis of aerosols of pollutants (Kellner, 1978).

Thermal analysis measures changes (e.g. phase changes, chemical changes) in tracer materials when heated or cooled, but requires a minimum sample size of 0.01 to 0.1 mg, and therefore is not particularly sensitive. Differential thermal analysis, thermogravimetry or thermo-X-ray analysis are alternative forms of thermal analysis should it need to be utilized.

Raman microprobe, unlike IR, can analyse individual aerosol particles and unlike X-ray fluorescence operates at atmospheric pressure. The principle is to measure shifts in wavelength of laser light when focused to a fine spot (as small as 0.1 μm). The scattered light wavelength can be greater or smaller than the incident wavelength; extent and direction of a Raman shift is characteristic of particle composition, while intensity is related to amount. Since visible wavelengths are involved, a microscope objective can serve to see samples and to focus a laser beam to a very fine spot for particle analysis. Compared to methods requiring electron microscopes, Raman microprobe is cheaper and operates at atmospheric pressure while offering comparable high sensitivity. A potential disadvantage is that it can be time consuming when analyses are to be made on individual particles. However, the method is convenient, relatively simple and like IR has been applied widely in studies of airborne pollutants.

Which technique is best will depend upon the actual tracer employed and sample quantities available for analysis together with the nature of any contaminating background materials. Methods applicable to analyses of single particles are among the most sensitive but are the most time-consuming because particles need to be scanned individually. On the other hand, more rapid analytical methods require comparatively large and background-free samples.

4.3 MICROBIOLOGICAL METHODS

In some instances it is inappropriate or impossible to substitute a biologically inert particle for a microbiological one. One example would be in determining the microorganisms dispersed by talking, coughing, sneezing, etc., or from bedmaking which re-aerosolizes microorganisms. Another is the release of microorganisms during pumping of milk (Dark, 1980). One approach to such a study is to collect aerosol samples with a high volume sampler and to use conventional microbiological assay procedures, e.g. colony counting. Such an approach while appropriate in some circumstances would be inappropriate where real-time monitoring is required as in sterile pharmaceutical preparation rooms, for example. In these types of circumstance automatic and rapid response are required.

Rapid monitoring or detection methods are either broad spectrum, responding to microbes generally or a given taxonomic group (e.g. bacteria, virus, yeast) without specific identification, or are specific, responding only to a particular microbe. The former are applicable to general air monitoring while the latter are restricted to monitoring isolation facilities, for example. Many of the known chemical, physical and physiological properties of microorganisms have been exploited to meet these aims, but a problem in common to all is the possibility of interference caused by dust and other background particles. The basic approach is to continuously sample the air space of concern and to deposit aerosols either into liquid or onto a surface. The next step is to assay collected materials. One direct method is to stain the sample and examine it microscopically, while others include pyrolysis followed by gas phase chromatography or analysis of ATP or haem content, for example. However, for viruses because of their much smaller mass compared to bacteria very large volumes of air may need to be processed.

In some applications very rapid response is not required, then classical methods (Noble, 1967) involving microbial growth are relevant. The general approach now is to culture microbial samples in liquid media and monitor growth by sensor. Suitable sensors measure light scattering, pH changes, CO_2 evolution, metabolite uptake, metabolic products, enzymatic activity (Strange, 1973) or electrical conductivity. The last property is invoked in the Malthus Microbiological Growth Analysers. Methods such as these detect growth within a few hours and are easier to apply than microculture techniques (Noble, 1967). Most are broad spectrum methods, although when electrical conductivity measurements are made carefully and continuously, resulting curves provide patterns characteristic of the cultured bacteria. Difficulties caused by innocula containing more than one microbial species may be reduced by including apposite antibiotics or antibodies.

A fairly rapid and specific process is immunofluorescence in which collected microorganisms are reacted with fluorescently-labelled antibodies. Those bind to homologous bacteria (or viruses, etc) so their uptake by as few as 500–5000 vegetative bacteria, for example, can provide in about 10 minutes their specific and rapid assay (Strange, 1973). For a tracer bacterium and homologous antibody, the method has much merit in terms of rapidity, specificity and sensitivity of response. Whether background particles interfere will depend upon particular circumstances. Very rapid but non-specific microbiological assay is possible with chemiluminescent-based reaction of luminol (5-amino-2,3-dihydro-1,4-phthalazindione) with haem iron of microorganisms (Oleniacz *et al.*, 1967, 1968). Of the microbial species tested in an automated system as few as 10^3 to 10^4 bacteria ellicited significant response. Comparable sensitivities are reported by Ewetz and Lundin (1973), but degrees to which background particles interfere depends on application. Nonetheless, chemiluminescence offers a relatively cheap but sensitive tracer assay procedure.

4.4 CONCLUSIONS

Potential biohazards associated with hospital, industrial and laboratory procedures involving infectious materials need to be established, monitored and controlled (Chapter 6). Required processes include sampling and assay of actual pathogens or tracers that may be non-living materials or non-pathogenic microorganisms. Depending on the nature of the aerosol, its monitoring can be achieved through different physical or microbiological assays. The most pertinent of them will depend on particular applications and facilities, with the most sensitive one likely to be microbiological.

REFERENCES

Corney, A. (1977). *Atomic and Laser Spectroscopy*, Clarendon Press, Oxford.

Dark, F. A. (1980). In *First International Conference on Aerobiology*, Federal Environmental Agency (Ed), Erich Schmidt Verlag, Berlin, pp. 387–392.

Ewetz, L. and Lundin, J. (1973). In *Fourth International Symposium on Aerobiology*, J. C. Ph. Hers and K. C. Winkler (Eds), Oosthoek, Utrecht, The Netherlands, pp. 23–26.

Goulden, P. D. (1978). *Environmental Pollution Analysis*, Heyden, London, Philadelphia.

Harper, G. J. (1980). In *First International Conference on Aerobiology*, Erich Schmidt Verlag, Berlin, pp. 420–431.

Kellner, R. (1978). In *Analysis of Airborne Particles by Physical Methods*, H. Malissa (Ed.), CRC Press, Florida, pp. 209–236.

McCrone, W. C. and Delly, J. G. (1973). *The Particle Atlas*. Vol. I. 2nd. ed., Ann Arbor Science, Ann Arbor, Michigan.

Malissa, H. (Ed.) (1978). *Analysis of Airborne Particles by Physical Methods*, CRC Press, Florida.

Noble, W. C. (1967). *Symp. Soc. Gen. Microbiol.*, **17**, 81–101.

Oleniacz, W. S., Pisano, M. A. and Rosenfield, M. H. (1967). In *Automation in Analytical Chemistry*, Technicon Symposium, 1966; Technicon, Ardsley, New York, pp. 523–525.

Oleniacz, W. S., Pisano, M. A., Rosenfield, M. H. and Elgart, R. L. (1968). *Environ. Sci. and Technol.*, **2**, 1030–1033.

Stenholm, S. (1978). *Progress in Atomic Spectroscopy*, W. Hanle and H. Kleinpoppen (Eds), Plenum Press, New York.

Strange, R. E. (1973). In *Fourth International Symposium on Aerobiology*, J. C. Ph. Hers and K. C. Winkler (Eds), Oosthoek, Utrecht, The Netherlands, pp. 15–23.

Chapter 5

Aerosol particle sizing

5.1 INTRODUCTION

A requirement to size particles cuts across a broad range of industrial, medical and research activities, some dating back to the turn of the century. Particles of interest may be residing on a surface, suspended in liquid as a colloid or suspended in air as an aerosol, and have a size in the range of 0.001 μm to 1000 μm. Of concern here are microbial particles having a size capable of invading a host through the respiratory tract. Consequently, particles between about 0.5 μm and 15 μm diameter are those primarily requiring analysis. Not surprisingly, particles with these sizes (and smaller) can remain airborne for considerable periods of time in turbulent air (see Chapter 8). The techniques of inertial classification, microscopy, light scattering and electrostatic mobility are those most frequently applied to sizing respirable aerosols (Cadle, 1975). Of these light microscopy is the most important and versatile. This chapter deals with the measurement of aerosol particle size, which represents a specialized topic of the general subject of particle sizing (Rideal, 1985).

5.2 AEROSOL DISPERSITY

Naturally occurring aerosols contain particles having a range of sizes (i.e. they are polydisperse) in contrast to some laboratory aerosols of a single particle size (i.e. monodisperse). Aerosols may comprise particles of the same shape (e.g. spheres) and are homogeneous, or contain particles of differing shapes and are heterogeneous. Generators such as the vibrating reed, Berglund-Liu and spinning disk (or top), described in Chapter 2, produce monodisperse aerosols, whereas two-fluid atomizers and dry powder disseminators give polydisperse aerosols.

For fundamental laboratory studies and instrument calibration, monodisperse aerosols are the most suitable as then interpretation of experimental data is simplified. Otherwise, when polydisperse aerosols are employed the actual distribution of particle sizes needs to be taken into account. Another situation when monodisperse aerosols would greatly simplify a problem is in crop spraying where downwind drift of spray has to be controlled. Without the monodisperse output of spinning disk atomizers, for example, control virtually would be impossible.

That aerosols often are polydisperse compounds the difficulty of deriving their particle size, while their heterogeneity of shape introduces so many more difficulties that in the past this latter factor more often than not was neglected. Alternatively, shape is allowed for by quoting an 'aerodynamic size', i.e. the size of a unit density sphere which behaves aerodynamically the same as a given irregular particle. But, nowadays with the advent of automated image analysis (see Section 5.6), shape can be quantified and described as a distribution, as is size.

5.3 DEFINITION OF SIZE

Inevitably, there are several different definitions of particle size with each being related to the principle invoked to measure it. One example is provided by real-time particle size analysers relying on light scattering. By measuring the intensity of light scattered by an aerosol particle, either over a narrow solid angle or a wide one, particle size is derived. But, scattered light intensity depends not only on the size of the particle, but also on its shape and its real and imaginary refractive indices, together with its structure.

In practice, the instruments are calibrated (often with polystyrene latex spheres of known size), then any particle eliciting a given scattered intensity is apportioned a diameter equal to that of the corresponding calibration particle. More precisely, it is an area rather than a diameter because light scattering intensity depends on the area of the scattering surface. These techniques, therefore, give a size-equivalent based on area and really are satisfactory only when the nature of test and calibration particles are the same.

Aerodynamic real-time particle size analysers such as the APS (TSI Inc.) provide aerodynamic particle size, as do inertial multistage devices. Microscopic analysis can provide a physical diameter but even then for droplets this may be their diameter after spreading, unless corrected for this effect by applying a spreading factor.

In some applications such as monitoring, the differences in size definitions may be unimportant, but nonetheless need to be recognized. In other instances the definitions are important especially when particles differ significantly from spherical shape and unit density. Extreme examples, perhaps, are the physical and other diameters of dandelion seeds and of pollen grains. For hygroscopic particles, particle refractive index will be relative humidity (RH) dependent while the reflectivity coefficient of non-hygroscopic particles can change with RH and amount of surface sorbed water. In an analogous way there can be changes in aerodynamic size too.

Whether in practice differences between the true physical diameter and that of light scattering or aerodynamic diameter are important will depend on application. But, generally, the aerodynamic behaviour and size are of concern in terms of physical behaviour of microbial aerosols whereas the actual physical size is more pertinent to microbial survivability. This view that aerodynamic diameter is the parameter most useful for describing the physical aspects of particle size is also that of Sern (1984).

5.4 PARTICLE SIZE DISTRIBUTIONS

Aerosols unless they are monodisperse will contain a range of particle sizes. This population of particles is made up of a number of individuals each of which can be assigned a size. Particle size distributions are representations of the frequency with which each size occurs. If for the moment it is supposed that the particles in our distribution have only integer values (e.g. 1, 2, 3....10 μm) of diameter, then the distribution may be Poisson, i.e.

$$f(d) = \frac{\lambda^d e^{-\lambda}}{d!} \quad (d = 0, 1, 2, 3, \text{etc.}) \tag{5.1}$$

where, $f(d)$ = the frequency (or probability) that the population will contain particles of diameter equal to d,

λ = an integer constant

$d!$ = d factorial (e.g. for $d = 5$, $d! = 5 \times 4 \times 3 \times 2 \times 1 = 120$)

Examples are shown in Figure 5.1.

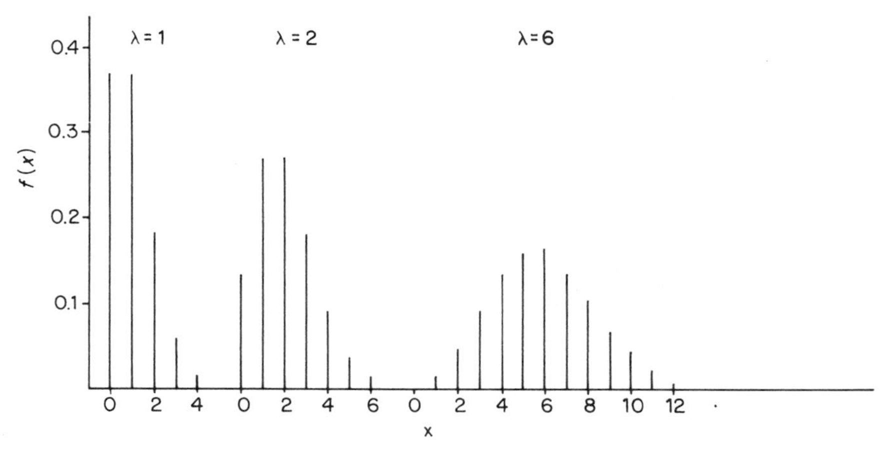

Figure 5.1 Examples of the Poisson distribution.

In practice, integer values of size are not usually observed except when a suspension of, for example, monodisperse spheres is atomized and there will be a discrete distribution with aerosol particles comprising one, two, three, four, etc., spheres. Another example of when this distribution is likely to be observed is when a suspension of microorganisms in water is atomized and there is likely to be a discrete distribution of microbes amongst the aerosol particles.

A more usual situation is for a continuous distribution of particle sizes, i.e. particle diameters can have any value between some minimum and some maximum value (e.g. 1.3157, 3.4131, 5.8765 μm, etc.). One of the more familiar distributions of this type is the normal distribution, i.e.

$$f(d) = \frac{1}{\sqrt{2\pi}\sigma} exp[-(d - \mu)^2/2\sigma^2] \qquad (5.2)$$

where, $f(d)$ = the frequency or probability of finding a particle of size d,

μ = the mean of the distribution

σ^2 = the variance of the distribution

σ = the standard deviation.

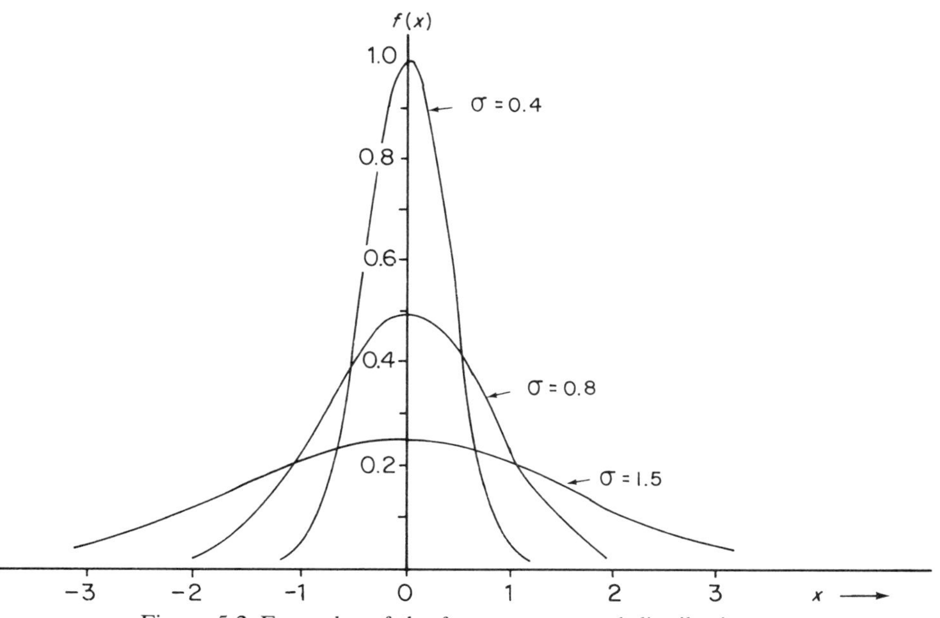

Figure 5.2 Examples of the frequency normal distribution.

A plot of the values of $f(d)$ against the corresponding values of d produces the bell-shaped curves of Figure 5.2, while the corresponding cumulative plots are given in Figure 5.3. For normal distributions the mean particle size (simple arithmetic mean), the median particle size (the size corresponding to a cumulative 50% of the particles) and the mode particle size (the size corresponding to the largest value of $f(d)$) are all the same. For distributions other than the normal distribution this equivalence between mean, median and mode does not hold; other distributions occur when there is a bias towards either small or large particles, i.e. when the distribution is skewed (Figure 5.4).

For a normal distribution plotted in cumulative form the median particle size is that corresponding to the 50% cumulative value while the standard deviation equals the difference between this value and that corresponding to the 15.87% or 84.13% cumulative values. While this is a convenient graphical method for deriving median particle size and standard deviation, the normal distribution strictly is not applicable to particle size analysis because it covers a size range of

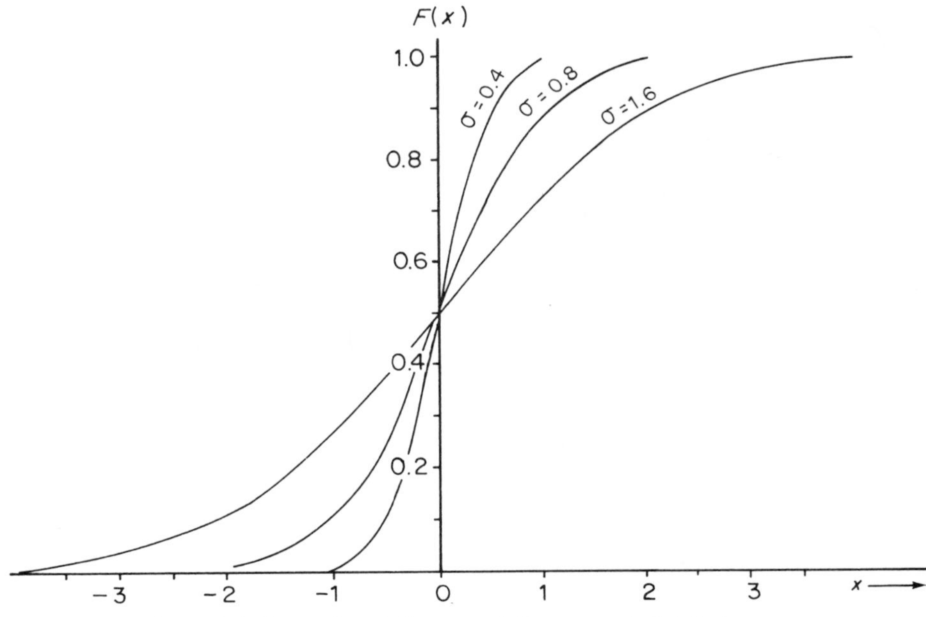

Figure 5.3 Examples of the cumulative normal distribution.

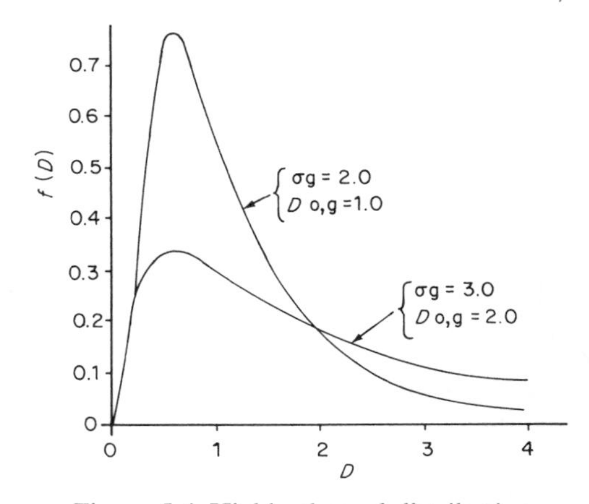

Figure 5.4 Highly skewed distributions.

$-\infty$ to $+\infty$. This means that it predicts a very small but definite probability that there are particles having negative diameters. The dilemma is alleviated somewhat if the logarithm of particle size (rather than particle size) is employed, when,

$$f(\mathrm{d}) = \frac{1}{\sqrt{2\pi}\ \ln \sigma_g} \left| exp[-(\ln d - \overline{\ln d})^2/2 \ln {}^2\sigma_g)] \right. \quad (5.3)$$

where, $\overline{\ln d}$ = mean of $\ln d$

σ_g = geometric mean deviation

and, $$\ln \sigma_g = \left[\frac{\sum_{j=1}^{j=n} (\ln d_j - \overline{\ln d_j})^2}{n - 1} \right]^{\frac{1}{2}} \tag{5.4}$$

n = number of particles

and, $\ln d_g = \overline{\ln d}$

d_g = geometric mean diameter.

Another advantage of the lognormal distribution for particle size analysis is that it compresses the particle size range while providing a more suitable weighting of particle sizes. (For comparison, the normal distribution puts equal importance on equal intervals of particle size.)

Unlike the normal distribution from which it derives, the lognormal distribution is skewed towards small diameters (Figure 5.4) as found for most naturally occurring aerosols. In cumulative form, the curves are similar to those for the normal distribution in that the cumulative percent value when plotted against geometric mean diameter yields an S-shaped curve also. Again the geometric mean diameter is the diameter corresponding to a cumulative value of 50%, while geometric mean deviation is equal to that diameter divided by that for the 15.87% cumulative value. Alternatively, the geometric mean deviation is equal to the diameter corresponding to 84.13% cumulative value divided by that for 50%.

Given that the distribution is truly lognormal, the data plotted cumulatively on log-probability paper yield a straight line. This result is because log-probability paper has the cumulative percent markings printed so that they are normally distributed about the 50% mark while the particle diameter scale is printed logarithmically. Data, without transformation, then may be readily plotted and geometric mean and geometric deviation easily obtained. In practice, it is surprising perhaps how often and well particle size data do conform to a log-normal distribution to give a straight line on log-probability paper. When this does occur, values for the geometric mean and geometric deviation provide a good description of the particle size distribution, except at its extremes or tails. For more accurate descriptions still, the moments of the distributions also can be provided.

So far in this discussion frequencies or particle number distributions have been considered, but often experimental data are in terms of particle mass. One example is data from Cascade impactor samples. In this case, cumulative particle mass is plotted as a function of stage d_{50} value (i.e. aerodynamic diameter) and mass median diameter and corresponding deviation found. When both mass and number data are available for the same sample and analysed similarly, it will be found that derived values for number and mass median and deviation will not be the same. This is because different particle properties were utilized in the analysis. Mathematically the two distributions are related and for the simple case of spheres the relationship between them is,

$$m(d) = n_p(d) \frac{\pi}{6} \times \rho_p \times d_p^3 \tag{5.5}$$

where, $m(d)$ = mass of aerosol particle of diameter d

ρ_p = particle density

d_p = equivalent spherical diameter

When comparing size distributions it is essential to know the particle property utilized because comparing number-size and mass-size distributions, for example, is not comparing like with like. A useful exercise is to take a simple number-size distribution and its mass-size transform calculated from equation 5.5. When such data are plotted as frequency versus size it is difficult to see any similarity between them. Analogous problems can arise when comparing size distributions obtained by say an instrument relying on light scattering with one obtained by a cascade impactor sampler. As the two devices measure different particle properties so the two distributions are fundamentally different and strictly cannot be compared meaningfully. That is, unless one of the distributions can be mathematically transformed to be expressed in a common particle property.

Another, perhaps more subtle, difficulty would be the comparison of a size distribution of a microbiological aerosol when derived by Andersen Biological Sieve Sampler and a Cascade Impactor measuring sampled mass, for example. Here the former is a number-aerodynamic size distribution and the latter a mass-aerodynamic size distribution. Consequently when giving details of a size distribution, it is imperative that the particle property being utilized is stated clearly.

Another important feature is that because aerosols consist of a large number of individual particles, their size distribution has to be estimated from a relatively small sample of them. A sample is ideally representative if all individual particles have an equal probability of appearing in it. Then, there is no bias in the sample towards any particular size. Even given isokinetic or stagnation point sampling (q.v.) it is still essential to minimize other factors causing bias, e.g. non-uniform mixing of the aerosol, sampler intake losses, etc., or to allow for them.

Having obtained a representative sample, analysed it and plotted the data as a lognormal distribution, for example, the question arises as to how well the actual distribution fits the theoretical distribution. Goodness-of-fit tests are required as there is little point in calculating mean and deviation, etc. values according to a lognormal distribution if the data conform to a totally different distribution, e.g. a Pearson type. Chi-squared and Kolmogorov-Smirnov tests are appropriate depending on the sample size. When it is 100 or less the former can be applied while for larger sample sizes the latter is applicable. A useful text for these tests and for distribution functions, etc., is that by Weatherburn (1968), while Deepak and Box (1982) give eight mathematical functions commonly used for representing aerosol particle size distributions.

From a practical point of view, having a distribution of particle sizes raises immediate difficulties in its determination. Because particles in each size class are not equally represented, a sample suitable for estimating the most frequent size will contain a very low percentage of the least frequent sizes. Conversely, a sample large enough to contain a significant number of the least frequent sizes may be overloaded by particles of the most frequent size. The most obvious outcome is that the 'tails' or extremes of a distribution are poorly defined. Samples of polydisperse aerosols taken by single stage impactor for microscopic analysis suffer badly from the above problem. In addition there is the difficulty that all particles because of their different sizes do not lie in the same focal plane. While manual focusing can accommodate the problem by frequent adjustment, it can become a serious source of error in automatic image analysers (q.v.) having autofocus. The use of a cascade impactor can alleviate this problem somewhat by providing a restricted range of sizes for each stage. But this then increases the number of samples requiring size analysis and may cause normalization difficulties for combining data from different stages to give a grand distribution. Whether, therefore, it is better to use a series of monodisperse aerosols or a polydisperse aerosol and size analyser depends on circumstances.

In addition to number-size and mass-size distributions, surface area-size distributions (strictly that given by light scattering) are appropriate. For particles of regular shape these may be mathematically interconverted given additional information, e.g. density, shape, etc. Even so, aerosol size distributions in practice quite often are highly skewed: a fact that is hidden when only median size and deviation are provided. Therefore, degree of skewness also should be quoted or even better still the actual distribution presented.

5.5 INERTIAL CLASSIFICATION

A feature of these and other similar devices is particle inertia (Chapter 1) and the impaction process. The basic configuration of an impaction jet is shown in Figure 5.5 with air being drawn through it by suction. The particles entrained in the air issuing from the jet are directed at the impaction surface. Larger particles having higher momentum than air molecules are unable to negotiate the changed direction of flow and collide with the impaction surface, i.e. particles cross the streamlines. Smaller particles having lower momentum are able to follow the streamlines and are not impacted. In cascade impactors a series of progressively finer jets are connected in series. As a result, successive stages have increasing air velocities and as the finer particles passing the first jet enter the second they are accelerated and attain a velocity approaching that of the air. With increased momentum the larger of them now are impacted at the second stage, and so on. In some devices, e.g. Andersen, Sierra, the final stage comprises a filter to trap particles too fine to be sampled by impaction stages.

The jets, or nozzles, can be any shape but usually are rectangular or round. As already indicated (Chapter 3) in the Andersen sampler each stage has a number of radially spaced circular jets in contrast to the Sierra with only one

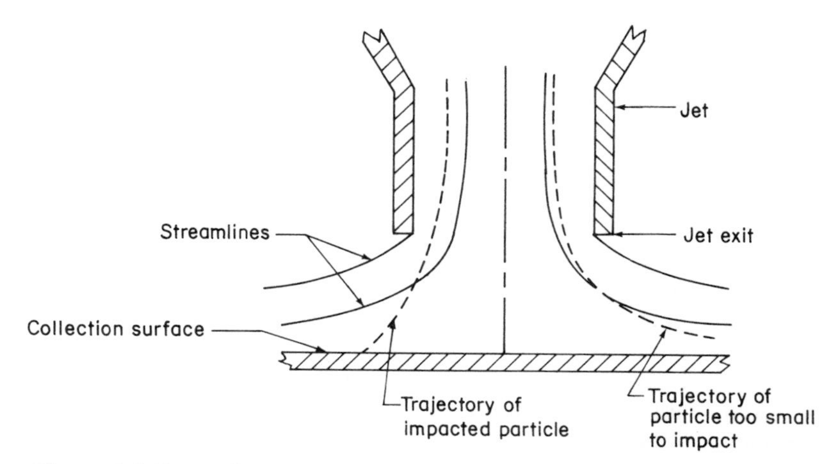

Figure 5.5 Streamlines and particle trajectories in an impactor jet. (After Rao, Ph.D. Thesis, *An Experimental Study of Inertial Impactors*, University of Minnesota.)

and a single rectangular (slit) jet per stage for the Cascade Impactor, for example. In these samplers, though, the most important parameter is the efficiency of collection of each stage as a function of particle aerodynamic diameter. Ideally each stage would collect all particles greater than a given size and none of the smaller ones. Such a performance corresponds to the ideal cut-off curve of Figure 5.5 and while it is approached in typical impactors, it is not actually achieved in practice owing to effects of the boundary layer near its walls (Marple and Willeke, 1979). The collection efficiency curves of impaction jets have received considerable experimental and theoretical study, the latter indicating that it is governed by fluid mechanics. The velocity flow field of the fluid (i.e. air) can be specified by Navier-Stokes equations, a solution to which has been provided by Marple and Liu (1974) and Marple *et al.* (1974). Their analyses indicate that the collection efficiency is a function of the Stokes number (*Stk*), the Reynolds number (*Re*) and the physical configuration of the jet, where,

$$Stk = \frac{\rho_p \, C \, V_0 D_p^2 / 18\mu}{W/2} \tag{5.5}$$

and, $$Re = \frac{\rho_a \, V_0 W}{\mu} \quad \text{(round jet)} \tag{5.6}$$

or, $$Re = \frac{\rho_a \, V_0 2W}{\mu} \quad \text{(rectangular jet)} \tag{5.7}$$

where, ρ_a = density of air

ρ_p = density of particle

C = Cunningham slip correction factor

V_0 = mean air velocity at jet throat

D_p = particle aerodynamic diameter

μ = air viscosity

W = jet width or diameter, as appropriate.

The square root of the Stokes number, $\sqrt{Stk}$, represents a dimensionless particle size and is widely used in relation to impactor efficiency.

Marple and Willeke (1979) provide sets of curves showing how the physical dimensions of the jet affect its collection efficiency. From their analyses several generalizations are possible. First, for $500 < Re < 25\,000$, all efficiency curves have the same shape. Second, the position of the efficiency curve in plots like that in Figure 5.6 are relatively independent of S/W, T/W and *Re*, except at very small values. S is the jet to impaction plate distance, W is the jet width or diameter and T is the jet throat length (Figure 5.5). These authors, therefore, are able to provide design charts for both round and rectangular jet impactors.

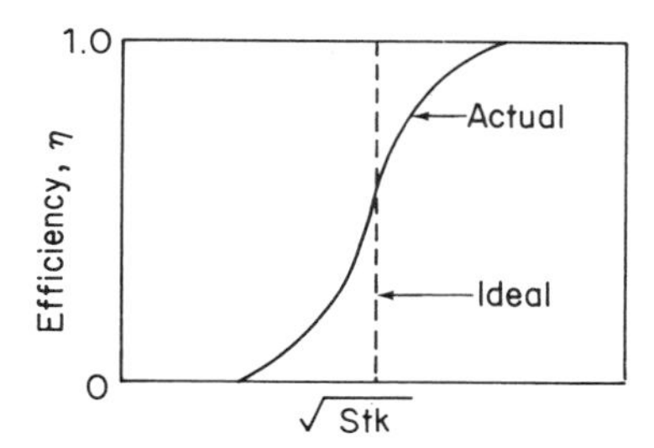

Figure 5.6 Comparison of actual and ideal efficiency curves for an impactor jet.

Even though inertial classifiers have been studied in depth, can sharply classify particles into ranges of particle size and are relatively simple devices, in practice they can all too easily lead to erroneous data unless their shortcomings are appreciated and taken into account. A major limitation is that a given impactor system is efficient over only a defined size range due to intake efficiency limitations, for example. In practice, their use often is limited to 0.08 to about 20 μm; even then not all models can cover this range. They can provide no reliable information for particles having sizes outside that operating range. Once the aerosol is sampled, the particles must be deposited onto the impaction plates only and not on instrument walls. On impaction, the particles must remain at their impaction points, otherwise they will be re-entrained and deposited on the 'wrong' stage for their size. If and when this happens, the indicated particle size distributions will be erroneous. While the application of adhesives to the impaction plates reduces particle bounce and particle blow-off following impaction, these processes are not always avoided (Rao, 1975; Wesolowski *et al.*, 1977). Furthermore, it is not always obvious that they have occurred and that the given size distribution is grossly in error. One aid is always to examine deposits microscopically to check that particles collected on a given stage do have physical sizes consistent with its calibrated or calculated size range. Unfortunately, though, this procedure is not always foolproof when determining the mass collected per stage, e.g. one 10 μm particle slipping to a stage sampling 1 μm particles, has the mass equivalent of 1000 × 1 μm particles. Hence, slipping of a few large particles can profoundly change the apparent particle size distribution with a shift to smaller sizes. Artefacts like these lead to discrepancies between the experimental mass median diameters determined on the same aerosol with three different cascade impactors (Knuth, 1979).

Another problem is that when aerosol concentrations are high or sampling times prolonged, stages can become overloaded with deposited aerosol. Thick deposits can modify sampling efficiency, particle bounce, particle blow-off, etc., as well as causing break-off in lumps which break up again at the next impaction stage.

Impaction size classifiers may seem simple but unless in their use attention to detail is given, together with the benefit of some operating experience, they are likely to provide erroneous particle size distribution data for the reasons outlined above. Given satisfactory operation, the next step is analysis of the data to provide size classification.

For illustrative purposes assume we are dealing with a seven-stage impaction device and a final stage of a filter, such as the Ultimate Cascade Impactor of May (1975). A normal approach is to assume that the lower size limit for particles collected by a given stage is the size corresponding to the d_{50} value or effective cut-off diameter. As mentioned in Chapter 3, this must introduce errors because some particles larger than the d_{50} value of a stage will deposit on the following stage, while some particles smaller than the d_{50} value deposit on this stage. (For this particular sampler, though, the efficiency curves for each stage (except number 1) are very steep and likely errors are small). Having determined the mass of deposited aerosol on each stage plus the filter by weighing, chemical analysis, including gas chromatography (Mitruka, 1977), the cumulative mass larger than the d_{50} value is found from the running total of the particle mass taken stage by stage from 1 to 7. The filter d_{50} is taken as zero and is treated as stage 8. The cumulative mass values for each stage conveniently are converted to percent by dividing each value by the total mass collected and multiplying by 100.

Cumulative percentage mass values then are plotted against their respective d_{50} values; as discussed in the previous section, common practice is to plot these data on log-probability paper. Most likely the data can be fitted reasonably by a straight line and the mass median diameter and geometric deviation calculated (see previous section). While this approach is convenient it tends to overestimate the true median diameter as discussed in Chapter 3 where ameliorative action of a kind is described. Another approach is to generate a differential rather than cumulative size distribution. For this, the percentage of the total mass for each stage is plotted (on linear graph paper) as

Table 5.1. Stage mid-point size values (Ultimate impactor)

Stage number	Cut-off values (μm)	Mid-point values (μm)
1	30	23
2	16	12
3	8	6
4	4	3
5	2	1.5
6	1	0.75
7	0.5	0.25
8	0.0	

a function of the midpoint of the size range for each stage. The value of the midpoint is not the same as the d_{50} value for a given stage. It is the arithmetic mean of the d_{50} values for successive stages as indicated by the entries of Table 5.1. The upper size limit for stage 1 can be estimated by determining microscopically the size of the largest particle sampled by this stage. Having plotted the data points, it is tempting to draw a smooth curve through them to indicate the shape of the size distribution. Unfortunately, the procedure can be misleading as relative small errors in the data can bring about quite marked changes in the shape of the distribution. A preferable procedure would be first to generate many more data points to more accurately define the distribution. Nonetheless the process still can be worthwhile for emphasizing the relatively small particle mass defining the tails or extremities of the distribution and their commensurate errors.

An inertial classifier of a different type is the cyclone which by suitable selection of cyclone geometry, diameter and flow rate, etc., provides for different cut-off diameters. Multistage cyclones, e.g. the six parallel cyclone classifier of Lippmann and Kydonieus (Chapter 3) inherently is capable of sampling much larger quantities of aerosol without the overloading that occurs with cascade impactors and, therefore, may provide more accurate particle size classification, especially in terms of particle weight. On the other hand, the cyclone efficiency-size curve does not have a sharp cut-off and inevitably must lead to a loss of precision in reconstructing particle size distributions. But when used, cyclone data may be analysed in analogous ways to those described for impactors.

While multiple cyclones offer certain advantages over cascade impaction devices, the aerosol centrifuge provides for a continuous gradation of particle size rather than 'discrete' sizes. The resolution of size provided by the aerosol centrifuge is difficult to better in terms of particle aerodynamic size, except for perhaps the TSI Aerodynamic Particle Sizer. Their main disadvantages are the relatively low sampling rate and poor commercial availability. In operation, aerodynamic particle size is obtained by measuring the distance of the deposit from one end of the collecting foil and making comparison with that for a previous calibration with polystyrene latex spheres, etc. Additional details are provided by Tillery (1979), Moss (1979) and Kasper and Berner (1981).

5.6 MICROSCOPIC AND IMAGE ANALYSIS

Aerosols collected by impactors, electrostatic and thermal precipitators are in a form amenable to microscopic analysis — a technique of great capability (McCrone and Pelly, 1973). For particles less than about 1 μm diameter the transmission and scanning electron microscopes are better suited than the light microscope with its resolution limit of about 0.3 μm; however, electron microscopes require sample exposure to a hard vacuum with a loss of sample water and evaporation of other volatile components. The alternative of providing sample replicas and analysing them can be helpful. For larger particles the light microscope is ideal. Its theoretical resolution limit is given by the equation,

$$\text{lim} = \frac{\lambda}{2\,NA} \tag{5.8}$$

where, λ = wavelength of light

NA = numerical aperture of the microscope objective.

The numerical aperture is governed by the refractive index of the medium between the lens and object. High refractive index immersion oil is used to achieve highest NA values and correspondingly lowest resolution limit. The depth of focus or depth of field of a lens also is inversely related to NA value. When used to size a polydisperse aerosol sample, choosing an objective with highest NA value will provide maximum resolution but minimum depth of focus, necessitating frequent microscope focus adjustment to accommodate different particle sizes. Having a range of differently sized particles in focus together requires a lens of relatively high NA value and low resolving power. In practice, a compromise of NA value can ease microscopic examination of particles. Contrast between particle and background can be increased by stopping down the microscope condenser but this simultaneously diminishes resolving power. It increases also with refractive index difference between particle and surrounding medium. The form of illumination plays a part with transmitted light usually providing better contrast than incident illumination.

As for other particle sizing techniques, unless the test aerosol is at a high concentration or is held in a chamber previously cleared of other particles, aerosol samples will be a mix of test and background (e.g. dust) particles. To derive correct size distributions test and background particles have to be differentiated. Depending on their nature, colour, polarizability, fluorescence, shape etc., can be invoked. Polarizing and fluorescent microscopes help in particle identification while phase contrast and differential interference contrast enhance particle visualization. For electron microscopy, X-ray dispersive analysers help in this task, but when operated frequently are time-consuming.

When microscopically sizing liquid droplets errors through their evaporation and spreading can be severe, but can be reduced by impacting onto MgO-covered slides and measuring the impact 'craters'. For low volatility droplets, spreading factors can be measured and applied.

For the most accurate measurements eyepiece micrometers are best as are size standards for electron microscopy. For less demanding work eyepiece graticules (e.g. Porton) are satisfactory. When large numbers of samples demand sizing microscopically, image-analysers virtually are essential. They provide additionally particle shape analysis. Basically, image analysers comprise a television camera suitably interfaced to a light or electron microscope. The TV image is converted into a mosaic of picture elements (or picture points or pixels) — usually minute squares. The pixel mosaic then is analysed electronically by hardware (e.g. logic circuits), by software (i.e. computer program), or a combination of both. Some analysers can integrate over several frames or scans of the TV picture, thereby, catering for very low light levels as found with fluorescence for example. Others provide for software-controlled image enhancement techniques, e.g. contrast enhancement, feature edge definition, etc. Such techniques, especially when coupled with that of a scanning light microscope achieve marked increases to microscope resolving power. The larger and more sophisticated image analysers can store pixel mosaics in memories, or image stores, allowing colour analysis, detection of differences between pictures, etc. Computer controlled systems having automatic stepping microscope stages and autofocus permit unattended analyses of very large numbers of samples.

Image analysers can measure size in a number of different ways. For example, most analysers measure projected area and circumference for each particle (or feature), together with Feret diameter. Feret diameter can be measured at any angle and is the distance between feature extremities at a particular angle (Figure 5.7). Of these measurements, size based on projected area usually is the most accurate as well as being independent of particle orientation. Circumference provides a poor estimate of size as its value depends on magnification factor and microscope resolving power. For

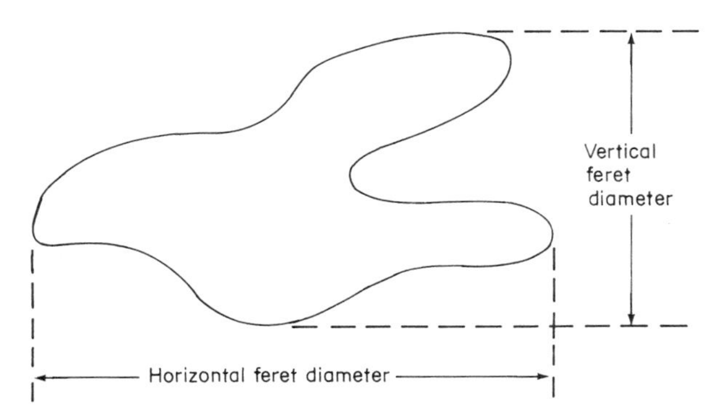

Figure 5.7 Horizontal and vertical Feret diameters.

instance, at low magnification a particle may appear to have a smooth circumference but as the magnification is increased it becomes more irregular because surface roughness is resolved. Values for circumference, therefore, increase with magnification even after allowing for magnification changes. While it may be argued that measured area would suffer similarly, the effect in practice is much less pronounced than for circumference and the error correspondingly smaller. Having measured particle area, values may be applied directly to compute size distributions, or the diameter of the equivalent projected sphere derived and utilized for the particle size distribution. Provided particle density is known, particle aerodynamic diameter can be calculated. Similarly, if required, a light scattering size could be derived by means of Mie theory. Given that most image analysers are built around powerful computers, data transformations like those above relatively are not too difficult to perform.

In addition to counting and sizing aerosol particles, image analysis enables particle shape to be described quantitatively. There are several ways in which this may be achieved, but one of the most generally applicable is to measure a number of different Feret diameters for each particle. For example, determine 12 Feret diameters at 30° intervals for each particle. Alternatively, find the horizontal Feret diameter for each TV scan line crossing the particle (Figure 5.8), i.e. a 'smashed' particle. A simple shape descriptor would be the mean and standard deviation of the Feret diameter set for each particle. More sophisticated shape descriptors involve complex mathematical treatments, e.g. Fourier analysis. One disadvantage of these types of analyses is the large number of measurements required for each particle. When a large number of particles also is involved, the entire process can be time-consuming. Even so, shape analysis is well developed and applied extensively (Beddow and Meloy, 1980; Beddow, 1984).

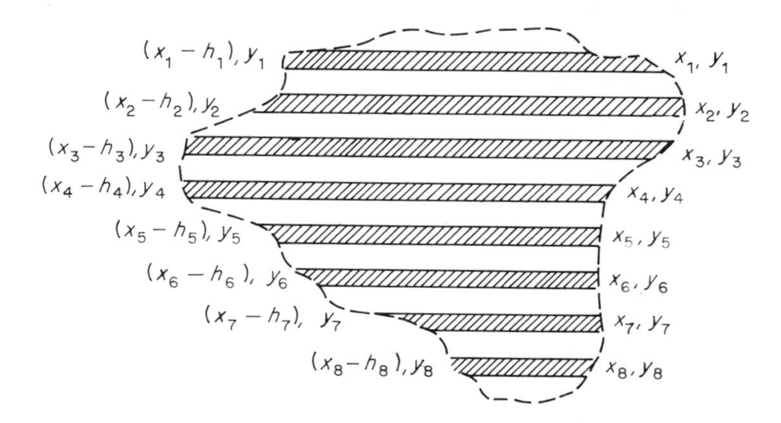

Figure 5.8 The x, y coordinates and horizontal Feret diameters (h) for a 'smashed' particle.

5.7 REAL-TIME PARTICLE SIZERS AND COUNTERS

Aerosols attenuate light, a familiar example being a reduction in visibility caused by fog. This attenuation is due to a combination of light absorption and light scattering. The process is a function of particle concentration, size, shape, refractive index and wavelength of light. Yellowing of lights caused by fog droplets provides an example of a wavelength effect, due in this instance to other light wavelengths being absorbed and scattered more than yellow light. A practical application is real-time aerosol particle sizing and counting. In order to understand their operation and some of the defects in performance of light scattering instruments, a brief account of light scattering follows.

There are three basic regimes of light scattering corresponding to when particles are much smaller than the wavelength of light (Rayleigh scattering), when particle size is comparable to wavelength (Mie scattering) and particle size is much greater than wavelength (corresponding to geometric optics). Considering Rayleigh scattering first, then for a single particle much smaller than the light wavelength (λ), the intensity (I_θ) of light scattered at an angle (θ) to the incident beam, over the angular element $d\theta$ is:

$$\frac{I_\theta}{I_0} = \frac{9\pi^2 v^2}{2R_0^2\lambda^4}\left(\frac{m^2 - 1}{m^2 + 2}\right)^2 (1 + \mathrm{Cos}^2\,\theta)\,d\theta \qquad (5.9)$$

where, I_0 = incident light intensity (unpolarized)

V = particle volume

R_0 = distance from particle to detector

m = particle refractive index

or in terms of an extinction coefficient, K, for a single particle of diameter, d,

$$K = \frac{4\pi^4}{3}\left(\frac{d}{\lambda}\right)^4\left(\frac{m^2 - 1}{m^2 + 2}\right)^2 \qquad (5.10)$$

The scattering coefficient decreases as the wavelength increases, which accounts for a blue (short wavelength) rather than a red (long wavelength) sky. Maximum scattered light intensity, though, is in the forward direction (i.e. narrow angles) and a smaller intensity backward. Between these limits the intensity changes smoothly with angle, θ. For Mie scattering by an individual particle, the mathematical functions are much more complex than those for Rayleigh scattering. Mie scattering is much more sensitive to changes in particle size, shape, and refractive index and for convenience results of scattering calculations often are expressed in terms of a size parameter, a, equal to $\pi d/4\lambda$.

Explitic equations for scattering in this regime were obtained by Mie who solved Maxwell's equations and it is these solutions that generally are referred to as Mie theory. For a single spherical particle, the intensity (I_θ) of light scattered with the electric vector, respectively, perpendicular and parallel to the scattering plane are given by:

$$\frac{I_\theta}{I_0} = \frac{\lambda^2}{4\pi^2 r^2} \, i_1(\theta) \quad \mathrm{Sin}^2\phi \tag{5.11}$$

and,
$$\frac{I_\theta}{I_0} = \frac{\lambda^2}{4\pi^2 r^2} \, i_2(\theta) \quad \mathrm{Cos}^2\phi \tag{5.12}$$

where, i_1, i_2 are intensity functions which mathematically are complex and involve derivatives of Legendre polynomials and Bessel functions,
θ, ϕ are polar angles for the scattered light waves,
r = particle radius.

Computer programs are available for carrying out calculations of light scattering according to Mie theory for particles of simple shape. They show that as for Rayleigh scattering, maximum intensity is in the forward direction and a smaller intensity backward, but between these two limits the intensity changes abruptly with angle. These changes are related in a complex way to particle size, shape and refractive index. In terms of the extinction coefficient as a function of the size parameter, it has the appearance of a sine wave progressively damped in intensity while increasing in frequency as the size parameter becomes larger. For a fixed wavelength then, the extinction coefficient demonstrates oscillatory behaviour as a function of particle diameter. That is, the same extinction coefficient occurs for particles of different sizes. For light scattering instruments the use of lasers enhances the problem as then the wavelength dependency is maximized and made more complex than that for a broadband source such as a hot filament when the effect is smoothed.

This account of light scattering while being very brief should be sufficient to indicate some potential difficulties in its application to particle size measurements. A much more thorough account of light scattering is provided by Kerker (1969). Light scattering aerosol analysers all work on the same basic principle, i.e. as particles flow through an illuminated volume, light scattered by a single particle into a solid angle is measured. The electrical pulses produced by the detector are fed to a pulse height analyser and thence to size channel counters. As indicated above, determination of actual particle size though is indirect because the response depends also on factors other than particle size, viz. particle shape, refractive index, solid angle, detector sensitivity, etc. A large number of theoretical and experimental investigations have been performed for many of the commercially available instruments. (For examples, see Lundgren *et al.*, 1979; Governmental Industrial Hygienists, 1978; Pinnick and Auvermann, 1979; Gerber and Hindman, 1982). That there should exist so many papers on the subject is indicative that light scattering aerosol analysers, in general, may not behave in practice quite as well as may be supposed. The basic problem is the complex nature of light scattering, as indicated above. In the author's experience if these instruments are to be used for other than crude particle size analysis, then first they should be calibrated with the test particles and then operated under the same conditions.

Alternatively, real-time aerodynamic size measurements can be made using laser Doppler velocimetry (e.g. Agarwal and Fingerson, 1979; Wilson and Liu, 1980; Fletcher *et al.*, 1980). One commercial version is the TSI 'Aerodynamic Particle Sizer' or APS. In this instrument the aerosol is introduced into a chamber through an accelerating nozzle and the velocity of the particle at the nozzle exit measured by a laser Doppler velocimeter. The velocity information for individual particles is recorded and processed by an integral microcomputer. Particle aerodynamic size and related parameters are displayed as histograms. Disadvantages are the size range limit of 0.6 to 15 μm diameter and the relatively long time (about 30 s) required to compute the data, for standard instruments. To order, modifications are made to alleviate these limitations.

Instruments in which particle electrical mobility is analysed operate for particles in the range 0.005 μm to 1 μm diameter and, therefore, are limited in their application to microbial aerosol sizing.

A difficulty in common to most real-time analysers is that of coincidence. As aerosol concentration increases the probability of there being more than one particle in the sensing volume correspondingly increases. The simultaneous presence of two or more particles when the instrument only anticipates one, usually is recorded as a single particle event. Particle numbers become underestimated and their size incorrectly assigned. Unfortunately, it is not always obvious that an instrument is suffering from coincidence, while manufacturers' estimates of when it becomes problematical cannot always be relied upon. Use of aerosol diluters can alleviate coincidence problems but then introduce problems of their own in terms of their efficiency as a function of particle size and shape.

Another and fundamental aspect of real-time particle analysers is that they take no account of the source of the counted particle. Consequently, analyser response may be due to a test particle, background dust, particles detaching from connecting tubing walls, etc. Whether interferences by particles other than test ones are important will depend upon application, but a frequent uncertainty is if the particles counted and sized are solely those of the test aerosol. Should this uncertainty be important in a given application remedial action may be to use gravimetric or microscopic methods. As for other aerosol samplers, account needs to be taken of whether or not the analyser intakes are obtaining their samples isokinetically, together with possible intake losses and particle losses in connecting tubing, if employed.

Light scattering aerosol analysers have an additional feature, in that identical particles need not produce identical signals and, therefore, are counted in different size channels. Part of this problem is due to instrument imperfections and another part arises from the nature of light scattering. Pinnick and Auvermann (1979) describe how monodisperse spherical latex particles of different sizes elicit the same detector response in light scattering aerosol particle analysers. This result is predicted by Mie theory.

When real-time analysers are operated with great care and consideration to detail, they can provide meaningful data. However, with the increasing use in them of microcomputers to control their function, a false sense of the validity of generated data can be only too readily engendered.

5.8 PARTICLE SIZES OF INHALED PARTICLES

For inhalation studies there is an additional particle sizing problem as mentioned in Section 2.5. On inhalation aerosol particles are exposed to atmospheres saturated with water vapour (100% RH). Consequently, water vapour rapidly condenses onto their surface. Should the aerosol particle be a solute, or contain a solute, it will dissolve in the condensed water to form a solution. Such solutions encourage further water vapour sorption and particle growth. The extent of increase in the size following inhalation depends on initial particle size, its hygroscopicity and the time elapsed between inhalation and deposition. It will be greatest for smallest particles which are highly hygroscopic.

As described more fully in Chapter 9, the point of particle deposition in the respiratory tract depends on particle size. Largest particles deposit in the nose and smallest in the alveoli. In order to correlate deposition point with size of hygroscopic particles their hydrated size is required.

There are several ways of estimating fully hydrated particle size. One simple method is to obtain the particle size of the dry aerosol particles and calculate their size assuming that they are at equilibrium with air at 99.5% RH. According to Druett (1967) this estimated size is satisfactory at least for glycerol droplets. An alternative is to determine the size distribution of particles in the air exhaled following aerosol inhalation. But, as retention efficiency by the lung is particle-size dependent, exhaled aerosol may not be truely representative of that inhaled and hydrated. A more direct method is to artificially hydrate particles using rehumidification chambers (like those described in Section 2.10) prior to size analysis. The most accurate method, though, is to use a series of monodisperse aerosols and measure the size of each one following their inhalation and exhalation, particle retention rarely being complete. In this way, ambiguities mentioned above, due to employing polydisperse aerosols are eliminated.

Having established hydrated particle size, correlations between hygroscopic particle and non-soluble particle (e.g. coal dust) retention by the respiratory tract then become possible.

5.9 CONCLUSIONS

Deriving aerosol particle size meaningfully and accurately is difficult. The first problem is in obtaining a truely representative sample of polydisperse aerosols. The next is to measure particle size, the definition of which is somewhat technique dependent. Then, there are artefacts associated with the sizing method employed, and sometimes with the data analysis, also. Microscopy is probably the best method, but is not without its difficulties. For inhalation work, particle hydration can occur and result in increased particle size. The task of aerosol particle sizing is full of pitfalls for the unwary.

REFERENCES

Agarwal, J. K. and Fingerson, L. M. (1979). *T.S.I. Quarterly*, **5**, 3–6.

Beddow, J. K. (Ed.) (1984). *Particle Characterization in Technology*, vol. II, *Morphological Analysis*, C.R.C. Press, Florida.

Beddow, J. K. and Meloy, T. P. (Eds) (1980). *Advanced Particulate Morphology*, C.R.C. Press, Florida.

Cadle, R. D. (1975). *The Measurement of Airborne Particles*, Wiley, New York, Chichester.

Deepak, A. and Box, G. P. (1982). In *Atmospheric Aerosols their Formation, Optical Properties and Effects*, A. Deepak (Ed.), Spectrum Press, Hampton, Virginia, pp. 79–109.

Druett, H. A. (1967). *Symp. Soc. Gen. Microbiol.*, **17**, 165–202.

Fletcher, R. A., Mulholland, A. W., Chabay, I. and Bright, D. S. (1980). *J. Aerosol Sci.*, **11**, 53–60.

Gerber, W. E. and Hindman, E. E. (Eds) (1982). *Light Absorption by Aerosol Particles*, Spectrum Press, Hampton, Virginia.

Governmental Industrial Hygienists (1979). Air Sampling Instruments, 5th edn, American Conference of Governmental Industrial Hygienists, Cincinnati, Ohio.

Kasper, G. and Berner, A. (1981). *J. Coll. Inter. Sci.*, **80**, 459–465.

Kerker, M. (1969). *The Scattering of Light*, Academic Press, New York, London.

Knuth, R. H. (1979). In *Aerosol Measurement*, D. A. Lundgren *et al.* (Eds), University Presses of Florida, Gainsville, pp. 108–116.

Lundgren, D. A. *et al.* (1979). *Aerosol Measurement*, University Presses of Florida, Gainsville.

Marple, V. A. and Liu, B. Y. H. (1974). *Environ. Sci. Technol.*, **8**, 648–654.

Marple, V. A. and Willeke, K. (1979). In *Aerosol Measurement*, D. A. Lundgren *et al.* (Eds), University Presses of Florida, Gainsville, pp. 90–107.

Marple, V. A., Liu, B. Y. H. and Whitby, K. T. (1974). *Trans. A.S.M.E. J. Fluid Engng.*, **96**, 394–400.

May, K. R. (1975). *J. Aerosol Sci.*, **6**, 413–419.

McCrone, W. C. and Pelly, J. G. (1973). *The Particle Atlas*, vol. I, 2nd edn, Ann Arbor Science, Ann Arbor, Michigan.

Mitruka, B. J. (1977). *Methods of Detection and Identification of Bacteria*, C.R.C. Press, Florida.

Moss, O. R. (1979). In *Aerosol Measurement*, D. A. Lundgren *et al.* (Eds), University Presses of Florida, Gainsville, pp. 24–28.

Pinnick, R. G. and Auvermann, M. J. (1979). *J. Aerosol. Sci.*, **10**, 55–74.

Rao, A. K. (1975). Ph. D. Dissertation. *An Experimental Study of Inertial Impactors*, University of Minnesota, Particle Trajectory Laboratory, publ. no. 269.

Rideal, G. (1985). *Lab. Equip. Digest.*, **23**(2), 91–97.

Sern, G. J. (1984). *T.S.I. Quarterly*, **10**(3), 3–12.

Tillery, M. I. (1979). In *Aerosol Measurement*, D. A. Lundgren *et al.* (Eds), University Presses of Florida, Gainsville, pp. 3–23.

Weatherburn, C. E. (1968). *A First Course in Mathematical Statistics*, Cambridge University Press, Cambridge.

Wilson, J. C. and Liu, B. Y. H. (1980). *J. Aerosol Sci.*, **11**, 139–150.

Wesolowski, J. J., John, W., Devor, W., Cahill, T. A., Feeney, P. J., Wolfe, G. and Flochinnia, R. (1977). In *X-ray Fluorescence Analysis of Environmental Samples*, T. G. Dzubay (Ed.), Ann Arbor Science Publishers, Ann Arbor, Michigan, pp. 121–131.

Chapter 6

Biohazard control: containment, ventilation and isolation

6.1 INTRODUCTION

Man, animals and plants are exposed continuously to airborne microorganisms and allergens unless specific measures are taken to prevent it. Some of the microorganisms potentially are pathogenic but whether they initiate disease in recipients depends upon numerous other factors, for example, ability to survive, infectivity, the host and numbers of microorganisms to which that host is exposed. The last is a function of the degree to which airborne microbes and hosts are kept separated. This chapter is concerned with measures available for achieving that separation; later chapters deal with those other factors involved in airborne transmission of disease, such as relative humidity, oxygen toxicity and repair mechanisms.

Laboratory, factory and hospital environments, as well as public transport, theatres and other public meeting places, provide opportunities for the airborne spread of disease through the Aerobiological Pathway. For the first three, complete separation of hosts and microorganisms can be achieved but at the expense of restricting mobility. For the last three, complete separation is impractical leaving ventilation as the only realistic means available for controlling airborne biohazards. Owing to the limited effects of ventilation which are reduced further by energy conservation, the airborne spread of disease is inevitable under most normal circumstances.

Such inevitability is demonstrated by infections acquired through laboratory accidents (Sulkin, 1960) and industry (Langmuir, 1961). According to Chatigny and Clinker (1969) every species of pathogenic microorganism studied in the laboratory has at one time or another caused infection of operators. These authors indicate that numbers of infections per million man years range from as high as 50 for research laboratories to 0.1–0.4 for clinical and public health laboratories, for the period 1930–1959. The corresponding hospital rate was 4.2 (Wedum, 1957).

Occasionally, laboratory epidemics have arisen because of accidents or procedures generating and spreading infectious aerosols. For example, in 1938, 94 persons in the Bacteriology Department of a State University contracted Brucellosis on aerosol leakage from a centrifuge. Details of this and other incidents are provided by Phillips (1965a, b). While centrifuges are notorious for generating aerosols, virtually all microbiological laboratory techniques do the same. Of most concern are those producing relatively high concentrations of respirable particles. One such is the familiar process of pipetting liquids when as many as 10 000 droplets (1–10 μm) are formed on forcibly expelling those remaining few μl in the pipette tip (Johansson and Ferris, 1946). Other examples are given in Table 6.1 taken from Chatigny and Clinker (1969), while Pike (1976) provides a summary and analysis of 3921 laboratory-associated infections, and Spendlove and Fannin (1982) discuss other cases.

With an increasing use of microorganisms in industry, microbial growth by large volume continuous and batch culture is becoming widespread. Potentially, large volumes of culture are massive sources of aerosols should an accident occur owing to an operating fault. Ashcroft and Pomeroy (1983) performed experiments to simulate accidents which might occur during large-scale fermentation. They found as much as 0.005% of the entire contents of the fermenter as aerosol following various equipment failures. Hence, for a fermenter holding say 10^{15} bacteria then $10^{15} \times 0.005 \times 10^{-2} = 5 \times 10^{10}$ bacteria could be aerosolized by an accident.

While accidental aerosols of infectious materials (including extracted RNA and DNA) can be a health hazard, or cause cross-contamination, as pointed out by Dimmick (1974) they can be a legal problem as well. The seriousness of these biohazards, though, is not altogether clear from the work of Chatigny and Clinker (Table 6.1) because their data are largely qualitative. The purpose of the work by Kenny and Sabel (1968) was to provide relevant quantitative data

Table 6.1. Aerosol generation by some microbiological laboratory operations. Reproduced by permission of Academic Press Inc.

Laboratory operation	Average number of colonies on air sampler plate (Minimum)	(Maximum)
Agglutination, slide drop technique	0.0	0.66
Animal injection	15.0	16.0
Centrifuge, broken tube, culture in cup	0.0	20.0
Centrifuge, broken tube, culture splashed	80.0	1800.0
One drop of *S. indica* falling 7.6 cm (3″)		
onto stainless steel	0.2	4.7
onto dry hand towel	0.0	0.35
onto towel wet with 5% phenol	0.0	0.05
Insert hot loop into culture	0.68	25.0
Insert cold loop into culture	0.0	0.22
Break ampoule lyophilized *S. indica*	1939.0	2040.0
Streaking agar plate with loop	7.0	73.0
Pipetting, innoculate culture	0.0	2.0
Use of blender with poorly fitting parts	77.0	1246.0
Opening screw cap bottle	0.0	45.0

by performing various operations in an enclosed chamber and obtaining aerial concentrations, as well as particle size distributions. But as the test organism was *Serratia marcescens* and the RH 50%, this organism quickly lost viability in the airborne state due to simultaneous desiccation and oxygen-induced death (see Chapters 10 and 11). Their estimates of biohazards, therefore, were too low.

This difficulty was recognized and rectified in the work of Dimmick *et al.* (1973) and Dimmick (1974) in which:

1. laboratory operations were performed in closed chambers of minimal practical volume;
2. a standard suspension of an aerostable bacterium was used;
3. the generated aerosols were sampled as quickly and efficiently as possible;
4. all chambers (of a size appropriate to the laboratory operation being evaluated) had a filtered air inlet, a sampling port, and provision for being 'air-washed';
5. Andersen stacked sieve samplers were used principally and in some instances samples were collected also by impingers and by membrane filters.

Flavobacterium sp., TI coliphage, poliovirus and EMC virus were test microbes, while the very hardy *Bacillus subtilis var. niger* spores were avoided because of problems arising from possible laboratory contamination.

Each laboratory operation was assessed in terms of a 'spray factor', i.e. the number of viable microbes/ml of the test fluid divided by the number of viable microbes in the sampled volume and divided by the time (min) for which the operation was carried out. The factor has the units ml/min and derived values are reproduced in Table 6.2. Dimmick gives the following example for their application.

Table 6.2. Spray factors for some microbiological laboratory operations. Reproduced by permission of the author

Operation	Spray factor
Blender, lid off	6×10^{-3}
Sonic homogenizer	
maximum aeration	1×10^{-4}
minimum aeration	5×10^{-7}
Pipetting vigorously	1×10^{-4}
Vortex mixer, overflow	8×10^{-8}
Drop spilled on zonal rotor	2×10^{-6}
Single drop of liquid dropped 1 m	2×10^{-6}

Suppose a bacterial suspension at 4×10^{10} bacteria/ml is to be sonicated for 5 min in a room of volume 6×10^4 litres, then from Table 6.2 the corresponding spray factor is 1×10^{-4} and the source strength is:

$$4 \times 10^{10} \times 1 \times 10^{-4} = \underline{4 \times 10^6 \text{ bacteria/min}}.$$

During sonication for 5 min, the total number of bacteria dispersed would be:

$$5 \times 4 \times 10^6 = \underline{2 \times 10^7 \text{ bacteria}}.$$

At equilibrium the aerial concentration would be:

$$\frac{2 \times 10^7}{6 \times 10^4} = \underline{3.333 \times 10^2 \text{ bacteria/l}}.$$

Given a breathing rate for man of 10 l/min and a lung retention factor of 0.3, then a person spending 5 min in that room would receive a dose of:

$$3.333 \times 10^2 \times 0.3 \times 10 \times 5 = \underline{5 \times 10^3 \text{ bacteria}}.$$

Dimmick also indicates how room ventilation and proximity to the source can be included, if required. To allow for ventilation, a rule of thumb is to reduce the estimated doses by one-third (i.e. multiply the value by 0.67) for every 10 changes of air/h of ventilation, provided exposure time is greater than 3 min. Under 3 min any correction for ventilation is of little significance. To allow for proximity to the source, whether in a ventilated room or not, then during the first 3 min exposure a person within 1 m of the source would effectively be in a volume of 10^3 l rather than 6×10^4 l in the example given above. The aerial concentration then would be:

$$\frac{2 \times 10^7}{10^3} \ \underline{\text{bacteria/l}}$$

and the dose:

$$\frac{2 \times 10^7}{10^3} \times 0.3 \times 10 \times 3 = \underline{1.8 \times 10^5 \text{ bacteria}}.$$

Trying to refine these calculations further is probably unrealistic because actual situations will vary somewhat from the set-up employed by Dimmick, as will the diffusion of the aerosols through the room (see Section 8.2). Also, because of those many factors known to affect survival and infectivity of airborne microorganisms (see Chapters 8–15), actual infectious concentrations may be difficult to predict. Even if this could be done accurately, values still would need to be compared with numbers of microorganisms (e.g. ID_{50}) known to initiate infection. Such doses have been established only for a few diseases. A practical approach, then, is to use the calculated doses to provide a 'worse case' that is, assume no loss of infectivity and that the ID_{50} is one bacterium, for example. Thus, any procedure likely to produce high doses should only be conducted within suitable enclosures. Better still is to perform all laboratory operations with microorganisms in safety cabinets, irrespective of whether they are likely to lead to high or low airborne doses. Then the probability of the Aerobiological Pathway causing unintentional infections should be minimized, as should problems of cross-contamination of cultures, etc.

6.2 AIR PURIFICATION

Whether containment, ventilation or isolation is invoked, a source of purified air will be required. This may be air that is recirculated, a continuous new supply or their combination, e.g. make-up air. Classes of airborne contamination, arising in outside or inside air, which need to be removed during air purification include:

1. microorganisms,
2. dusts (including rust and other metal particles),
3. pollens and other microflora,
4. hair, faeces and animal dander,
5. mists and fogs (including oil and water),
6. vapours and gases (including combustion products).

In addition to a necessity for their removal, it is usual to deliver purified air in a respirable condition with respect to oxygen concentration, relative humidity (RH), temperature, and these days, negative ion content even. As far as air quality is concerned, cleaning of air for indoor use generally demands the highest possible degree of purification. In contrast, cleaning air for discharge outdoors as waste usually requires less stringent measures.

Opinions differ widely as to permissible concentrations of air contaminants while protecting human health (Witheridge, 1958) and these levels are frequently revised. Even so, required extent of purification depends on application as the following example shows. In animal or human environments, trace levels of ozone engender significant physiological response, whereas in an apparatus for sampling inert aerosol particles onto clean surfaces, ozone levels may be inconsequential. The former problem can be exacerbated through day-to-day variations in the composition of outside air, and in variations between locales. Providing a precisely defined and controlled purified air supply, therefore, is not always easy.

Of all the properties of air supplies that confront the designers of purification equipment, temperature requires the greatest attention (Witheridge, 1958). When coupled with pressure and relative humidity, properties of the air itself, e.g. viscosity, density, diffusivity, chemical activity, are affected. So, too, are any airborne particles, the durability of construction materials and sizes of the purification plants. Pressure influences design through material strength and leakage, and in this regard maintenance of correct pressure differentials can be crucial in controlling direction of flow of any leaks. For example, maintenance of a negative pressure differential in a cabinet containing a centrifuge is essential if possible aerosol leaks to the surroundings are to be prevented.

6.2.1 Purification processes

In general, purification processes may be considered under the following headings.

Aerosol removal

Settling chambers: removal by gravity.
Cyclones and inertial separators: removal by impaction.
Filters: particle laden air flows through irregular channels in cloth, paper, glass fibre, etc. Particulates through inertial, electrostatic, diffusion, and gravity forces collide with and are trapped by the filter elements even though the air passage sizes greatly exceed particle dimensions.
Electrostatic precipitators: particles are charged in an electric field and deposited on an electrode.
Scrubbers: centrifugal, packed tower, spray, venturi, etc. particles and liquids are brought into close contact causing particle removal by deposition.

Aerosol inactivation

Heating: sterilization or incineration.
Radiation: electromagnetic or particulate radiation — UV, IR, α, β, γ.
Chemicals: vapour phase disinfection by ethylene oxide, formaldehyde, β-propriolacteone, etc.

Gases and vapours

Adsorption: removal by solid adsorbents in fluidized beds and towers.
Absorption: removal by liquid absorbents in scrubbers.
Cooling and compression: removal by condensation.

Air ions, charged particles

Precipitation: charged plates.
Neutralization: radioactive source, generation of ions of opposite sign.
Most of these techniques are common to industrial plants and other large scale installations and they are discussed together with their relative advantages and disadvantages, and equipment and operating costs, etc., in engineering handbooks and other publications (e.g. Witheridge, 1958; Whitby *et al.*, 1961; McPhee, 1966; Danielson, 1967; Chatigny and Clinker, 1969; Baturin, 1972; Scales, 1972; Coulston and Korte, 1972; Nelson, 1972; Ascott, 1979; Strauss, 1978). For ease of reference the very useful summary table of Chatigny and Clinker (1969) is reproduced in Table 6.3 (a), (b), while that of Baturin (1972) is given in Table 6.4.

Of the techniques for removing particles, filtration is widely applied when the Aerobiological Pathway is involved. The basis of operation of filters (Darlow, 1973) is that being composed of cylindrical fibres of small diameter, when air passes through them, the streamlines diverge around those fibres. Particles of diameter of 0.5 μm and above, having a density greater than that of air, follow a trajectory towards the fibre to which they adhere by Van der Waals forces if collision occurs. For particles too small to be affected by inertial

Table 6.3. (a), (b) Characteristics of air and gas cleaning devices. Reproduced by permission of Buffalo Forge Co., Buffalo, N.Y.
(a)

Name of device		Description of device
General class	Specific type	(for each specific type or variation thereof)
Odour absorbers	Shallow bed	Activated charcoal beds in cells or cartridges, molecular sieve.
Air washers	Spray chamber	One or two coarse spray banks followed by bent plate eliminators.
	Wet cell	Wetted glass or synthetic fibre cells followed by bent plate eliminators.
Electro. precip., low voltage	Two stage, plate	Ionizing (+) wires followed by collecting (−) plates.
	Two stage, filter	Ionizing (+) wires followed by filter (−) cells.
Air filters, viscous coated	Throwaway	Deep bed of coarse glass, vegetable or synthetic fibres in cells.
	Washable	Deep bed of metal wires, screens or ribbons in cells.
Air filters, dry fibre	5–10 micron	Porous mat of 5–10 micron glass or synthetic fibres pleated into cells.
	2–5 micron	Porous mat of 2–5 micron glass or synthetic fibres pleated into cells.
Absolute filters	Paper	Porous paper of <1-micron glass, ceramic or other fibres pleated into cells.
Industrial filters	Cloth bag	Bags made of natural or synthetic fibre fabrics.
Electro. precip., high voltage	Single stage, plate	Ionizing (−) wires between parallel collecting (+) plates.
Dry inertial collectors	Settling chamber	Straight horizontal chamber—some with shelves.
	Cyclone	Chamber with provisions for spiral flow.
	Impingement	Alternate stages of nozzles and baffles.
Scrubbers	Cyclone	Cyclone collector with coarse radial sprays.
	Impingement	Impingement collector with wetted baffles.
	Fog	Cyclone collector with fine tangential sprays.
	Multi-dynamic	Power driven normal and reverse flow fan stages with coarse sprays.
	Venturi	Venturi with coarse sprays at throat.
	Submerged nozzle	Nozzle partially submerged in water.
Incinerators	Direct	Combustion chamber with supplemental fuel firing.
	Catalytic	Combustion chamber with catalyst plus supplemental fuel.
Gas absorbers	Spray tower	Vertical-up airflow chamber with downward sprays.
	Packed column	Tower with counter-currently wetted Rashig rings, Berl saddles, etc.
	Fibre cell	One or more stages of co-currently wetted fibre cells.
Gas adsorbers	Deep bed	Activated charcoal beds in regenerative-recovery equipment, molecular sieve, activated alumina, silica gel.

(b)

Name of device	Optimal size particle	Limits of gas temperature	Usual face velocity		Usual air resist.	Usual efficiency
General class	microns	°F	fpm	through	WG	% by Wt
Odour adsorbers (1)	(Molecular)	0–100	50–120	bed	< 0.3	< 95
Air washers (1, 2)	>20	40–700	300–500	chamber	< 0.4	< 25
	> 5	40–700	200–350	cells	< 0.7	> 25
Electro. precip., low voltage (2)	< 1	0–250	275–500	plates	< 0.3	< 90
	< 1	0–180	200–300	cells	< 0.2	> 50
Air filters, viscous coated (2)	> 5	0–180	300–500	cells	< 0.1	< 25
	> 5	0–250	300–500	cells	< 0.1	< 25
Air filters, dry fibre (2)	> 3	0–180	5–25	mat	< 0.3	> 50
	> 0.5	0–180	5–25	mat	< 0.5	< 95
Absolute filters (5)	< 1	0–1800	4–6	paper	< 1	> 99.95
Industrial filters (3)	> 0.3	0–180[a]	1–30	fabric	> 4	> 99
Electro. precip., high voltage (3)	< 2	0–700	180–600	plates	< 1	< 95
Dry inertial collectors (3)	>50	0–700	300–600	chamber	< 0.1	<50
	>10	0–700	2000–4000	inlet	< 2	< 80
	>10	0–700	3000–6000	nozzles	< 4	< 80
Scrubbers (3, 4)	>10	40–700	2000–4000	inlet	> 2	< 80
	> 5	40–700	3000–6000	nozzles	> 2	< 80
	< 2	40–700	3000–4000	inlet	> 2	< 99
	< 1	40–700	2000–3000	inlet	(up to 4″ developed)	< 99
	< 2	40–700	12000–24000	throat	>10	< 99
	> 2	40–700	2000–4000	nozzles	> 2	< 90
Incinerators (4, 5)	any	2000	500–1000	chamber	< 1	< 95
	(Molecular)	1000	500–1000	chamber	> 1	< 95
Gas absorbers (4)	(Molecular)	40–100	300–800	tower	< 1	< 95
	(Molecular)	40–100	500–1000	bed	<10	< 95
	(Molecular)	40–100	200–300	cells	< 4	< 95
Gas adsorbers (4)	(Molecular)	0–100	20–120	bed	<10	<100

[a] 500 °F if glass fibre

Removable contaminants

(1) Maladors, gases
(2) Lints, dusts, pollens, tobacco smoke
(3) Dusts, fumes, smokes, mists
(4) Gases, vapours, maladors
(5) Special. Bacteria, radioactive or highly toxic fumes etc

Table 6.4. Effects on man of various concentrations in air (in mg/l) of poisonous gases and vapours [Taken from Baturin, *Fundamentals of Industrial Ventilation*, Pergamon Press, Oxford, 1972.] Reproduced by permission of Pergamon Journals Ltd.

		For 0.5–1 h exposure			For many hrs exposure	
Substance	Rapidly fatal	Fatal (rapidly or after some time)	Illness, possibly fatal (Gess)	Tolerable without immediate or retarded after-effects	Lower limit of dangerous concentration (Gess)	Tolerable for 6 h with no noticeable after-effect
Chlorine	2.5	0.1–0.15	0.04–0.06	0.01	0.001	0.003–0.005
Bromine	5.5	0.22	0.04–0.06	0.022	0.001	0.005
Hydrogen chloride	5.5	1.8–2.6	1.5–2.0	0.06–0.13	0.01	0.013
Hydrogen sulphide	1.2–2.8	0.6–0.84	0.5–0.7	0.24–0.36	0.1–0.15	0.12–0.18
Sulphur dioxide	—	1.4–1.7	0.4–0.5	0.17–0.64	0.02–0.03	0.06–0.1
Ammonia	—	1.5–2.7	2.5–4.5	0.18	0.1	0.06
Nitrous gases	—	0.6–1.0	—	0.2–0.4	—	(0.2)
Hydrogen phosphide	—	0.56–0.84	0.4–0.6	0.14–0.26	0.1 (death follows in 6 hr)	—
Hydrogen arsenide	5.0	0.05	0.02	0.02	0.01	0.01(?)
Carbon monoxide	—	2–3	2–3	0.5–1.0	0.2	0.1
Carbon dioxide	360	90–120	60–80	60–70	20–30	10
Phosgene	—	0.02–0.1	0.05	—	—	—
Petrol	—	30–40	25–30	10–20	5–10	10
Benzene	—	20–30	—	10	5–10	5–10
Chloroform	—	200	—	30–40	—	20–30
Carbon tetrachloride	—	400–500	150–200	60–80	10	—
Carbon bisulphide	—	15	10–12	3.5	1–1.2	—
Prussic acid	0.3	0.12–0.15	0.12–0.15	0.05–0.06	0.02–0.04	0.02–0.04
Nitrobenzene	—	—	—	1.0–1.5	—	0.3–0.5
Aniline	—	—	—	0.5	—	0.15–0.2

forces, diffusion, electrostatic and Brownian motion cause particle collision with filter fibres. Particles about 0.2 to 0.3 μm diameter and of about unit density are those least prone to inertial impaction, diffusion and Brownian motion, and are those most likely to penetrate fibre filters. Even so, virtually 100% filter efficiencies are achievable while offering little resistance to air flow owing to the large air channels of fibre filters. Such operation is fundamentally different to that of a sieve in which the pore size has to be smaller than the aerosol particles (to be removed) consequently offering high resistance to air flow. Fibres used in constructing filters include those of glass, asbestos, paper, cotton, synthetic plastics and ceramic materials. Absolute or HEPA (high efficiency particulate air) filters are those most often installed for the removal of airborne microorganisms (Sivinski, 1968). Preferably, they are fitted with a prefilter to remove large particulates, thereby reducing the burden on the absolute filter. Owing to a relatively fragile nature, air filters can be easily damaged in transit and during installation. Testing filters both before and after installation therefore is essential, especially so when pathogenic microbes need to be trapped. Various test methods are available, based on comparisons between upstream and downstream concentrations.

Challenge or test aerosols employed commonly are those of NaCl or DOP (dioctylphthalate) with particle sizes in the range 0.1 to 1 μm diameter. Monodisperse aerosols over a range of sizes provide the most stringent challenges, but owing to demands of time and cost, routine testing of filters is much more limited in scope (Darlow, 1973). The wide availability of real-time aerosol particle counters, though, has eased this situation. According to Chatigny and Clinker (1969) the most frequent causes of failures of installed air filters occur in the ductings or mounts into which they are fitted. Faults such as leakage through cracks or badly matting surfaces are common. Although when properly installed and operated with a prefilter, commercial standard HEPA filters have an operational life-time of two years or more depending on how dirty is their working environment.

Whether filters are required with efficiencies as high as 99.99994% or as high as 99.5%, say, depends on application. Because of greater costs and difficulties in testing, use of extremely high efficiency filters may not be warranted. Instead it may be preferable to employ two lower efficiency filters connected in series. Another consideration then would be the pressure drop across the combination compared to a single filter except on installation in high-pressure pipelines when tolerance to pressure drop is greater than for a ventilation system, for example. But whenever high humidity conditions need to be tolerated moisture-resistant glass fibre filters are virtually essential.

As high efficiency filters demonstrate an approximately direct relationship between pressure drop and flow rate, a differential pressure gauge, or manometer, connected across a filter indicates its flow rate and when the filter resistance rises due to particulate burden.

Purification of air through removal of toxic gases and vapours is achieved by their adsorption in the pores of activated charcoal, activated alumina, silica gel or molecular sieves. Charcoal is activated by selective oxidation of constituents through controlled exposure to a set temperature. Activated charcoal is produced from numerous materials including coconut shell. It is usual to impregnate it with metal salts (e.g. Cr, Cu), thereby increasing its ability to catalytically degrade those toxic gases which are not adsorbed. Most condensable toxic gases consequently are removed either by adsorption or degradation (e.g. by Purafil, activated alumina impregnated with potassium permanganate) although carbon monoxide is an exception. It first has to be oxidized to CO_2 by a bed of platinum catalyst or Hopcalite, a mixture of copper and manganese oxides.

Silica gel principally is for removing water vapour. It is relatively cheap, strongly sorbs water vapour while readily being regenerated by heating. Molecular sieves comprising sodium or calcium aluminium silicates have a broader role. They have high porosity, pores of uniform size and large internal surface area, with access to these internal sorption sites being controlled by pore size. Molecules with diameters appreciably larger than those of the pores being unable to reach the interstices pass through a bed of molecular sieve, whereas smaller molecules are retained. Molecular sieves are regenerated by heating or through displacement with a low molecular weight gas. Having two sorption beds in parallel permits one to be regenerated while the second is operational. Alternating their roles provides for continuous purification even though absorbants have finite capacity.

Preferably, all purified air supplies are continuously monitored, using methods akin to those described in Chapter 4 or by Patty (1958), e.g. infrared monitors, in-line gas phase chromatography or mass spectrometry. Combustible gases and vapours pose an additional problem due to explosion hazards (Jones, 1958; Baturin, 1972), as do combustible dusts (Hartmann, 1958).

In most instances little attention is paid to the control of air ion content, but as described in Chapter 13 it can affect the survival of airborne microbes as well as influence degree of human comfort.

6.2.2 Air sterilization processes

An alternative to placing several HEPA filters in series to ensure complete removal of pathogenic microbial aerosols is to pass the effluent air through an air burner. Such devices — powered by electricity, gas or oil — operate at 300+ °C. The ratio of (temperature/residence time) provides some estimate for effectiveness of air sterilization. Chatigny and Clinker (1969) indicate that 300 °C for 6+ seconds ensures a good safety margin even for killing bacterial spores. But, due to their high exhaust temperatures, air burners are not generally applied and are suited better to the treatment of waste prior to discharge outdoors.

In certain circumstances dry heat sterilization under vacuum is the method of choice, as in the cabinet complexes deployed at the initial handling of returned lunar samples. The interiors of these cabinets needed to be sterile to prevent sample contamination by terrestrial microorganisms, etc. The method is to evacuate to about 1×10^{-6} torr thereby desiccating any microorganisms and making them highly susceptible to temperatures around 100 °C (Miller, 1970). The more familiar option of steam sterilization has some limitations imposed by the high temperature and pressure required as well as by condensed water vapour. These do not seem to apply when sterilizing large aerosol holding facilities like those described by Hood (1971, 1973).

For general air sterilization many laboratories, factories, hospitals, schools, etc., exploit short wavelength ultraviolet light (2537Å) (Mazzarella and Flynn, 1969). UV germicidal action is well established, although as discussed in Chapter 14 some microorganisms repair UV-induced damage. Of more concern, perhaps, is ensuring that UV light intensity is maintained. Simple visual examination can be dangerous to the eyes (Guth and Lindsay, 1958) as well as being unreliable in that UV and visible output are not necessarily correlated. Without doubt, lamps should be tested routinely, but the method of test can depend on the installation and the number of lamps (Thomson *et al.*, 1970). Difficulties, though, still can accrue when shadows are cast and areas become shielded from the UV. On the other hand, if done properly, UV-sterilization can be effective, convenient, readily available and with minimal clean-up procedures to follow (*Handbook of Environmental Control*, 1975).

Particulate radiation (i.e. α, β, γ) compared to UV has limited application for air sterilization, whereas for sterilization of laboratory apparatus such as petri dishes, pipettes, γ-irradiation is common because of the high penetrating ability of γ-rays due to their high energies. Powerful radiation sources (e.g. high energy cobalt) are required to deliver the high dosage necessary to ensure sterilization. But these impose their own safety requirements because of radiation hazards which are reduced through shielding by lead, concrete or steel (Curtiss, 1958). Other forms of radiation such as X-rays and infrared do not seem to be effective in air sterilization; nonetheless, like γ-rays, they can pose a radiation hazard for man (Curtiss, 1958).

In some instances it is practicable to spray or vaporize aerial disinfectants. Ozone was one of the first tested and Elford and Van den Ende (1942) found that at concentrations tolerated on breathing, it was effective only against 'naked' (i.e. unprotected) airborne bacteria, and then only between 60 and 90% RH. Protected airborne bacteria (e.g. bacteria covered by saliva) were inactivated only at ozone concentrations deemed to damage lung tissue. For bacteria deposited on surfaces even higher ozone concentrations were required for their inactivation. Consequently, ozone as a disinfectant is not really satisfactory and becomes limited to those situations when it will not be injurious to man or his interests. Ozone biohazards may still arise unintentionally when, for example, UV lamps are operated in confined spaces and a build-up of the gas takes place through the action of UV on oxygen in the air.

As one alternative to ozone, Williamson and Gotaas (1942) studied the germicidal action of aerosols of ethanol, chlorine, chloramine, sodium hypochlorite and resorcinol-glycerine mixtures. Whereas these aerosols were not particularly effective, Elford and Van den Ende (1942) observed that those of sodium hypochlorite and of hypochlorous acid, like ozone, were most effective at 70–90% RH but caused respiratory irritation and corrosion of certain metals. Formaldehyde vapour behaved in an analogous fashion being effective at 86% RH but not at 32% RH (Beeby *et al.*, 1967). On the other hand, less obnoxious aerial sterilizing agents can be derived through reactions between ozone and olefins (Dark and Nash, 1970; de Mik *et al.*, 1977), but these still work best at high RH. According to Nash (1951, 1962) all good aerial disinfectants work best at about 80% and have low vapour pressure so that they condense readily onto aerosol particles. These same features arise also for somewhat special aerial pollution products, OAF, as described in Chapter 12.

The general conclusion of an editorial article in the *British Medical Journal* (1949): '...that air disinfection is practical only in special circumstances. There is no prospect whatever of the widespread adoption of any of these methods in public buildings and conveyances; such problems are almost insuperable owing to the great size of the task or to the variable conditions for which provision would have to be made' seems equally true today. Even ethylene oxide, which at one time showed much promise, is now considered carcinogenic. Perhaps it is unrealistic to expect to find a chemical vapour which, for example, can denature the proteins and nucleic acids of microorganisms yet be totally innocuous to similar moieties of higher life forms. Consequently, aerial disinfectants are apt only for those situations where they would not endanger human health, e.g. disinfection of sealed rooms. Even so, the effectiveness of other techniques for air purification may make them redundant as epitomized by Polaris submarines and space vehicles where air is successfully recycled and artificially replenished over considerable periods of time (Watkins, 1970).

6.3 CONTAINMENT

6.3.1 Laboratory systems

Here, containment is taken to mean enclosing aerosols of hazardous materials to prevent their release into the environment. Unless contained, their effects can be widespread in a manner dependent on their nature (Edmonds, 1979). Equipments range from simple hoods to sophisticated sealed safety cabinets requiring that work be performed through attached thick rubber gloves.

The first major steps towards establishing internationally recognized criteria in microbiological safety cabinets was the World Health Authority's classification of types I, II and III. Later, the US National Sanitation Foundation Standard 49 and British Standard BS 5726 followed suit. The Classes I, II and III refer to levels of protection afforded and to the hazard of materials for which they are appropriate.

Class I hoods are open-fronted cabinets (Figure 6.1) through which room air is drawn at a velocity of 20–35 m/min. After passing through a pre-filter and a HEPA filter of 99.999% efficiency for 0.3 μm particles, the air is exhausted either to the room or outdoors. Some manufacturers provide a solid state

Figure 6.1 Class I open front safety hood. (From Chatigny and Clinger, in *An Introduction to Experimental Aerobiology*, R. L. Dimmick and Ann B. Akers (Eds), Wiley-Interscience.) Reproduced by permission of John Wiley & Sons Inc.

temperature compensated air velocity indicator, together with a motor control circuit to provide constant air intake velocity under varying conditions — both external and internal.

Other features can include high or low level flow alarms, UV lights, a formalin vaporizing unit for sterilization, a timer, and an hours-run counter. As a Class I system is intended to protect a worker against inhalation whilst handling hazardous materials, anti-blow back (non-return) valves are essential to this and all other systems. Preferably, air is sucked rather than blown through the filters so that hazardous aerosols are never in air at a higher pressure than ambient. Otherwise there is always a potential hazard due to leakage. The cabinet itself may be of stainless steel or of mild steel coated with white epoxy resin, aluminium sheet or 6 mm thick welded PVC which is flame retardant. Radiused corners are advantageous. Front vizors can be of toughened glass, transparent plastic or polycarbonate, while interior lights should provide even illumination. Depending on manufacturer, systems are designed to be bench or floor mounted. Class I safety cabinets are appropriate for handling category 1 aetiological materials or (depending on classification system) medium to low-risk pathogens, category B1.

Class II design incorporates laminar flow and its development was given impetus from the space programme. Class II systems are suitable for category 2 materials or category B2, C, low-risk pathogens, under certain circumstances. These circumstances are related to the routing of the laminar flow of air.

Unfortunately, the design of Class II systems has not been standardized and in some, perhaps better termed laminar flow clean air work stations, the air is discharged horizontally *towards* the worker. This type of design would offer that worker no protection from microbiological aerosols, rather it is likely to increase exposure to them. Such systems are more appropriate to keeping non-hazardous materials free from contamination during their manipulation.

In other designs for Class II, a vertical downflow containment hood provides an ultra-clean work area offering protection both to the materials being manipulated and to the operator (e.g. Figure 6.2). Such would seem preferable to horizontal laminar flow systems, which were originally devised by Whitfield *et al.* (1962) and were extended to clean rooms and areas required

Figure 6.2 Class II open front safety hood with access ports. (From Chatigny and Clinger, as Figure 6.1.) Reproduced by permission of John Wiley & Sons Inc.

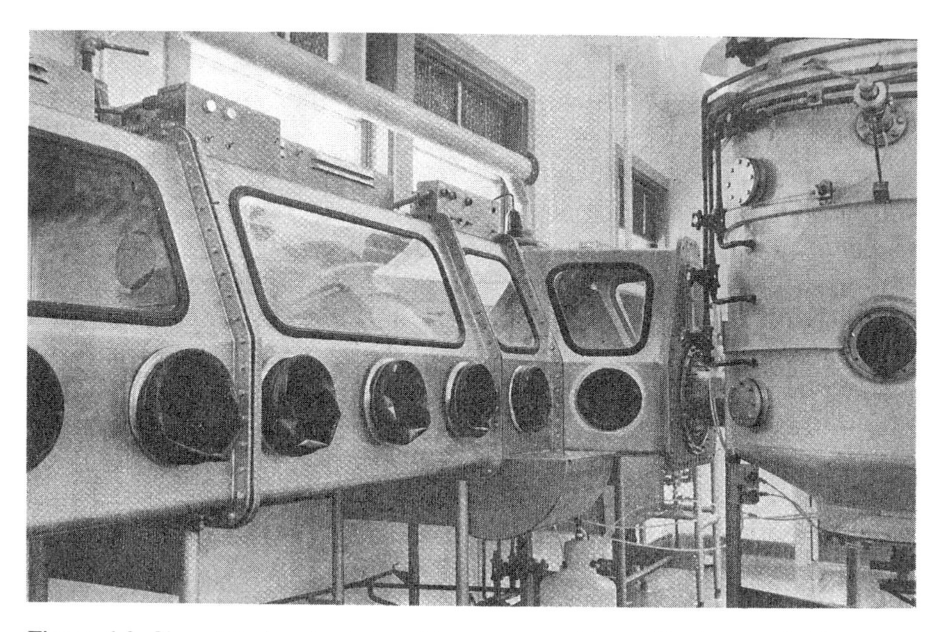

Figure 6.3 Class III closed safety cabinet system. (From Chatigny and Clinger, as Figure 6.1.) Reproduced by permission of John Wiley & Sons Inc.

for tissue culture, assembly of spacecraft, satellites, and to the electronic and pharmaceutical industries. But, for microbiological work, *horizontal* laminar flow Class II systems probably are best avoided.

In most other aspects Class II systems are similar to those for Class I. But the reliance placed by both classes on air inlet velocity and volume for their efficiency is a potential weakness as that velocity may be exceeded by local air currents caused by room ventilation or by the movement of people. The latter can suck air from a properly operating Class I cabinet into a room in a quantity large enough to cause a biohazard. This effect can occur because a person walking at 2 mph creates air currents having a velocity up to about 55 m per minute. Likewise, the act of rapidly withdrawing one's hands can create an outflow. According to Chatigny (1974) and to Chatigny and Clinker (1969) design values of 45 m per minute are required for these hoods, but even then following usage filters become partially laden with concomitant increased pressure drop and decreased flow. The advisability of having low-flow alarms fitted to cabinets cannot be over stressed.

Class III cabinets (e.g. Figure 6.3) are hermetically sealed with air inlet and outlet being controlled and with manipulations taking place through fitted rubber gloves. Safety is enhanced by the cabinet interior pressure being at below ambient (i.e. under negative pressure) and by having an airflow/pressure monitoring and alarm system. Materials are introduced into or removed from the cabinet through a front port, dunk tank carrying disinfectant, double ended sterilizer or transfer hatch. They also feature internal lighting to a level of 1000 lux. As with all Class III systems, they are intended for use with category 3 or A high-risk pathogens and recombinant DNA, where complete isolation of work and operator is required.

Their inlet air is drawn into the work area through a HEPA filter and then exhausted to the atmosphere through a prefilter and a final HEPA filter bank. Other features include many of those for Class I and II cabinets plus hinged access door/viewing panel with quick release fasteners, and gas, air and electrical fittings. That for gas (for bunsen burners, etc.) preferably incorporates a solenoid valve that allows flow only when the exhaust fan is operating.

A correct negative pressure differential is essential for maintaining safety while ensuring that the chamber pressure is not too low to make the rubber gloves rigid and difficult to work with. Automatic control of the exhaust fan is an asset in this regard and also when the Class III hood is operated in Class I mode for which the rubber gloves are removed. Commensurate with a change in classification must be the category of microorganisms being handled. Class III operation provides maximum user protection, with Class I next and Class II least, but because no system can provide perfect protection, or may fail, Class III systems preferably should be operated in a ventilated room dedicated for that purpose and having restricted access.

Since the ventilation rate of these cabinets is about 400 to 600 cubic feet per minute, their operation can modify air currents within the room containing them, as well as modifying room ventilation. This is especially so if the cabinet exhaust is vented outdoors rather than returned to the room. One alternative is to make the hood part of the room ventilation system and run it continuously. This approach can be expensive because the hood HEPA filter will require frequent replacement since it will become laden quickly by room and replacement outside air. Returning vented air to the room following its purification, therefore, can be advantageous. Under some circumstances this may not always be possible, that is, when the temperature and humidity requirements for the room and cabinet interior are not compatible.

Positioning of Class I and II systems should be to minimize interference caused by people moving past them, while causing minimum disturbance of room airflow and ventilation. System construction preferably should be of metal or reinforced plastics, e.g. fibre glass resin, which is unaffected by chemical disinfectants and by UV light. In the long term, cabinets made from stainless steel have advantages of mechanical strength, durability and economy. To be most effective their size and sophistication should be tailored to the task (e.g. to contain a centrifuge, rotating drum aerosol holding facility, etc.) and to the likely biohazards.

6.3.2 Testing safety cabinets

That safety cabinets must be safe is obvious but how to ensure their continuing operation in this state is less so, other than to frequently test them. These tests are of three types: those carried out by the manufacturer on completing the cabinet, type testing to satisfy standards and routine testing following installation. They all more or less rely on the same procedures. Types of tests are determining the percentage penetration of the HEPA filters, the operator protection factor and the air velocities and flow patterns. Efficiency of filtration usually is established by one or more of the following techniques. One is to release an aerosol of microorganisms (Board and Lovelock, 1973) (e.g. *Bacillus subtilis* var. *niger* spores, T1 coliphage) inside the cabinet and to sample it (e.g. by impinger) within and outside the cabinet. Alternatively, samples are collected upstream and downstream of the filter. Comparison of corresponding samples permits cabinet leakage or filter penetration to be measured. Conveniently, aerosols may be generated by Collision spray or other small two-fluid atomizers, but unfortunately, this approach is difficult to standardize for all those reasons given in later chapters dealing with survival. In addition, a trained aerobiologist and good laboratory facilities are required, while the release of large numbers of microorganisms as aerosols inevitably leads to various problems of contamination. Furthermore, when low levels of test organisms need to be detected problems are likely due to background contamination, while in susceptible persons problems due to allergenic reactions can be precipitated.

Another approach is to disseminate sodium chloride while another is to use dioctyl phthalate (DOP) aerosols, as described, respectively, in BS 3928 and US MIL STD 282. The advantage of the sodium chloride method is that assay is in real time by the sodium flame test (Murphy, 1984), but suffers from the disadvantage of requiring frequent calibration which is relative humidity dependent. The DOP method is to generate with a Dautrebande atomizer a fine aerosol of relatively non-volatile and non-toxic DOP, having 98% of particles less than 3 μm diameter and 11% less than 0.3 μm. A light scattering particle photometer is set to sample the challenge airstream carrying the aerosol and the reading taken as 100%. A corresponding downstream reading gives a direct value for percentage filter penetration. A major disadvantage is that the analyser responds equally well to contaminating particles and to DOP particles passing through the filter or leakage from ducting, etc. However, the method is commonly applied because of its ease of use, portability and reliability.

A much better method, which also is utilized in determining operator protection factors, is the KI or potassium iodide discus test. It owes its reproducibility largely to generation of test aerosols by means of a spinning disc onto which is fed at a controlled rate an alcoholic solution of 1.5% KI. A monodisperse aerosol (plus some much smaller satellite particles) is formed of size determined by angular velocity of the disc perimeter. Samples collected by membrane filters inside and outside the cabinet provide a figure for cabinet leakage. This may be expressed in terms of operator protection factor, that is, the ratio of the exposure to airborne contamination when generated on the open bench to the exposure from the same dispersal when generated inside a cabinet. The factor should not be less than about 10^5 as in US Standard NSF 49.

Two other tests are the measurement of air velocity through the front opening of Class I cabinets, for example, and the use of smoke to see air-flow patterns within the cabinet and around its immediate surroundings, as well as to check laboratory ventilation. Another check is to see that the laboratory is at negative pressure to associated corridors.

Tests like those described above should be conducted when a cabinet is installed, moved, when new filters are fitted and also when routinely serviced at least twice a year. Other checks are required, for example, examination of seals, gaskets, pipe work, ducts, etc. together with control gear.

Frequent testing of cabinets together with their good care and maintenance will help to keep them safety cabinets. While in the past the use of microorganisms in such a programme produced delayed results, nowadays, the advent of rapid microbiological methods, e.g. light scattering (Autobac), bioluminescence (Lumac), electrical impedence (Bactmatic, Malthus) overcomes that difficulty while offering great specificity.

6.3.3 Aerosol holding chambers

These include rotating drums (Goldberg *et al.*, 1958; Dimmick and Wang, 1969), large volume spheres (Hood, 1971, 1973) and other similar facilities. Rotating drums owing to their rotating seals have a potential leakage problem which is minimized by operating them at negative pressure. Even then it is prudent to mount them in a safety cabinet. An extensive facility incorporating eight drums in a Class III system has been described by Chatigny and Clinker (1969) and by Goldberg (1970).

Large aerosol chambers, effectively pressure vessels, can be tested hydrostatically for leaks. In essence, a simple water manometer connected to the vessel is monitored after it has acquired negative or positive pressure with respect to the ambient air and then is sealed. Provided the temperature remains constant, changes in manometer readings give a measure of leakage rate. This rate will be that also for particles of respirable sizes which if appreciable requires the actual point or points of leakage to be found with halogen leak testers, for example.

Aerosol holding facilities may be tested also in an analogous manner to laboratory safety cabinets using tracer aerosols.

6.3.4 Rooms and buildings

Through their relatively large sizes, checking leakage into or out of rooms and buildings can be difficult because amounts of tracer sampled are likely to be small and confused with the natural aerosol. One successful method, due to its high sensitivity, is microaerofluorometry (Goldberg, 1968). Another promising method is that of using halogenated hydrocarbons and an electron-capture detector (Foord, 1973). While microbe tracers provide for sensitivity, their potential disadvantages for testing laboratory cabinets apply in this context as well.

6.3.5 Animal holding facilities

Determining the aerosol infectivity of microbial aerosols necessitates exposing animals to them under controlled conditions. Afterwards, these animals are maintained in individual cages which may need to be well separated to avoid cross-infection. Alternatively, animals may be kept in individual ventilated cages. In addition, personnel should wear protective equipment, either ventilated plastic suit or surgical gown plus respirator, together with rubber gloves.

Air from rooms housing exposed animals or from ventilated cages should be filtered then preferably passed through an air-burner before release to the atmosphere. UV lamps operating in these rooms and in their associated air-locks (for entry and exit of personnel) also help to limit the spread of air contagion. For similar reasons, materials leaving such rooms preferably do so through an autoclav and thereby are sterilized.

Such containment rooms have been termed tertiary barrier systems and may be considered the final line of protection provided access is limited to essential workers. By means of pressure differentials between rooms, as well as adequate ventilation, contamination control is enhanced as long as the air flow is directed inwards to those rooms containing the infected animals. Finishes in the rooms should allow easy cleaning, e.g. monolithic surfaces which withstand disinfectants. Polyurethane and epoxy wall and ceiling finishes and waterproof fixtures are recommended by Chatigny and Clinker (1969).

To conclude this section on containment, the following quotation taken from Dimmick's paper of 1974 provides, perhaps, the best basis for controlling biohazards in the laboratory, '... and the best judge of the real hazard, in terms of potential infection, is the responsible scientist at the bench, who should know as much about the particular microbe he is using as any other expert. Safety begins at home.'

6.4 AIR CURTAINS AND DOUCHES

In the previous section on containment and in those sections to follow on ventilation and isolation there is a common problem for personnel entering or leaving the rooms concerned. As indicated already, air-locks and pressure gradients can help contain air contagion. Two other potentially helpful techniques are air curtains and air douches. Air curtains are literally curtains of air (Figure 6.4). Warm filtered air is blown upwards, from the side, or downward, to interact with the cooler air of the room exterior (Baturin, 1972). The generated air curtain provides pneumatic resistance to the flow of cooler air through the doorway. In this way, the temperature of the room is

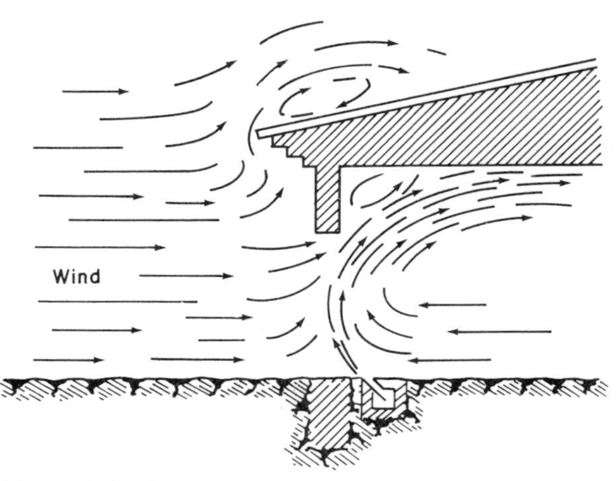

Figure 6.4 Air curtain at a doorway. (From Baturin (1972), *Fundamentals of Industrial Ventilation*, Pergamon.) Reproduced by permission of Pergamon Journals Ltd.

maintained while reducing the entry of airborne material. Air curtains seem appropriate for operating theatres and for patient and animal isolation rooms.

An air douche is effectively being 'washed' by a directed flow of filtered air. On exiting an animal holding room, for example, preferably through an air-lock, the incorporation into it of an air wash can reduce the transfer of materials from that room to the exterior. Depending on the extent of the biohazard an air douche rather than a shower may suffice, thereby significantly reducing the time required for personnel to exit such a facility. Baturin (1972) provides design and construction details for both air curtain and air douche systems.

6.5 TRANSMISSION IN HOSPITALS

Hospitals admit infected people for treatment and isolation, while simultaneously creating centres with a high infection risk, e.g. burns units. At the same time many opportunities are afforded for the spread of infection because of shared staff and equipment. Furthermore, common microbes to be found in hospitals usually have become resistant to many antimicrobial agents so that any infections they may cause can become difficult to treat.

Two types of airborne infection well recognized in hospitals are those of tuberculosis and infections of the respiratory tract (e.g. group A streptococcus) and those generated from the body surface. Considering the latter first, skin squalms or scales are constantly shed. Clark and Cox (1973) indicate that as many as seven million skin scales per minute are liberated by the abrasive and bellows action of clothing fabrics and bedclothes. Detached skin scales bearing microorganisms then may be aerosolized directly or by movements and bedmaking, for example, and lead to contamination of hospital wards and operating theatres (*Handbook of Environmental Control*, 1975). Showering increases the rate of shedding, presumably because of a loosening of skin scales (Spears *et al.*, 1965), as does physical activity (May and Pomeroy, 1973). These latter authors report also rates of shedding of viable bacteria (number/min) in the range about 250 for a clothed female to 17 650 for a naked male. By wearing tightly fitting neoprene sponge rubber dresses output from the perineal region was prevented with a concomitant marked reduction in the numbers of bacteria dispersed from males. Should this prove to be a general finding, the wearing of this garment may be advantageous in operating theatres, for instance. In this regard, the work of Blowers *et al.* (1973) also indicates that males disperse many more *Staphylococcus aureus* from the perineum than do women. On the other hand, Ayliffe *et al.* (1973) find that heavy staphylococcal disperses (men more often than women) usually have high counts of this bacterium in the nose, on the fingers, face and hair. In contrast, Lidwell (1973) studying patients over a period of 58 weeks found that women dispersed more *Staphylococcus aureus* than did males. An excellent collection of data of these kinds will be found in *Handbook of Environmental Control* (1975).

During talking, coughing and sneezing, air velocities may reach 50 m/s in mucous lined respiratory passages (Chausse and Magne, 1916) and cause liberation of droplets. During sneezing, according to Jennison (1942) as many as 40 000 droplets can be dispersed, which quickly evaporate to give particles mainly in the size range 0.5–12 μm (Duguid, 1946). Consequently, microorganisms associated with respiratory disease and with peridontal diseases enter the Aerobiological Pathway and spread more rapidly than by personal contact and communal practices. According to Laurell and Hambreus (1973) dispersal from the clothing of nursing staff also is significant. In general, any process or procedure which causes shear forces, or thermal gradients, will result in the generation of aerosols, while the amounts of pathogens so dispersed will be a function of the source in an analogous way to that described by Dimmick (1974) for laboratory manipulations.

Considering the relative ease with which aerosols are generated in hospitals, it is somewhat surprising that the airborne route is not implicated much more in studies like those of Laurell and Hambreus (1973), Jameson (1973), Gould (1970), Brachman (1970) and Ayliffe (1970). It may be that airborne pathogens rapidly lose infectivity, their numbers are low or their hosts have immunity, etc. There could be many reasons, but in studies like those above, an area which has received scant attention is the effects of the artificial sampling methods on the viability and infectivity of recovered microorganisms. As discussed at length in Chapters 2, 8, 9 and 10, the precise method employed for sampling microbial aerosols can play a vital role in determining observed viability/infectivity. It seems expedient, therefore, that in any study concerned with the airborne transmission of disease, exceptional care and attention should be paid to the sampling techniques. Apropos is that by adopting this approach, Thomas (1970a, b, c) increased the frequency with which he detected viable microorganisms in natural aerosols. To emphasize the point, Dimmick (1974) states: 'The (sampling) problem is very real but is usually ignored in field studies...'.

To prevent the general airborne spread of infection in hospitals and in other public places would require sophisticated and expensive equipment. A more practical solution in hospitals is to provide ventilation and isolation areas so that especially vulnerable patients (e.g. burn, transplant) need not be exposed unnecessarily to risks of airborne infections. A description of these systems follows in the next two sections.

6.6 VENTILATION SYSTEMS

Approaches to the provision of microbe-free rooms or enclosures are exclusion of contaminated air from without, minimizing dispersal of microorganisms from potential sources from within, and the rapid and complete removal of any contagion as it is generated. Unfortunately, ventilating a room with clean air only dilutes rather than replaces that already present. One solution, somewhat analogous to that in safety cabinets, is to use laminar-flow, either horizontal or vertical, with velocities of about 30 m/min. While with this system turbulence is not eliminated entirely, it is usually small enough not to impair system efficiency. Unidirectional laminar flows can provide the high levels of hygiene appropriate for surgery (McDade *et al.*, 1968; Charnley, 1968, 1970; Lidwell, 1973; Gould, 1970; Gould *et al.*, 1973), for individual patient care (Burke, 1967, 1970; Michaelson *et al.*, 1968; Foord, 1973), for rooms housing several patients (Lidwell, 1973; Dietrich, 1973) and for sterile rooms in the pharmaceutical industry (Sykes, 1970).

Lidwell (1973) provides experimental data which show that laminar rather than turbulent flow ventilation is from 10^2 to 10^3 times more effective in preventing airborne bacterial transfer provided the velocity is greater than 0.18 m/s (35 f/min.). At lower velocities performance was degraded at least in part by thermal gradients, while higher velocities up to 0.5 m/s (100 f/min.) produce no noticeable sensation for patients. On the other hand, due to enhanced body cooling, air temperatures have to be 1 or 2 °C warmer than when unventilated, while fan noise can be troublesome. Humidity control is required also. In general, there seems little doubt that laminar flow is more effective than conventional ventilation for reducing biohazards (*Handbook of Environmental Control*, 1975).

Air supplies are either crossflow or downflow, requiring, respectively, perforated end walls or floors, together with appropriate exhaust slots at ceiling or floor level. Of the two, horizontal flow is the more prevalent especially for rooms with several patients as their care is made easier by a direction of flow parallel to the long axis of the beds. While nursing is not quite as easy as in an open ward, it is easier than for single room isolation systems.

The consensus of opinion is that with this type of ventilation external airborne particles are excluded, that there is an extremely rapid removal of any airborne contagion generated within the room and that there are no transfers to the patients from areas downstream or to the sides. Several patients, therefore, may safely be nursed within a single room and yet be isolated one from another and from staff dealing with other patients. In units of this kind, though, contact with nursing staff still can be responsible for considerable transfer of microorganisms to patients. The wearing of sterile textile gowns reduces but does not give full protection against such transfers between patients via nurses clothing, although the use of plastic aprons, cleaned with detergent, is likely to help. Overall, the technique is highly effective with downflow being slightly better than crossflow (*Handbook of Environmental Control*, 1975).

As predicted by Lidwell (1973), over the years, ventilation of operating theatres has seen an increasing use of laminar air-flow rooms. This system when compared with others produced lowest bacterial counts around the immediate vicinity of the wound during operations (Wanner, 1973), that is, provided laminar flow functions properly and is regularly maintained and checked. Another particular point to watch is that humidifiers themselves are sometimes prolific sources of contamination, e.g. bacteria, moulds, fungi, etc.

Given a good laminar-flow ventilation system, the incidence of infection perhaps surprisingly is not significantly less than that for conventionally ventilated theatres, largely due to nursing contacts (Wanner, 1973; Sattel, 1974) or because of infections derived from the patients own tissues (Gould *et al.*, 1973). According to Laufman (1973) the appropriate application of fundamental surgical, technical and hygienic measures rather than the implementation of very expensive laminar-flow systems is sufficient for achieving asepsis.

Of systems installed in operating theatres, the question arises as to whether downflow from the ceiling or cross-flow from one wall is better. For rooms, Van der Waaij and Van der Wal (1973) found cross-flow more convenient as well as more effective in patient protection. However, when providing laminar-flow air only in the region of an operating table, as opposed to an entire room, downflow bounded by an outer zone created with air jets (rather than an air curtain) is probably best (Bossers *et al.*, 1973). Yet, as the actual area to be protected during surgery is relatively small, the best system of all probably is to generate a continuously expanding bubble of sterile air at the site of surgery. A protective barrier then is formed between the operating room team and the wound as it continuously drives all airborne contagion from the wound area. Such an air-bath unit was described by Westwood *et al.* (1973) in which sterile air is flexibly ducted to a plenum. Thereafter, the sterile air is fed into a diffuser which surrounds the wound and comprises open-weave fabric incorporated into the final sterile surgical drape. At least in laboratory tests the system provides complete exclusion of airborne contamination.

6.7 ISOLATION SYSTEMS

During chemotherapy of patients suffering from acute leukemia, from congenital immune deficiency syndrome, under cancer chemotherapy, during immunosuppressive treatment for organ transplants and other diseases, patients are highly susceptible to infection. Prophylactic therapy is to prevent infection by exogenous microorganisms as well as by microbes carried by the patient, i.e. it is to achieve a germ-free state. Several approaches have been tried including: laminar air flow rooms, the plastic isolator of the 'Life Island' type, the Ulm isolated bed system, and combinations thereof. All have their relative advantages and disadvantages, with the laminar air flow room being the most debatable in terms of the ease of breaking 'the barrier'. Each one requires high financial investment and an increase in personnel for each patient compared to conventional care. It is pertinent to ask, therefore, whether the use of the protected environment and a prophylactic antibiotic programme may be expected to extend remission.

Herman and Hart (1973) treated patients in an isolater system, the 'Life Island'. Isolators of this type are designed to provide a complete barrier between the inner compartment, in which is the patient, and his surroundings. Glass fibre filters (AAF type) and HEPA filters are employed for the removal of bacteria and viruses in the air supply to the inner compartment. Materials entering it first pass into an airlock fitted with UV lights as a final sterilization of surfaces and a disinfection of the associated air. Sterile tubes for infusions, transfusions, gas inhalation, etc., enter via special connectors fitted through the walls of the isolator and a row of glove ports on each side of the isolator allow for handling the patient.

Initial sterilization of the patient compartment is achieved by spraying 3% peracetic acid for 30 min followed by ventilation for 24 h to remove the acetic acid vapour (Wendt *et al.*, 1973). Ethylene oxide sterilization also is appropriate provided its explosive and mutagenic properties are taken into account. Following isolator sterilization the patient may enter it but in so doing breaks the mechanical barrier for about 30 s during which time reliance is placed on the over pressure of the inner compartment to prevent ingress of contamination.

The patient himself is surface sterilized daily during the 2 to 3 days prior to his entry, by whole-body washing with ampholyte soap. However, transition areas between skin and mucous membranes pose problems and call for antibiotics and chemotherapy. Oropharyngeal and nasopharyngeal areas are sterilized frequently with hexetidine-derivatives, antibiotics and antimycotics, while decontamination of the intestinal tract is by antibiotics (Klastersky, 1973). Difficulties still can arise especially from microflora of the oral cavity and ear, and from antibiotic resistance. Problems also can develop due to leaks in the plastic barriers, although the internal over-pressure, if maintained, minimizes these. Even then, damaged integral gloves may convey infection as can pharmaceutical products which sometimes carry microorganisms although usually of a non-pathogenic nature.

Burke (1970) reports that treating burns patients in plastic ventilated bed-isolators significantly reduces wound infections due to *Pseudomonas aeruginosa*, *Staphylococcus aureus* and *Escherichia coli*, acquired extramurally and by auto-infection. Their techniques included aerosol sampling with settle plates, Andersen samplers and a large slit sampler. Unfortunately, the last sampler gave erroneous results as its use created a negative pressure within the compartment and the entry of unfiltered air. Herman and Hart (1973) using an analogous isolator in conjunction with the treatment of 22 severely burned or leukemia patients observed a 100 to 1000-fold reduction in airborne microflora in the isolator compared to an average hospital ward or to a private room. Of the bacteria that were airborne in the isolator, relatively high concentrations, originating from the patient himself, arose during bedmaking or other periods of high activity.

Another method for assessing the efficiency of isolation procedures is by typing the patients microflora, but as pointed out by Noble (1973), there is then the problem of the sampling process (e.g. what body area should be swabbed). Given that sampling is adequate, an isolation of a microorganism originally not carried by the patient implies a failure of the quarantine procedures. Should then the microorganism be of a type that cannot be matched with another normally associated with the hospital environment, it may have been carried undetected by the patient. Due to the large number of strains of *Staphylococcus aureus* and *Pseudomonas aeruginosa* found in hospital environments, for these species the method is applicable only when completely different strains are isolated from patient and environment (Noble, 1973). Fortunately, and in contrast to *Pseudomonas aeruginosa*, *Staphylococcus aureus* only occasionally is the aetiological agent of infections in patients with impaired immunity, such as those suffering from Acute Myeloid Leukemia and kidney transplant patients. Rather, most infections originate from species of Enterobacteriaceae and by typing them Van der Waaij *et al.* (1973) observed that, perhaps surprisingly, isolation did not appear to reduce their colonization in immune deficient patients. In these particular conditions, the incomplete isolation of patients was traced to Enterobacteriaceae contamination of food, water and medicines supplied to them. But, how general a problem this may be, is not known.

Following treatment in isolators, patients before release have to be recolonized: a process which is dose dependent. (Such dose dependency also is of importance in man for the degree of the required isolation which is less in an individual having resistance to colonization.) Recolonization may be best accomplished with natural pathogen-free faecal flora because the anaerobic compliment plays a major role in promoting colonization resistance (Dietrich, 1973).

6.8 CONCLUSIONS

Control of biohazards in public places such as stores, theatres, lecture halls, public transport, etc., in practice is possible only on a very limited scale and the airborne spread of numerous diseases (e.g. respiratory, periodontal, etc.) by aerosols is inevitable. While the problem is serious enough and leads to much human discomfort, it would be much worse if the causative organisms frequently engendered lethalities in the population. When restrictions on movement are possible as in laboratories, factories, hospitals, etc., very much better control of biohazards is achievable. Purified and conditioned air supplies coupled with laminar flow and isolation techniques can very significantly reduce the spread of and the contamination by airborne microbes. Much pertinent experimental data have been compiled in the *Handbook of Environmental Control* (1975), while recent safety recommendations are to be found in HMSO *Categorization of Pathogens According to Hazard and Categories of Containment* (1984) and *Biosafety in Microbiology and Biomedical Laboratories* (1984), US Dept. of Health and Human Services.

REFERENCES

Ascott, R. (1979). *Toxic Chemical and Explosive Facilities*, American Chemical Society, Washington.

Ashcroft, J. and Pomeroy, N. P. (1983). *J. Hyg. (Camb.)*, **91**, 81–91.

Ayliffe, G. A. J. (1970). In *Third International Symposium on Aerobiology*, I. H. Silver (Ed), Academic Press, London, New York, p. 91.

Ayliffe, G. A. J., Babb, J. R. and Collins, B. J. (1973). In *Fourth International Symposium on Aerobiology*, J. C. Ph. Hers and K. C. Winkler (Eds), Oosthoek, Utrecht, The Netherlands, pp. 435–437.

Baturin, V. V. (1972). *Fundamentals of Industrial Ventilation*, 3rd edn, transl. O. M. Blunn, Pergamon Press, Oxford, New York. Sydney, Braunschweig.

Beeby, M. M., Kingston, D. and Whitehouse, C. E. (1967). *J. Hyg. (Camb.)*, **65**, 115–130.

U.S. Dept. of Health and Human Services (1984). *Biosafety in Microbiology and Biomedical Laboratories*, U.S. Govt Printing Office, Washington, DC.

Blowers, R., Hill, J. and Howell, A. (1973). In *Fourth International Symposium on Aerobiology*, J. F. Ph. Hers and K. C. Winkler (Eds), Oosthoek, Utrecht, The Netherlands, pp. 432–434.

BMJ Editorial (1949). *Br. Med. J.* **2**, 641.

Board, R. G. and Lovelock, P. W. (1973). *Sampling — Microbiological Monitoring of Environments*, Academic Press, New York, London.

Bossers, P. A., Crommelin, R. D. and van Gunst, E. (1973). In *Fourth International Symposium on Aerobiology*, J. F. Ph. Hers and K. C. Winkler (Eds), Oosthoek, Utrecht, The Netherlands, pp. 591–594.

Brachman, P. S. (1970). In *Third International Symposium on Aerobiology*, I. H. Silver (Ed.), Academic Press, London, New York, p. 87.

Burke, J. F. (1967). *Hospital Practice*, **2**, (2) February.

Burke, J. F. (1970). In *Third International Symposium on Aerobiology*, I. H. Silver (Ed.), Academic Press, London, New York, pp. 157–166.

Charnley, J. (1968). *Hospital Management, Modern British Operating Theatres, Supplement*, Sept. p. 44.

Charnley, J. (1970), In *Third International Symposium on Aerobiology*, I. H. Silver (Ed.), Academic Press, London, New York, pp. 191–198.

Chatigny, M. A. (1974). In *Developments in Industrial Microbiology*, **15**, American Institute of Biological Sciences, Washington, D.C., chap. 6, 48–55.

Chatigny, M. A. and Clinger, D. I. (1969). In *An Introduction to Experimental Aerobiology*, R. L. Dimmick and Ann B. Akers (Eds), Wiley-Interscience, New York, London, pp. 194–263.

Chausse, P. and Magne, H. (1916). *Arch. Med. exp.*, **27**, 213–251.
Clark, R. N. and Cox, R. N. (1973). In *Fourth International Symposium on Aerobiology*, J. F. Ph. Hers and K. C. Winkler (Eds), Oosthoek, Utrecht, The Netherlands, pp. 413–426.
Coulston, F. and Korte, F. (1972). *Environmental Quality and Safety*, vols I, II, III, Academic Press, New York, London.
Curtiss, L. F. (1958). In *Industrial Hygiene and Toxicology*, vol. I, *General Principles*, 2nd edn, F. A. Patty (Ed.), Interscience, New York, London, pp. 743–788.
Danielson, J. A. (Ed.) (1967). *Air Pollution Engineering Manual*, Public Health Service, Pub-999-AP-40, Cincinnati.
Dark, F. A. and Nash, T. (1970). *J. Hyg. (Camb.)*, **68**, 245–252.
Darlow, H. M. (1973). In *Fourth International Symposium on Aerobiology*, J. F. Ph. Hers and K. C. Winkler (Eds), Oosthoek, Utrecht, The Netherlands, pp. 516–519.
Dietrich, M. (1973). In *Fourth International Symposium on Aerobiology*, J. F. Ph. Hers and K. C. Winkler (Eds), Oosthoek, Utrecht, The Netherlands, pp. 532–533.
Dimmick, R. L. (1974). In *Developments in Industrial Microbiology*, **15**, American Institute of Biological Sciences, Washington, D.C., chap. 5, 44–47.
Dimmick, R. L. and Wang, L. (1969). In *An Introduction to Experimental Aerobiology*, R. L. Dimmick and Ann B. Akers (Eds), Wiley-Interscience, New York, London, pp. 164–176.
Dimmick, R. L., Vogl, W. F. and Chatigny, M. A. (1973). In *Biohazards in Biological Research*, Cold Spring Harbor Laboratory, Cold Spring Harbor, New York, pp. 242–266.
Duguid, J. P. (1946). *J. Hyg. (Camb.)*, **44**, 471–479.
Edmonds, R. L. (Ed.) (1979). *Aerobiology, The Ecological Systems Approach*, Dowden, Hutchinson and Ross, Stroudsburg, Pennsylvania.
Elford, W. J. and Van den Ende, J. (1942). *J. Hyg. (Camb.)*, **42**, 240–264.
Foord, N. (1973). In *Fourth International Symposium on Aerobiology*, J. F. Ph. Hers and K. C. Winkler (Eds), Oosthoek, Utrecht, The Netherlands, pp. 450–454.
Goldberg, L. J. (1968). *J. Appl. Meteorol.*, **7**, 68–72.
Goldberg, L. J. (1970). In *Third International Symposium on Aerobiology*, I. H. Silver (Ed.), Academic Press, London, New York, p. 268.
Goldberg, L. J., Watkins, H. M. S., Boerke, E. E. and Chatigny, M. A. (1958). *Am. J. Hyg.*, **68**, 85–93.
Gould, J. C. (1970). In *Third International Symposium on Aerobiology*, I. H. Silver (Ed.), Academic Press, London, New York, pp. 62–75.
Gould, J. C., Bone, F. J. and Scott, J. H. S. (1973). In *Fourth International Symposium on Aerobiology*, J. F. Ph. Hers and K. C. Winkler (Eds), Oosthoek, Utrecht, The Netherlands, pp. 572–575.
Guth, S. K. and Lindsay, E. A. (1958). In *Industrial Hygiene and Toxicology*, vol. I, *General Principles*, 2nd edn, F. A. Patty (Ed.), Interscience, New York, London, pp. 743–788.
Handbook of Environmental Control (1975). v. *Hospital and Health Care Facilities*, C. P. Straub (Ed), CRC Press, Cleveland, Ohio.
Hartmann, I. (1958). In *Industrial Hygiene and Toxicology*, vol. I, *General Principles*, 2nd edn, F. A. Patty (Ed.), Interscience, New York, London, pp. 549–578.
HMSO (1984). *Categorization of Pathogens according to Hazard and Categories of Containment*. HMSO, London.
Herman, L. G. and Hart, L. J. (1973). In *Fourth International Symposium on Aerobiology*, J. F. Ph. Hers and K. C. Winkler (Eds), Oosthoek, Utrecht, The Netherlands, pp. 536–539.
Hood, A. M. (1971). *J. Hyg. (Camb.)*, **69**, 607–617.
Hood, A. M. (1973). In *Fourth International Symposium on Aerobiology*, J. F. Ph. Hers and K. C. Winkler (Eds), Oosthoek, Utrecht, The Netherlands, pp. 149–151.
Jameson, B. (1973). In *Fourth International Symposium on Aerobiology*, J. F. Ph. Hers and K. C. Winkler (Eds), Oosthoek, Utrecht, The Netherlands, pp. 465–469.

Jennison, M. W. (1942). In *Aerobiology*, publ. no. 14, American Association for the Advancement of Science, Washington, DC.
Johansson, K. R. and Ferris, D. H. (1946). *J. Infect. Dis.*, **78**, 238–252.
Jones, G. W. (1958). In *Industrial Hygiene and Toxicology*, vol. I, *General Principles*, 2nd edn, F. A. Patty (Ed.), Interscience, New York, London, pp. 511–548.
Kenny, M. T. and Sabel, F. L. (1968). *Appl. Microbiol.*, **16**, 1146–1150.
Klastersky, J. (1973). In *Fourth International Symposium on Aerobiology*, J. F. Ph. Hers and K. C. Winkler (Eds), Oosthoek, Utrecht, The Netherlands, pp. 542–544.
Langmuir, A. D. (1961). In *Conference on Airborne Infection*, W. McDermott (Ed.), Williams and Wilkins, Baltimore, pp. 356–358.
Laufman, H. (1973). In *Fourth International Symposium on Aerobiology*, J. F. Ph. Hers and K. C. Winkler (Eds), Oosthoek, Utrecht, The Netherlands, pp. 575–580.
Laurell, G. and Hambreus, A. (1973). In *Fourth International Symposium on Aerobiology*, J. F. Ph. Hers and K. C. Winkler (Eds), Oosthoek, Utrecht, The Netherlands, pp. 462–464.
Lidwell, O. M. (1973). In *Fourth International Symposium on Aerobiology*, J. F. Ph. Hers and K. C. Winkler (Eds), Oosthoek, Utrecht, The Netherlands, pp. 534–536.
May, K. R. and Pomeroy, N. P. (1973). In *Fourth International Symposium on Aerobiology*, J. F. Ph. Hers and K. C. Winkler (Eds), Oosthoek, Utrecht, The Netherlands, pp. 426–432.
Mazzarella, M. A. and Flynn, D. D. (1969). In *An Introduction to Experimental Aerobiology*, R. L. Dimmick and Ann B. Akers (Eds), Wiley-Interscience, New York, London, pp. 437–462.
McDade, J. J., Whitcomb, J. G., Ryptia, E. W., Whitfield, W. J. and Franklin, C. M. (1968). *J. Amer. Med. Assoc.*, **203**, 125–130.
McPhee, C. W. (Ed.) (1966). *ASHRAE Handbook of Fundamentals*. Amer. Soc. of Heating, Refrigeration and Air-conditioning Engineers, Inc., New York.
Michaelson, G. S., Halbert, M. M., Sovenson, S. D. and Vesley, D. (1968). Development of an open isolation system for the care of low resistance hospital patients. *Report, School of Public Health*, University of Minnesota, Minneapolis.
de Mik, G. and de Groot, I. (1977). *J. Hyg. (Camb.)*, **78**, 175–187.
de Mik, G., de Groot, I. and Gerbrandy, J. L. F. (1977). *J. Hyg. (Camb.)*, **78**, 189–198.
Miller, W. S. (1970). In *Third International Symposium on Aerobiology*, I. H. Silver (Ed.), Academic Press, London, New York, p. 96.
Murphy, C. H. (1984). *Handbook of Particle Sampling and Analysis Methods*. Verlag Chemie International, Deerfield Beach, Florida.
Nash, T. (1951). *J. Hyg. (Camb.)*, **49**, 382–399.
Nash, T. (1962). *J. Hyg. (Camb.)*, **60**, 353–358.
Nelson, G. O. (1972). *Controlled Test Atmospheres*, Ann Arbor Science, Ann Arbor, Michigan.
Noble, W. C. (1973). In *Fourth International Symposium on Aerobiology*, J. F. Ph. Hers and K. C. Winkler (Eds), Oosthoek, Utrecht, The Netherlands, pp. 544–546.
Patty, F. A. (1958). In *Industrial Hygiene and Toxicology*, 2nd edn, F. A. Patty (Ed.), Interscience, New York, London, pp. 174–210.
Phillips, G. B. (1965a). *J. Chem. Educ.*, **42**, A43–A48.
Phillips, G. B. (1965b). *J. Chem. Educ.*, **42**, A117–A130.
Pike, R. M. (1976). *Health Lab. Sci.*, **13**, 105–114.
Sattel, W. (1973). In *Fourth International Symposium on Aerobiology*, J. F. Ph. Hers and K. C. Winkler (Eds), Oosthoek, Utrecht, The Netherlands, pp. 570–572.
Scales, J. W. (1972). *Air Quality Instrumentation*, vols I, II, Instrument Society of America, Pittsburgh.
Sivinski, H. D. (Ed.) (1968). *NASA Contamination Control Handbook*, SANDIA Laboratory Report No. SC-M-68-370, SANDIA Corp., Alburquerque, New Mexico.

Spears, R., Bernard, H., O'Grady, F. and Shooter, R. A. (1965). *Lancet*, **i**, 478.
Spendlove, J. C. and Fannin, K. F. (1982). In *Methods in Environmental Virology*, C. P. Gerba and S. M. Goyal (Eds), Marcel Dekker, New York, Basel, pp. 261–329.
Straub, C. P. (Ed.) (1975). *Handbook of Environmental Control. V. Hospital and Health Care Facilities*, CRC Press Inc., Cleveland, Ohio.
Strauss, W. (Ed.) (1978). *Air Pollution Control. III. Measuring and Monitoring Air Pollutants*, Wiley, New York, Chichester,
Sulkin, S. E. (1960). In *Conference on Airborne Infection*, W. McDermott (Ed.), Wilkins and Wilkins, Baltimore, pp. 203–209.
Sykes, G. (1970). In *Third International Symposium on Aerobiology*, I. H. Silver (Ed.), Academic Press, London, New York, pp. 146–156.
Thomas, G. (1970a). In *Third International Symposium on Aerobiology*, I. H. Silver (Ed.), Academic Press, London, New York, p. 266.
Thomas, G. (1970b). *J. Hyg. (Camb.)*, **68**, 273–282.
Thomas, G. (1970c). *J. Hyg. (Camb.)*, **68**, 511–517.
Thomson, W. K., Malysheff, C. and O'Connell, D. C. (1970). In *Third International Symposium on Aerobiology*, I. H. Silver (Ed.), Academic Press, London, New York, p. 97.
Van der Waaij, D. and Van der Wal, J. F. (1973). In *Fourth International Symposium on Aerobiology*, J. F. Ph. Hers and K. C. Winkler (Eds), Oosthoek, Utrecht, The Netherlands, pp. 584–587.
Van der Waaij, D., Vossen, J. M., Kal, H. B. and Speltie, T. M. (1973). In *Fourth International Symposium on Aerobiology*, J. F. Ph. Hers and K. C. Winkler (Eds), Oosthoek, Utrecht, The Netherlands, pp. 546–548.
Wanner, H. U. (1973). In *Fourth International Symposium on Aerobiology*, J. F. Ph. Hers and K. C. Winkler (Eds), Oosthoek, Utrecht, The Netherlands, pp. 568–570.
Watkins, H. M. S. (1970). In *Third International Symposium on Aerobiology*, I. H. Silver (Ed.), Academic Press, London, New York, pp. 9–53.
Wedum, A. G. (1957). In *Fourth National Conference on Campus Safety*, Safety Monographs for Colleges and Universities, Joint Project of Purdue University and the National Safety Council, Chicago, Illinois, pp. 15–20.
Wendt, F., Grüning, B., Linzenmeier, G., Scholz, N. and Brittinger, G. (1973). In *Fourth International Symposium on Aerobiology*, J. F. Ph. Hers and K. C. Winkler (Eds), Oosthoek, Utrecht, The Netherlands, pp. 540–542.
Westwood, J. C. N., Criddle, E., Satar, S. A., Synck, E. J. and Neals, P. (1973). In *Fourth International Symposium on Aerobiology*, J. F. Ph. Hers and K. C. Winkler (Eds), Oosthoek, Utrecht, The Netherlands, pp. 594–596.
Whitby, K. T., Lundgren, P. A., McFarland, A. R. and Jordan, R. C. (1961). *J. Air Pollut. Control Assoc.*, **11**, 503–515.
Whitfield, W. J., Mashburn, J. C. and Neitzel, W. E. (1962). A new principle for airborne contamination control in clean rooms and work stations. ASTM Special Tech. Publication No. 342. Publ. by Amer. Soc. for Testing and Materials, Philadelphia, Pennsylvania.
Williamson, A. E. and Gotaas, H. B. (1942). *Ind. Med. Surg.*, **11**; *Ind. Hyg.*, Section **3**, Section **1**, 40.
Witheridge, W. N. (1958). In *Industrial Hygiene and Toxicology*, vol. I, *General Principles*, 2nd edn, F. A. Patty (Ed.), Interscience, New York, London, pp. 285–341.

Chapter 7

Field techniques

7.1 INTRODUCTION

Most field techniques and equipments are derived from their laboratory counterparts. Where they differ most, perhaps, is in terms of scale of events. For instance, generating an aerosol in a laboratory chamber may involve a few micrograms of materials while to disinfest even an acre of land may require as much as 5 kg of a particular biological insecticide. Similarly, when sampling aerosols in the field, air speed and direction may abruptly change between wide limits whereas under laboratory conditions such factors usually change little. Events like these, and their random nature, often lead to a lack of control which makes work in the field more difficult than in the laboratory. Yet, in the case of crop spraying, for example, unless tight control is exercised over aerosol downwind drift, there is a likelihood that a crop can be unintentionally treated, with sometimes extensive and expensive ramifications. This chapter deals with techniques appropriate for field use.

7.2 AGRICULTURAL AEROSOLS

In the context of this book there are basically two kinds of agricultural aerosol, namely, that generated unintentionally during irrigation, dung spreading, etc. and that generated purposely for the application of microbial insecticides. The former situation together with the question of microbial survival in water, wastewater and sewage is receiving increasing attention. Of topical interest is *Legionella pneumophilia* which in pneumonic form may produce fatalities through Legionnaire's disease. Some outbreaks of this disease have been associated with air-conditioning, cooling towers, evaporative condensers and other man-made water installations (Tobin *et al.*, 1981; Arnow *et al.*, 1982; Dennis *et al.*, 1984). In each case, exposures to aerosols generated by these installations were implicated, while the work of Baskerville *et al.* (1981, 1983a, b) and of Hambleton *et al.* (1983) clearly demonstrate the viability, virulence and infectivity of airborne *L. pneumophilia*. Consequently, the modern practices of spray irrigation, etc., must increase the probability of acute human, animal and plant disease initiated by bacteria and related microorganisms. Those caused by viruses such as polio, coxsackie, reo, adeno, infectious hepatitis and oncogenic viruses, likewise become more probable through the treatment of wastewater and by spray irrigation. Given that a microorganism withstands being airborne to some extent, then any practice which results in its generation as an aerosol must produce a biohazard. Its likely severity, though, depends on the numerous factors which affect the Aerobiological Pathway while improvements in field sampling techniques and procedures (e.g. Fannin *et al.*, 1976; Rao, 1982; Spendlove and Fannin, 1982) lead to its better quantitation.

In contrast to unintentional aerosols, the application of microbial insecticides involves deliberate spraying or dust spreading, of, for example, nuclear polyhedrosis virus for controlling cotton bollworm and *Bacillus thuringiensis* against insect larvae of the order ***Lepidoptera*** (e.g. Gypsy moth). As an illustration, the latter example will be considered in more detail. *Bacillus thuringiensis* (Bt) is a spore-forming bacterium that concomitantly with sporulation produces a parasporal body. Berliner in 1911 isolated the organism from *Anagasta kuhniella* and named it *B. thuringiensis* in 1915 after the town of Thuringia in Germany. This original strain was lost but Mattes (1927) reisolated it, while Angus (1954) demonstrated that the pathogenicity for the silkworm was initiated by the parasporal body. This inclusion is a proteinaceous δ-endotoxin or crystal which when ingested does not dissolve or have any effect until it reaches the mid-gut. There it is exposed to a slightly alkaline environment whereupon the crystal dissolves and inhibits peristalsis so the food bolus does not move, feeding stops and the gut pH reduces. This *modus operandi* is common to all susceptible species and is described in detail by Fast (1974).

Toxicity of Bt towards human volunteers, animals, birds and beneficial insects has not been demonstrated (Fisher and Rosner, 1959) while a variety, *Kurstaki* serotype III (of higher infectivity than the serotype I Berliner strain), forms the basis of commercial preparations derived from large scale batch fermentation which is quality controlled. The Bt vegetative cell is about 1 μm by 5 μm while the spore and parasporal body are each about 0.5 to 1 μm diameter. The parasporal crystal is a single glyco-protein of molecular weight about 1.2×10^5, with the carbohydrate largely consisting of glucose and mannose. Under alkaline conditions, or in the presence of certain enzymes, the crystal dissolves and becomes activated through the breaking of sulphydryl bonds in the glycoprotein thereby liberating the biologically active subunits.

One of the major attractions of Bt as an insecticide is that it is naturally occurring while being selective against larvae of the insect order ***Lepidoptera***, i.e. a host specific pathogen. It is registered, sold and used in more than 40 countries as an insecticide against about 150 different species of caterpillar. Probably because of the way it acts, as far as is known, none of these species has developed resistance to it, unlike chemical pesticides.

Persistence in the field depends upon many factors including soil pH, sunlight, rainfall, but, generally, Bt remains active for about 7 days after dissemination. Other factors include manufacturers' additives such as wetting agents, stabilizers, adjuvants, and in some cases the type of foliage onto which it is sprayed. Spraying is by any spray system delivering droplets in the range 100–400 μm diameter. Depending on areas to be treated, hand-operated, power-operated air-carrier or aircraft-carried sprays are appropriate (Matthews, 1979; Haskell, 1985). In each, liquid is pumped or gravity-fed to one, or more nozzles, of which there are several kinds (Matthews, 1979; Haskell, 1985). They range from simple orifices to spinning discs, but at the throughputs required in the field none produce monodisperse aerosols, which would be ideal for controlled drop application (Johnstone, 1978a). Such applications are a function of droplet size and meteorological factors, as well as the nature of the target (Johnstone, 1978b) and particle charge (Jones and Hopkinson, 1979).

In contrast to chemical pesticides, drift of droplets containing Bt will neither affect adjacent crops, as it is non-phytotoxic, nor other life-forms in that area due to its host specificity. On the other hand, Bt applications are best timed to coincide with peak early larval instar activity. They, also, should uniformly cover the foliage, with minimal waste, as Bt must be ingested to be effective.

7.3 FIELD AEROSOL SAMPLING AND SIZING TECHNIQUES

Virtually all aerosol sampling and sizing techniques described for laboratory investigations can be applied in the field, under appropriate conditions. But, being designed primarily for handling respirable sized particulates, they are not always suitable for the larger ones associated with microbial insecticides. A good adaptation though, is to coat glass slides with MgO and to measure the diameters of craters made by drops impacting onto the surface (Andrews *et al.*, 1983). Such slides may be placed on the ground when collection is through gravitational settling or to collect airborne droplets by mounting a slide at each end of a centrally pivoted and rotated arm. An alternative is a Rotorod sampler in which U-shaped rods are coated with a suitable adhesive and rotated at 2400 rev/min.

Ground recoveries may be assessed by placing filter papers on it especially when the droplets are coloured. However, the stains tend to be irregular in shape, and the smaller ones difficult to see. A superior method is that described by King and Johnstone (1973) in which a suitable dye is deposited on fixed glazed photographic bromide paper. When spray droplets impact on its surface the dye is displaced to give a clear spot. By careful matching of dye and droplet, it is possible to have a spot/droplet diameter ratio of up to 10:1, making the technique applicable to even 5 μm diameter droplets.

For sampling droplets reaching leaves and other parts of plants, papers and slides may be fixed to upper and lower surfaces. However, as pointed out by Matthews (1975), the main problem then is that small droplets may not be sampled whereas they may adhere to plant hairs. Large droplets, on the other hand, may shatter on impact. In these instances, a very soft grease is better and should the collected droplets be likely to evaporate a light oil layer may be best. Here the difficulty is to achieve a viscosity which permits droplet penetration but prevents droplet coalescence.

For droplets of low vapour pressure, fine fibres akin to those of the microthread technique (q.v.) but stronger may be used to catch airborne droplets. When utilized as synthetic angora wool this sampling method is both practical and effective (Andrews *et al.*, 1983).

For sizing droplets collected onto filter papers or sampling papers or kromekote cards, allowance has to be made for droplet spread. One way is to multiply the actual size by a spreading factor, which may be determined in the laboratory with monodisperse droplets. Microscopy is the sizing method of choice for spray droplets (100–400 μm diameter), a technique made easier by employing image shearing eyepieces and eyepiece graticules (Johnstone and Huntingdon, 1970).

Field sampling of another kind is that required for monitoring downwind biohazards created by water, wastewater, sewage, etc. management. Objectives here may be aerosol concentration–time profiles as a function of particle size, location, wind speed and direction, season, atmospheric stability, source, etc. Obtaining such data involves a considerable investment in time, effort and equipment. Even then, there is the complication of ensuring that the collection methods will provide samples of microbes at sufficiently high concentrations to give statistically significant data. Unless large volumes of air are sampled, the most likely result is a negative one owing to the relatively low aerosol concentrations found downwind of relatively small sources (see Chapter 8). As already indicated (Chapter 3) difficulties can arise with high volume samplers so the problem is not trivial especially when account is taken of the few bacteria, viruses, etc. required to initiate some diseases (Chapter 15).

7.4 CONCLUSIONS

Field equipments and techniques are based on laboratory counterparts and have to work satisfactorily even though conditions outdoors are not well controlled. Practices of wastewater treatment and spray irrigation, etc., are likely to cause an aerosol biohazard, the quantitation and evaluation of which is benefitting from improved sampling procedures. Replacing chemical pesticides with naturally occurring microbial insecticides produces many advantages owing to their host specificity.

REFERENCES

Andrews, M., Flower, L. S., Johnstone, D. R. and Turner, C. R. (1983). *Trop. Pest Managem.*, **29**, 239–248.

Angus, T. A. (1954). *J. Invertebr. Pathol.*, **11**, 145–146.

Arnow, P. M., Chou, T., Weil, D., Shapiro, E. N. and Kretzschner, C. (1982). *J. infect. Dis.*, **146**, 460–467.

Baskerville, A., Broster, M., Fitzgeorge, R. B., Hambleton, P. and Dennis, P. J. (1981). *Lancet*, **ii**, 1389–1390.
Baskerville, A., Dowsett, A. B., Fitzgeorge, R. B., Hambleton, P. and Broster, M. (1983a). *J. Pathol.*, **140**, 77–90.
Baskerville, A., Fitzgeorge, R. B., Broster, M. and Hambleton, P. (1983b). *J. Pathol.*, **139**, 349–362.
Berliner, E. (1911). *Z. Ges. Getreidew.*, **3**, 63–70.
Berliner, E. (1915). *Z. Angew. Entomol.*, **2**, 29–56.
Dennis, P. J. L., Wright, A. E., Rutter, D. A., Death, J. E. and Jones, B. P. C. (1984). *J. Hyg. (Camb.)*., **93**, 349–353.
Fannin, K. F., Spendlove, J. C., Cochran, K. W. and Gannon, J. W. (1976). *Appl. Environ. Microbiol.*, **31**, 705–710.
Fast, P. G. (1974). In *Developments in Industrial Microbiology*, American Institute of Biological Sciences, Washington, DC, vol. 15, chap. 19, pp. 195–198.
Fisher, R. and Rosner, L. (1959). *J. Agric. Food Chem.*, **7**, 686–688.
Hambleton, P., Broster, M. G., Dennis, P. J., Henstridge, R., Fitzgeorge, R. and Conlan, J. W. (1983). *J. Hyg. (Camb.)*., **90**, 451–460.
Haskell, P. T. (Ed.) (1985). *Pesticide Application — Principles and Practice*, Oxford University Press, Oxford.
Johnstone, D. R. (1978a). In *Symposium on Controlled Drop Application*, pp. 35–42.
Johnstone, D. R. (1978b). In *Symposium on Controlled Drop Application*, pp. 43–57.
Johnstone, D. R. and Huntingdon, K. A. (1970). *J. Agric. Engng. Res.*, **15**, 1–10.
Jones, C. D. and Hopkinson, P. R. (1979). *Pestic. Sci.*, **10**, 91–103.
King, W. J. and Johnstone, D. R. (1973). Papers for sampling spray deposits of ultra-low volume carrier solvents. *Miscellaneous Report No. 9*. Centre for Overseas Pest Research, London.
Mattes, O. (1927). *Sitzungsber. Ges. Befoerder. Ges. Naturw. Marburg*, **62**, 381–417.
Matthews, G. A. (1975). PANS, **21**, 213–225.
Matthews, G. A. (1979). *Pesticide Application Methods*, Longman, London, New York.
Rao, C. (1982). In *Methods in Environmental Virology*, C. P. Gerba and S. M. Goyal (Eds), Marcel Dekker, New York and Basel, pp. 1–13.
Spendlove, J. C. and Fannin, K. F. (1982). In *Methods in Environmental Virology*, C. P. Gerba and S. M. Goval (Eds), Marcel Dekker, New York and Basel, pp. 261–329.
Tobin, J. O'H., Swann, R. A. and Bartlett, C. J. R. (1981). *Br. Med. J.*, **282**, 515–517.

Chapter 8

Aerial transport

8.1 INTRODUCTION

The earth's atmosphere swarms with microorganisms: some are harmless others spread disease among man, animals and plants. Because of its continuous movement and mixing, air transports microbes on local and global scales (Edmonds, 1979). For particle sizes of concern here, diffusion is the primary mechanism for their aerial dispersals — a subject which forms the basis of this chapter.

8.2 AIR MOVEMENTS INDOORS

Air flow in rooms varies widely according to their design, manner of operation, heating, ventilation, size, shape, furnishings and the activities of occupants. Changes in weather influence air flow particularly where building insulation and ventilation is poor. Even so, like global air movements, those indoors owing to surface friction are turbulent, with convection causing additional vertical circulation. Turbulent air flows result in microbial aerosols rapidly spreading through buildings unless strict precautions, as described in Chapter 6, are taken to prevent it.

The action of the wind is to exert a pressure predominantly on windward faces, diminishing with angle of impingement, while faces to leeward experience a suction. Consequently, changes in wind speed and direction modify movements of indoor air. By considering these factors it is possible to estimate rates of air flow through rooms of a building in given circumstances (Daws, 1967).

8.3 ATMOSPHERIC LAYERS

The atmosphere usually is thought of as being in layers as illustrated in Figure 8.1 in which altitude is given on a logarithmic scale. The three vertical panels represent different weather conditions, while the boundaries between the various layers in practice are not so definite. The region from ground level to about 10 km is the troposphere and comprises the laminar boundary layer, turbulent boundary layer, frictional layer and convective layer. Above the

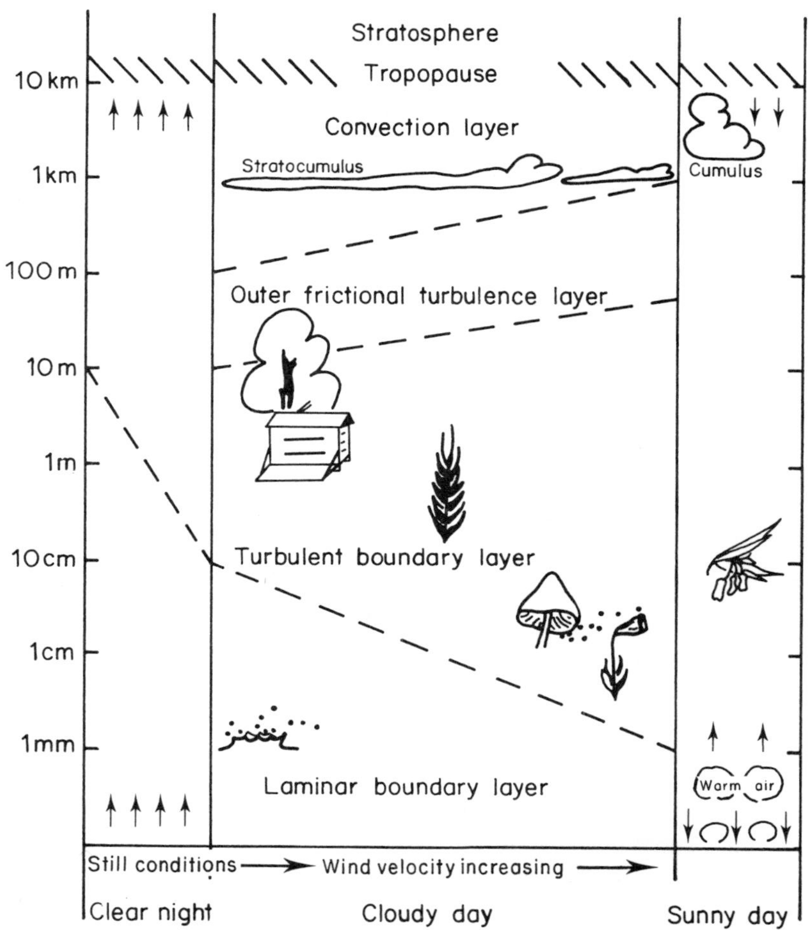

Figure 8.1 Diagrammatic representation of the layers of the atmosphere (with logarithmic vertical scale). (From Gregory (1973), *The Microbiology of the Atmosphere*, Leonard Hill.) Reproduced by permission of Blackie, Glasgow and London

troposphere is the stratosphere extending to the outer limit of our atmosphere. For the aerial transport of microorganisms the troposphere is mainly of concern and events occurring in it will be considered next.

8.4 AIR MOVEMENTS OUTDOORS

Surface friction generates turbulence in outside air up to heights of about 1 km (Tyldesley, 1967), thereby forming a frictional layer within which vertical fluctuations are up to about 1 m/s. These velocities greatly exceed the terminal velocities of small particles (Table 8.1) so the disorganized vertical air motion reduces particle fall-out and aerosols become spread vertically.

Table 8.1 Terminal velocities for particles of density 1.1 g/cm^3

Diameter (µm)	Terminal velocity (cm/s)
20.0	1.3
10.0	3.3×10^{-1}
2.0	1.3×10^{-2}
1.0	3.3×10^{-3}
0.2	1.3×10^{-4}

Horizontal spread of particles liberated from small sources is by the action of the wind.

Above about 1 km turbulence decreases but still is sufficient to suspend aerosol particles and cause their diffusion. In contrast, very close to the ground air movement is laminar rather than turbulent, thereby forming a laminar boundary layer only 1 or 2 mm thick even on a windy day. On clear nights with light winds though, a boundary layer several metres deep develops. It has only quasi-laminar flow characteristics because the smooth air flow is sporadically disrupted by turbulent activity. Resulting rates of diffusion are an order of magnitude greater than those for true laminar flow, but smaller than the day time (turbulent) diffusion rates.

Turbulence of the turbulent boundary layer is modified by buoyancy effects produced by interactions of random frictional turbulence with organized vertical motion caused by vertical temperature gradients. The particular condition occurring when air temperatures increase with vertical height is known as a temperature inversion and is often formed by cooling at the ground on clear nights. Inversions can form also at heights of hundreds or even thousands of metres by large-scale atmospheric motion, or radiation from cloud or haze. Inversions are very effective barriers to diffusion either from above or from below them. The path traced by an airborne particle due to air movements is termed particle trajectory and it is an irregular line observable directly or estimated from the known winds. The use of a zero-lift balloon (i.e. one that stays at a fixed height) provides a good estimate of downwind drift and a particle's jerky motion. Deviations from a smooth path result from wind gusts — usually taken as random events — which can be measured using the technique of Jones and Hutchinson (1976).

With powerful forces of the atmosphere operating, it should not be too surprising to find that particles can be transported over considerable distances under suitable conditions. Details of some of the more outstanding examples follow now. Craigie (1945) and Peturson (1958) demonstrated that stem rust caused by *Puccinia graminis tritici* travelled from the Missisippi Valley (where the disease was endemic) to the central and northern regions of Canada when the meteorological situation created a strong airflow to those regions. Tree pollens also can travel over considerable distances as indicated by Tyldesley (1973) who found them to be carried from Scandinavia to the Shetland Isles, a distance of about 400–500 km. For this investigation Tyldesley employed back-trajectory analysis as used to determine probable pollutant sources. Essentially the technique evaluates mean surface wind speeds from appropriate synoptic weather charts and predicts the position of the pollutant at progressively earlier times. Unfortunately, calculating back-trajectories in this way suffers from two major weaknesses. First, no account is taken of vertical components of atmospheric motion and, second, diffusion of material due to atmospheric turbulence is ignored. On the other hand, within these limitations it is a useful technique which can be computer automated (Battalino *et al.*, 1982).

An approach which does take account of vertical component as well is to follow air movements on isentropic surfaces, i.e. surfaces on which the potential temperature of the air is a constant (Bartlett, 1973). It was used by Stevenson (1969) for her analysis of the case when red dust fell in Southern Britain on 1 July 1968. She showed that most probably the dust was injected into the atmosphere through dust storms in the southern Sahara. The red dust then travelled northwards arriving over Britain at an altitude of 5 or 6 km. Simultaneous local rain showers then caused dust deposition following aerial transport of about one million tons of dust over a distance of about 4000 km.

Extensive use also has been made of calculated back-trajectories to help explain movements of spores and pollens (Hirst and Hurst, 1967; Edmonds, 1979). For prolific materials such as these, vertical concentration profiles can be measured (rather than estimated) from samples collected by plane. These values, at least for unstable atmospheric conditions, indicate a positive correlation with temperature profile, confirming the isentropic method and that gravitational settling of spores of all sizes is insignificant compared to their convective ascent. Rain, though, tends to reduce vertical spore and pollen concentrations due to 'wash-out' as for the Sahara dust in the example given earlier. Hirst and Hurst (1967) and Gregory (1973) provide additional details, while other examples of long-distance aerial transport including that of disease producing organisms are to be found in books by Gregory (1973), Edmonds (1979), Pedgley (1982) and papers by Bovallius *et al.* (1980), Zadoks (1973) and Edmonds and Benninghoff (1973).

One of the more recent and thorough investigations of the long-distance airborne transport of bacterial spores is due to Bovallius *et al.* (1978). These workers proved that spores sampled in the centre of Sweden originated from the Black Sea 1800 km away. Simultaneously at two sampling stations in Sweden 40 km apart, brown-red coloured snow fell and samples were taken from the air and from snow profiles. Bacteria collected from each were similar but distinguishable from the local flora on the basis of several properties (Table 8.2). In the snow profile, bacterial concentrations above, in and below the coloured layer were determined to be 1, 70–120 and <1 bacteria/ml, respectively. On further examination, 36 colonies selected randomly from the

Table 8.2 Distinguishing features of local flora and of 'red snow' flora (Taken from Bovallius *et al.*, 1980). Reproduced by permission of the author.

	Mean local airborne bacterial flora	Mean 'Red Snow' airborne bacterial flora
Total number of bacteria/m^3	39–74	144–245
% outgrowth in two days	10–20	85–90
% less than 6.0 μm in size	41–44	80–85
% spore formers	20–30	85–90
% outgrowth at 37°C	30–50	70–80

air and snow samples were type classified. All bacteria were *Bacillus* species, with about 35% *megaterium*, 15% *pumilius*, 10% *mycoides*, 30% *firmus/lentus* and 10% *subtilis/licheniformis*.

The authors conducted a computer-based horizontal back-trajectory analysis with the altitude set at 1.5 km owing to the meteorological conditions prevailing at the relevant time. An area of low pressure covered central Europe and a strong high pressure was over eastern parts of European Russia. Two steep pressure gradients between them resulted in strong northwesterly winds of 21–36 m/s. Trajectories back-calculated from the Swedish Baltic coast in the area of the brown-red snowfall, coincided with the occurrence of a sandstorm in the area north of the Black Sea. Between the two regions the ground was snow covered and could not have contributed dust. The time required for the 1800 km journey was about 36 h during which time bacterial flora suffered exposure to direct sunlight, which must have adversely affected UV-sensitive species. Temperature of exposure was −5 °C to +2 °C. That only spore-forming bacteria were observed in samples of the brown-red snow layer, therefore, may have resulted from lethal effects of UV or from sub-zero temperature, or their combination.

Independent corroborative evidence of the aerial transport of *Bacillus* species over this 1800 km emerged from chemical analyses of the inorganic particles of the brown-red snow. The dust particles comprised a mica-like substance and a complex clay having a high iron content, that is, the same composition as loess. It had also an organic content of 9–12%, including pollen grains (Aartolahti and Kulmala, 1969). The pollen species, as well as the ratios of tree pollens, bush pollen and herb pollen, were indicative of the source being of southern origin (Lundquist and Bengtsson, 1970; Aartolahti and Kalmala, 1969). Given the snow cover north of the site of the sandstorm, together with sandlike particles, a clay found only in Asia, Europe or North America, and pollen species of southern origin, together with the analyses of bacterial species, it is difficult to find an explanation other than that proposed, viz. aerial transport of bacteria over a distance of 1800 km.

Turning now to the aerial transport of animal diseases such as foot-and-mouth disease (FMD), Newcastle disease, Q-fever, rinderpest disease, coccidiomycosis, histoplasmosis, infectious laryngotracheitis, anthrax, avian coccidiosis, etc.,

FMD aerial transport was first suggested as long ago as 1938 (Hugh-Jones, 1972), but proving that these and other animal diseases are transported in the atmosphere over tens of kilometres is difficult (Hugh-Jones, 1972; Sellers *et al.*, 1973; Gregory, 1973; Pedgley, 1982). However, from the physical point of view, particles containing microorganisms are expected to behave like bacilli, pollen grains, plant pathogens and even fine Sahara and Black Sea dust. Consequently, the proven aerial transport of these particles provides a good model for those organisms causative of animal diseases, although one difference is likely to be the scale or quantity of material released. Certainly, the red Sahara dust source mentioned above was about a million tons and phytopathogenic fungal spores above an infected crop can attain concentrations of millions of spores/m^3 (Zadoks, 1973). In comparison, an animal with FMD represents a relatively small source of about 10–100 ID_{50}/min (Sellers *et al.*, 1973). Therefore, the probability of artificially collecting statistically significant quantities of FMD for positive identification, say 10 km downwind, must be very small. On the other hand, an infectious dose for an animal can be expected at this distance downwind (Sellers and Parker, 1969; Norris and Harper, 1970). Consequently, obtaining a positive correlation between the two events is unlikely to be easy.

The overall conclusion must be that microbial aerosols can travel over very large distances given suitable meteorological conditions. What is in doubt, perhaps, as for their spread in buildings, is whether or not they will be infective at their landing site and providing that answer is the aim of later chapters of this book. As explained in the following section, hopefully, it soon will become possible to forecast where and when infections due to airborne contagion are likely to arise.

8.5 THEORETICAL MODELS

Downwind concentrations of particles from small sources can be estimated using mathematical models (either empirical or theoretical) which take account of meteorological parameters such as diffusion coefficients, windspeed, stability category, etc. The stability of the atmosphere generally is a function of wind speed and insolation and Pasquill (1961) assigns six different stability categories. Stability is important in determining the way in which particles diffuse downwind. Unfortunately, diffusion models become less reliable as distance downwind of the source increases since it is assumed that properties of the atmosphere remain essentially uniform in time and space. Once the distances involved are about 10 km or more, or the time of travel exceeds about 1 h, then the assumption represents a poor approximation.

According to Bartlett (1973), the concentration at a distance downwind of a continuous point source of strength Q at ground level is approximately given by,

$$\chi = \frac{Q}{\pi\sigma_y\sigma_z\bar{\upsilon}}\, exp\left[-\tfrac{1}{2}\left(\frac{y^2}{\sigma_y^2}+\frac{h^2}{\sigma_z^2}\right)\right] \tag{8.1}$$

where, h = height of plume above the ground,
y = distance from axis of the plume,
$\bar{\upsilon}$ = mean wind speed
σ = standard deviation of the plume in the horizontal (y) and vertical (z) planes.

The quantities σ_y and σ_z are functions of the distance downwind and of the stability of the atmosphere. The importance of atmospheric stability in determining downwind diffusion is likely to be a major one as the equation predicts a 20-fold difference in concentration downwind from the source for moderate instability (Pasquill category B) and for slight stability (category E).

Equation 8.1 represents an approximate theoretical equation for predicting effects of atmospheric turbulence on downwind diffusion of materials injected into the atmosphere. In contrast, Benarie (1976) offers a simple empirical relationship (equations 8.2 and 8.3) for a continuous source at ground level,

$$\frac{D}{D_0} = 7.3 \times x^{-2} \quad \text{(day)} \tag{8.2}$$

or,

$$\frac{D}{D_0} = 25 \times x^{-3} \quad \text{(night)} \tag{8.3}$$

where, D = dosage downwind,

D_0 = emitting dosage,

x = distance downwind,

while Wu (1977) provides an analogous relationship, viz.

$$\chi = \frac{C\,Q}{\bar{\upsilon}} \tag{8.4}$$

where, χ = particulate aerosol concentration,

Q = source strength per unit area,

$\bar{\upsilon}$ = mean wind speed

C = 892.

The equations due to Benarie (8.2 and 8.3) were based on studies of gaseous pollutants. Even so, they are equally applicable to aerosol particles (less than about 20 μm diameter) because in turbulent air both particles and gaseous pollutants diffuse similarly.

Equations given above take no account of possible loss of viability/infectivity of microorganisms during their downwind travel. Lighthart and Frisch (1976) try to remedy this deficiency by incorporating into their diffusion equation a simple exponential term for biological decay, viz.

$$\frac{\chi}{Q} \cdot \frac{\bar{\upsilon}}{exp(-\lambda x/\bar{\upsilon})} = \frac{1}{2\pi\sigma_y\sigma_z}\, exp\left[-\frac{h^2}{2\sigma_z^2}\right] \tag{8.5}$$

where, χ = number of particles/m^3,

Q = number of particles emitted/s,

$\bar{\upsilon}$ = mean air speed (m/s),

σ_y, σ_z = diffusion constants in the horizontal and vertical planes,

h = height of source above ground (m),

λ = rate of loss of viability (s^{-1}) for various specific conditions,

x = distance downwind from the source (m).

Teltsch *et al.* (1980) and Bovallius *et al.* (1980) also provide analogous expressions which take biological decay into account.

As will be shown in later chapters, describing loss of viability or infectivity in terms only of exponential decay, i.e.

$$V_t = V_0 exp(-Kt) \tag{8.6}$$

where, V_t = viability (or infectivity) at time t,

V_0 = viability (or infectivity) at time zero,

K = decay rate constant,

is a gross over-simplification. Even under otherwise constant conditions (e.g. fixed RH, temperature, oxygen concentration, etc.) the relationship between V_t and time is much more complex than that given by equation 8.6. However, taking account of all parameters known to affect loss of viability, or infectivity, will not necessarily give a more accurate prediction. That approach would lead to an equation with 20 or more independent constants, values of which have been derived in most cases for only one or two strains of microorganisms so far. In addition, in practice many of the constants would change irregularly with environmental factors during downwind travel. A very practical approach, therefore, is that of Bovallius *et al.* (1980) who rank bacteria into categories of toughness or ability to survive. (Tough microbes are those such as spores.) To each category is assigned a decimal reduction time (DRT), i.e. the time for a 90% reduction in viability. For small or point sources the DRT downwind diffusion is used to compute viable concentrations as a function of distances downwind from 1 to 10 km. In addition, the model takes account of the fraction of the aerosol deposited onto surface materials such as grass. Required deposition values were determined experimentally by comparing concentrations of airborne *Bacillus thuringiensis* spores just above the ground with amounts of spores deposited per unit time at the same distance downwind from the source. Deposition values (cm/s) were obtained at different downwind distances.

Results of their computations for a source strength of 8×10^3 bacteria/m^3

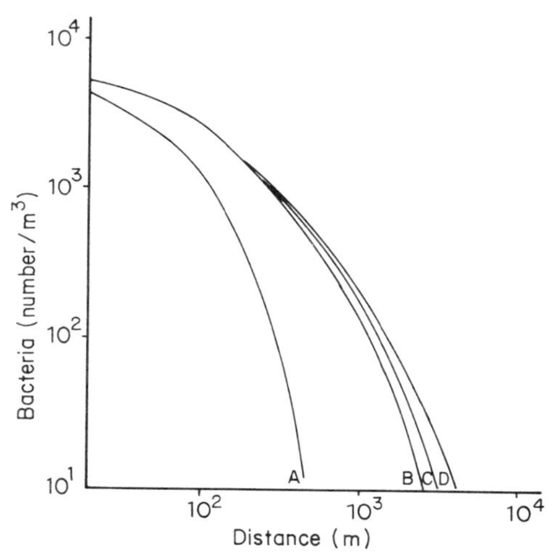

Figure 8.2 Variation in concentration of airborne bacteria with distance, decimal reduction time (DRT) (see text) and dry deposition V_d as calculated by a plume model for dispersal from a small continuous source during neutral conditions. Windspeed 5 m/s; bacterial size 2 μm and density 1 g/cm^3.

A : V_d = 0.16 cm/s : DRT = 1 min
B : V_d = 0.16 cm/s : DRT = 0.5 h
C : V_d = 0.16 cm/s : DRT = ∞
D : V_d = 0 : DRT = ∞

(From Bovallius and Ånäs (1980), *Annals N.Y. Acad. Sci.*, **353**, 186.) Reproduced by permission of the author.

(a value typical of sewage treatment plants, for example) are given in Figure 8.2. They show that for neutral conditions dispersed bacteria in particles 2 μm diameter theoretically will not be transported downwind further than about 3 km. The main factor influencing that downwind distance is the prevailing meteorological condition, although very small DRT values (e.g. 1 min) will significantly further limit downwind dispersal. In contrast, the value of dry deposition value has a much smaller influence.

For estimating long-distance downwind concentrations from large scale sources, different types of air pollution models (Scriven and Fisher, 1975; Draxler and Elliot, 1977; Högström, 1964) have been used. In the box-model of Scriven and Fisher, transport is assumed to take place between ground level and an inversion layer with complete vertical mixing between. The model was modified by Bovallius *et al.* (1980) to include viability loss and dry deposition as for their Gaussian-plume model, described above. In addition, the long-range transmission model takes account of wet deposition by precipitation. The modified box-model of Bovallius *et al.* (1980) for a large scale source assuming a homogenous concentration of about 10^3 bacteria/m^3 in a large volume of air, predicts that 2 μm diameter particles of unit density will travel more than 1000 km from their source. In this case, dry deposition becomes an important factor as does wet deposition by rain and snow. However, ability to survive is the most important factor of all because of the extended times involved in travelling such long distances. In the event that dispersed microbes rise above the inversion layer, the described models no longer apply because, due to the jet airstreams and low turbulence, aerosol dilution is reduced and conditions for long-range transport favoured. Wet deposition then becomes a dominating factor as is sunrise when jet streams tend to decay and completely disappear as the inversion breaks down (Izumu, 1964).

A probable situation, therefore, is that of material launched into the atmosphere during the day being spread reasonably uniformly by turbulent diffusion. With the onset of darkness and the development of an inversion layer, material below that layer relatively will be restricted in movement, whereas that above the inversion layer will be entrained in the jet stream. Then it will be transported over substantial distances during darkness without contacting any surfaces. At sunrise, as the inversion breaks down the material spreads through the lower atmosphere and part of it deposits onto the ground. Alternatively, rain or snow can bring material through the lower atmosphere to the surface of the earth.

8.6 COMPARISONS BETWEEN PRACTICE AND THEORY

Industrial and agricultural activities such as waste water treatment plants, waste water irrigation and cooling towers, can cause large numbers of microorganisms to become aerosolized. Some may be pathogenic e.g. *Legionella* species (Marshall 1980; Cordes and Fraser, 1980), or cause spoilage, with their concentrations at the launch site being in the range 10^4–10^5 bacteria/m^3 (Hickey and Reist, 1975; Andersson *et al.*, 1973; Wanner, 1975; Sorber *et al.*, 1976). Some examples of observed downwind dispersal distances for bacteria from sewage treatment plants, waste water irrigation and cooling towers are given in Table 8.3. The data imply that bacterial counts significantly greater than background levels were rarely found downwind at distances greater than about 1 km. Such distances are very similar to those of Figure 8.2 calculated for a plume model by Bovallius *et al.* (1980). Other bacteria detected in the vicinity of these sources are *Aerobacter*, *Streptococcus*, *Proteus* and *Pseudomonas* (Hickey and Reist, 1975; Dart and Stretton, 1977; Högström, 1964). On the basis of the above, it seems that severe biohazards from relatively small sources are of low probability at distances much greater than about 1 km, although Parker *et al.* (1977) indicate that that probability still can be significant. What is more, the biohazard could be a lethal one as exemplified by Legionnaire's disease liberated from cooling towers (Marshall, 1980).

Table 8.3 Observed downwind dispersal distances for bacteria from sewage treatment plants and waste water irrigation (Taken from Bovallius *et al.*, 1980) Reproduced by permission of the author

Bacteria measured	Longest recorded Down-wind distance (m)	Source of bacteria
Total number	90	STP
	270	STP
	80	STP
	100	STP
	100	STP
	200	STP
	400	SI
	198	SI
	400	SI
	650	SI
Coliforms	90	STP
	1290	STP
	20	STP
	700	STP
	150	STP
	200	STP
	350	SI
	152	SI
Klebsiella aerogenes	90	STP
Salmonella	60	SI

STP: sewage treatment plant; SI: spray irrigation.

For longer distance transmission of microorganisms or at least detection of it, large sources seem necessary. Densely populated areas, wind erosion of soil, large water surfaces including lakes, rivers and seas are appropriate. Bacteria from such land-based sources have been detected as far as 160 km out to sea (Zobell, 1942), while those arising from the sea itself have been found up to 54 km inland (Zobell and Mathews, 1936). Emphasizing, perhaps, the importance of survivability for long-distance travel mentioned above, observations in Sweden show a greater prevalence of spore-forming bacteria in air masses travelling over the sea (57%) than in air masses originating over land (7%) (Roffey *et al.*, 1977). But, as for bacteria, the number of fungal spores also decreases over the ocean with increasing distance from land.

In and above densely populated areas, concentrations of 10^3–10^4 bacteria/m^3 have been found to be significantly higher than those normally recorded over agricultural land or open sea. At higher altitudes up to 3 km, values range from 1–3000 bacteria/m^3 (Anderson *et al.*, 1950; Trådgård, 1977), while at 6.8 km above the North Pole a few bacteria and fungi have been found, indicative of long-distance travel in jet streams (Polunin and Kelly, 1952). At these heights sampling from balloons indicated concentrations of 0.14 bacteria/m^3 (all *Bacilli*) (Meier, 1936) and 10^{-2}–10^{-3} bacteria/m^3 at 9–27 km (Greene *et al.*, 1964). Unfortunately, as the balloons were not sterile the concentration values must be in doubt because of possible sample contamination. On the other hand, by means of a rocket, microorganisms were detected at heights of 48–77 km (Imshenetsky *et al.*, 1978). In this instance the bacteria were of the genera *Mycobacterium* and *Micrococcus*, with five out of the six species having chromogenic pigments and possibly, therefore, protection against ultraviolet light. In this study, though, no *Bacillus* species were sampled.

For intermediate scale aerial transport, excellent agreement between experiment and theory has been obtained for 2.3 μm diameter tracer particles travelling over about 20 km (Perkins and Vaughan, 1961).

8.7 CONCLUSIONS

That microorganisms can spread by aerial transport through rooms, buildings, cities, continents and throughout the atmosphere, now is difficult to refute. In contrast, at the turn of this century, the opposite view was held (e.g. Chapin, 1912) and was still quite prevalent in the late 1930s (e.g. Chope and Smillie, 1936). This difference of opinion and the better understanding of today is largely due to the developing subject of Aerobiology. Some implications of the ease with which microbial aerosols travel throughout the atmosphere are beginning to emerge as are the distances involved and the widespread nature of the effects. Yet, a general appreciation by the public of the importance of the Aerobiological Pathway seems somewhat lacking.

REFERENCES

Aartolahti, T. and Kulmala, A. (1969). *Terra*, **81**, 98–104.

Anderson, D. T., Mitchell, R. B., Dorris, H. W. and Timmons, D. E. (1950). US air Force School of Aviation, Medicine, Project 21-02-118, Report No. 4, Randolph Field, Texas.

Andersson, R. B., Bergstrom and Bucht, B. (1973). *Vatten*, **2**, 117–123.

Bartlett, J. T. (1973). In *Fourth International Symposium on Aerobiology* J. F. Ph. Hers and K. C. Winkler (Eds), Oosthoek, Utrecht, The Netherlands, pp. 385–391.

Battalino, T. E., Helvey, R., Walcek, C. J., Rosenthal, J. and Gottschalk, J. (1982). In *Atmospheric Aerosols, Their Formation, Optical Properties, and Effects*, A. Deepak (Ed.), Spectrum Press, Hampton, Virginia, pp. 125–137.

Benarie, H. M. (1976). *Atmos. Environ.*, **10**, 163–166.

Bovallius, Å., Bucht, B., Roofey, R. and Ånäs, P. (1978). *Appl. Environ. Microbiol.*, **35**, 1231–1232.

Bovallius, Å., Roffey, R. and Henningson, E. (1980). *Annals. N.Y. Acad. Sci.*, **353**, 186–200.

Chapin, C. V. (1912). *The Sources and Modes of Infection*, 2nd edn, Wiley, New York.

Chope, H. D. and Smillie, W. G. (1936). *J. Ind. Hyg. and Toxicol.*, **18**, 780–792.

Cordes, L. G. and Fraser, D. W. (1980). *Med. Clinics of N. America*, **64**, 395–416.

Craigie, J. H. (1945). *Scientific Agriculture*, **25**, 285–401.

Dart, R. K. and Stretton, R. J. (1977). In *Microbiological Aspects of Pollution Control* Elsevier, New York, pp. 29–51.

Daws, L. F. (1967). *Symp. Soc. Gen. Microbiol.*, **17**, 31–59.
Draxler, R. R. and Elliot, W. P. (1977). *Atmos. Environ.*, **11**, 35–40.
Edmonds, R. L. (Ed.) (1979). *Aerobiology, The Ecological Systems Approach*, Dowden, Hutchinson and Ross, Stroudsburg, Pennsylvania.
Edmonds, R. L. and Benninghoff, W. S. (1973). *Aerobiology Program. US Comp. Int. Biol. Prog.*, Report No. 3.
Greene, V. W., Pederson, D. A., Lundgren, D. A. and Hagberg, C. A. (1964). In *Proceedings of Atmospheric Biology Conference*, H. M. Tsuchiya and A. H. Brown (Eds), University of Minnesota, Minneapolis, pp. 199–211.
Gregory, P. H. (1973). *The Microbiology of the Atmosphere*, Leonard Hill, Aylesbury.
Hickey, J. L. S. and Reist, P. C. (1975). *J. Water Pollut. Control Fed.*, **47**, 2758–2773.
Hirst, J. M. and Hurst, G. W. (1967). *Symp. Soc. Gen. Microbiol.*, **17**, 307–344.
Högström, U. (1964). *Tellus*, **16**, 205–251.
Hugh-Jones, M. E. (1972). *US/IBP Aerobiology Program Handbook*, No. 2, W. S. Benninghoff and R. L. Edmonds (Eds), University of Michigan, Ann Arbor, Michigan, pp. 69–75.
Hugh-Jones, M. E. (1973). In *Fourth International Symposium on Aerobiology* J. F. Ph. Hers and K. C. Winkler (Eds), Oosthoek, Utrecht, The Netherlands, pp. 399–404.
Imshenetsky, A. A., Lysenko, S. V. and Kazakov, G. A. (1978). *Appl. Environ. Microbiol.*, **35**, 1–5.
Izumu, Y. (1964). *J. Appl. Met.*, **3**, 70–82.
Jones, C. D. and Hutchinson, W. C. A. (1976). *J. Atmos. Terrest. Phys.*, **38**, 485–494.
Lighthart, B. and Frisch, A. S. (1976). *Appl. Environ. Microbiol.*, **31**, 700–704.
Lundquist, J. and Bengtsson, K. (1970). *Geologiska Föreningens i Stockholm Forhandlingar*, **92**, 288–301.
Marshall, E. (1980). *Science*, **210**, 745–749.
Meier, F. C. (1936). *Nat. Geo. Soc. Stratosphere Series*, No. 2, 152–153.
Norris, K. P. and Harper, G. J. (1970). *Nature (Lond.)*, **225**, 98–99.
Parker, D. T., Spendlove, J. C., Bondurant, J. A. and Smith, J. H. (1977). *J. Water Poll. Control Fed.*, **49**, 2359–2365.
Pasquill, F. (1961). *Met. Mag.*, **90**, 33–49.
Pedgley, D. E. (1982). *Windborne Pests and Diseases, Meteorology of Airborne Microorganisms*, Ellis Horwood, Chichester.
Perkins, W. A. and Vaughan, L. M. (1961). In *Conference on Airborne Infection*, W. McDermott (Ed.), Williams and Wilkins, Baltimore.
Peturson, B. (1958). *Can. J. Plant Sci.*, **38**, 16–28.
Polunin, N. and Kelly, C. D. (1952). *Nature (Lond.)*, **170**, 314–316.
Roffey, R., Bovallius, Å., Ånäs and Konberg, E. (1977). *Grana*, **16**, 171–177.
Scriven, R. A. and Fisher, B. E. A. (1975). *Atmos. Environ.*, **9**, 49–58.
Sellers, R. F. and Parker, J. (1969). *J. Hyg. (Camb.)*, **67**, 671–677.
Sellers, R. F., Barlow, D. F., Donaldson, A. J., Herniman, K. A. J. and Parker, J. (1973). In *Fourth International Symposium on Aerobiology* J. F. Ph. Hers and K. C. Winkler (Eds), Oosthoek, Utrecht, The Netherlands, pp. 405–412.
Sorber, C. A., Bausum, H. T., Schaub, S. A. and Small, M. J. (1976). *J. Water Pollut. Control Fed.*, **48**, 2367–3479.
Stevenson, C. M. (1969). *Weather*, **24**, 126–132.
Teltsch, B., Shuval, H. I. and Tadmor, J. (1980). *Appl. Environ. Microbiol.*, **39**, 1191–1197.
Trådgård, H. C. (1977). *Grana*, **16**, 139–143.
Tyldesley, J. B. (1967). *Symp. Soc. Gen. Microbiol.*, **17**, 18–30.
Tyldesley, J. B. (1973). In *Fourth International Symposium on Aerobiology*, J. F. Ph. Hers and K. C. Winkler (Eds), Oosthoek, Utrecht, The Netherlands, pp. 396–398.
Wanner, H. U. (1975). *Zbl. Bakt. Hyg. I. Abt. Orig. B.*, **161**, 46–53.
Wu, D. (1977). *J. Air. Pollut. Cont. Assoc.*, **27**, 1207.

Zadoks, J. C. (1973). In *Fourth International Symposium on Aerobiology*, J. F. Ph. Hers and K. C. Winkler (Eds), Oosthoek, Utrecht, The Netherlands, pp. 392–396.
Zobell, C. E. (1942). In *Aerobiology*, S. Moulton (Ed.), American Association of Science, No. 17, Washington, D.C., pp. 55–68.
Zobell, C. E. and Mathews, H. N. (1936). *Proc. Nat. Acad. Sci. USA*, **22**, 567–572.

Chapter 9

Take-off and landing

9.1 INTRODUCTION

Microorganisms together with particles of dust, pollen, spores, etc., are distributed throughout the atmosphere. Some are there because of a specific liberation process (as for fungal spores) while others (dust, microorganisms, etc.) have become airborne through chance. Those chance processes resulting in take-off, together with those involving particle deposition (or landing) at a site, provide the topics of this chapter.

9.2 TAKE-OFF PROCESSES

9.2.1 Talking, coughing and sneezing

During these actions expired air carries with it materials stripped from surfaces of the respiratory and oral tracts. Predominantly, the droplets are mucus but microorganisms are likely to be within them. The particle size distribution is difficult to predict as it depends on exhaled air velocity, fluid viscosity and the path taken, e.g. whether through the nose or mouth.

Of the three processes, sneezing is the most vigorous and according to Tyrrell (1965) the act generates particles both from the nose and the mouth, while a cough results in a much smaller number of particles. There is, however, a much higher likelihood that those so liberated arise from the mucosa of the lower respiratory tract rather than from the buccal cavity. Apart from this direct dispersal, coughing also may contaminate the saliva with lower respiratory tract microflora which thereby become airborne subsequently through talking or sneezing. Sneezing generates a million or so droplets up to 100 μm diameter plus several thousand larger particles (Duguid, 1945), formed predominantly with saliva from the front of the mouth. Much less frequently, nasal secretion is expelled and even then, owing to its high viscosity, atomization is poor. However, during an upper respiratory infection nasal secretion is increased with concomitant reduction in viscosity and increased likelihood of effective aerosolization.

Talking produces a few droplets per word and is less effective in aerosol dispersal than coughing. Larger droplets fall to the ground where they dry and can be aerosolized by walking, etc. As the number of viable microorganisms per millilitre of saliva is only likely to be about 10^6–10^7/ml, most generated droplets comprise saliva only. As the concentration increases, during an infection, for example, then most particles will contain microorganisms.

One of the best studied microorganisms in terms of take-off processes is *Staphylococcus* as it is present in the nose of between 30 and 50% of adults and on the skin, including that of the perineum. Lidwell (1967) in his review on this subject suggests hand contact may lead to *Staphylococcus* dispersal on skin scales.

9.2.2 Dispersal from surfaces

All surfaces exposed to the atmosphere are contaminated by particles, some of which will be, or will contain, microorganisms. Any factor which causes a force to act on the particles is likely to move or even disperse them. Therefore, the act of walking causes aerosolization of particles from the floor while at the same time microbe-bearing particles, such as skin scales, will be deposited. According to Carson (1966) walking on a floor contaminated with 8 tracer bacteria/cm^2 produced an aerial concentration in a single bed ward of about 180 tracer bacteria/m^3 of sampled air. Such concentrations depend on the floor surface, with waxed and polished floors retaining particles better than unwaxed surfaces (Jones and Pond, 1966). Even so, sweeping floors with a dry brush is highly effective for dispersal, whereas a brush wetted with water, oil, etc., retains particles. But the most effective method is to vacuum-clean floors. Other actions such as bedmaking and closing curtains will increase numbers of airborne particles, as will maintenance work such as hammering, for example. Similarly, air movements through doors and windows are likely dispersal factors. Outdoors, the action of wind and rain produces aerosols and, similarly, the movement of animals, vegetation and vehicles. Whether generated aerosols present biohazards will depend on circumstances as indicated in the previous chapter. In the case of an infected animal, the ground may become contaminated from its saliva, excrement or infected tissues. On drying, this contamination is more readily aerosolized. Infected plants may shed microorganisms directly into the air, or via ground contamination, through the action of the wind, rain, etc. Raindrops can produce splashing or wash particles off leaves and onto the ground where subsequently they are dispersed into the air.

In contact with the ground and other surfaces is a laminar flow boundary layer which, as indicated in the previous chapter, can be a millimetre thick, or much more, especially at night. Only when there is strong turbulence will the eddies be sufficiently powerful to break through this boundary layer and lift microorganisms into the troposphere. Raindrops, on the other hand, owing to their momentum penetrate the boundary layer even when falling vertically. Of the particles liberated, most will quickly fall to the ground with only those less than about 20 μm diameter remaining airborne. The efficiency of dispersal seems to be low in most cases, perhaps only 10%. Exceptions do occur as exemplified by Sahara dust falling in Southern England (see Chapter 8) and by the liberation mechanisms of fungi (Ingold, 1967).

Liquid surfaces provide sources of airborne microorganisms through bursting of bubbles, wave action, waste-water treatment, spray irrigation, etc., which like those from solid surfaces can pose biohazards for man, animals and plants (Rao, 1982; Spendlove and Fannin, 1982). The likelihood of infection, though, is a function of their aerial transport (Chapter 8), survival and landing site.

9.3 LANDING ON SURFACES

Sedimentation under gravity is an important factor only for those particles having relatively large terminal velocities, i.e. several cm/s. These correspond to aerodynamic particle diameters of tens of microns and so this mechanism is important for bacteria, viruses, etc., only when attached to or contained within much larger airborne particles (e.g. skin scales, saliva drops).

Impaction onto obstacles also is important for particles. For obstacles of simple shape such as spheres and cylinders, the efficiency of impaction is related to a parameter P given by,

$$P = \frac{S}{R} = \frac{uv}{gR} \tag{9.1}$$

where, S = particle stopping distance,

R = radius of sphere or cylinder,

u = wind speed,

v = particle terminal velocity,

and, g = acceleration due to gravity.

Values of impaction efficiency as a function of P have been calculated by Davies and Peetz (1956) for large (10^3) obstacle Reynolds number, given by,

$$Re = \frac{2Rv\rho_a}{\eta_a}$$

where, ρ_a = density of air,

η_a = viscosity of air.

Chamberlain (1967) provides experimental data for the efficiency with which *Lycopodium* spores and other particles impact on cylinders, twigs and leaves. Values range over more than an order of magnitude. He concludes that for particles 10 μm diameter and larger impaction on to natural obstacles is a principal mechanism outdoors. For particles 1–5 μm diameter, impaction is of low efficiency, whereas interception by fine structures (e.g. surface hairs, spiders webs, etc.) is of high efficiency. Similar arguments seem to apply for grass which when wet shows an enhanced collection. Although for smaller particles of 1–2 μm diameter deposition efficiencies are similar for sticky and non-sticky surfaces but greater than expected from Stoke's settling velocity (Bovallius and Ånäs, 1980). The higher values arise probably because eddy diffusion assists the passage of small particles across the boundary layer. Another factor enhancing deposition is the action of raindrops (Edmonds, 1979). For particles 20–30 μm diameter raindrop capture efficiency is high but for particles 5 μm diameter or less it falls off sharply. One explanation for the latter is that as raindrops fall the pressure wave on their leading surface is sufficient to push small particles out of their path. Then, collision frequency is low. Raindrop action of a different kind is when microbial particulates are carried into the rain forming area of the atmosphere to become condensation nuclei.

Indoor convection currents are comparable in velocity to fine particle sedimentation velocities and carry particles to surfaces. Wall stains above radiators and hot water pipes are familiar examples of the turbulent deposition of particles in rising warm air.

Landing sites of especial importance in the present context are the airways of animals. On each inspiration air is taken into the respiratory system together with any fine particles suspended in it. For humans with a resting breathing rate around 10 l/min, very large volumes of air are processed relatively rapidly, during which the particulate matter of the air is likely to be deposited to some extent on surfaces exposed in the respiratory tract. These landing sites are flat surface cells or squamous epithelia, or the columnar ciliated epithelia with their hairlike projections. They are likely to be bathed in liquid while the airways will be saturated with water vapour. Various factors dictate whether and where particles deposit, and include the actual structural and physiological features of the respiratory tree. These are considered in the following section.

9.4 RESPIRATORY SYSTEM

9.4.1 Regions

For convenience, the respiratory tract as a landing site (Reid, 1973) may be considered in terms of the following three regions (Figure 9.1).

1. The upper respiratory portion comprising the nasopharynx and/or mouth.
2. The conducting air passages of the larynx, trachea and large bronchi by means of which air taken in through the nose or mouth is uniformly distributed during inhalation, and conversely during exhalation.
3. The respiratory gaseous exchange region formed by the bronchi, alveolar sacs and alveoli.

The nasopharynx region in humans, with a volume of about 20 ml, has two nasal cavities with ciliated epithelia and separated and supported by bone, cartilage and connective tissue. Only the posterior two-thirds has goblet cells and serous and mucous glands while the anterior third is partially covered with

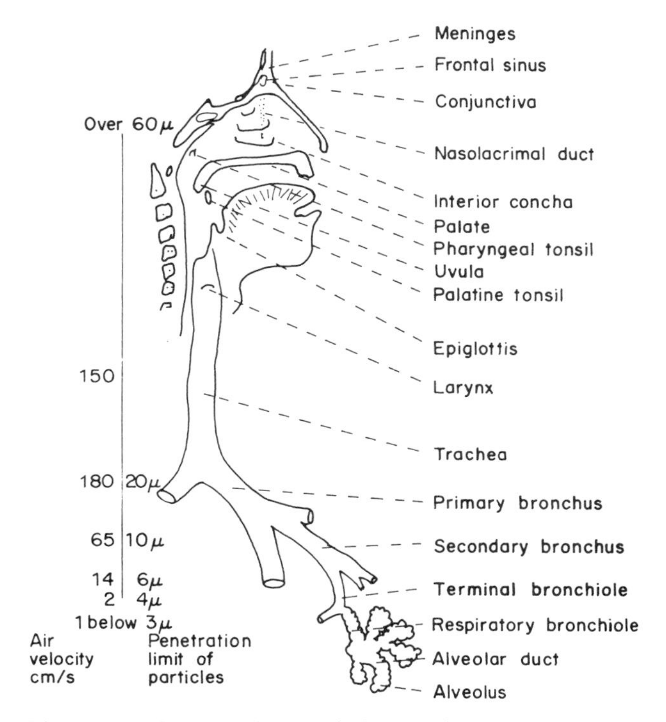

Figure 9.1 Characteristics of the respiratory system of man. (From Pappagianis (1969), in *An Introduction to Experimental Aerobiology*, R. L. Dimmick and Ann B. Akers (Eds), Wiley-Interscience.) Reproduced by permission of the Colston Research Society.

hairs. During breathing, in addition to trapping larger aerosol particles, the nose helps to control the humidity and temperature of air before it enters the pharynx. At this point, the nasal cavity becomes narrowed and makes a 90° turn downwards. (In the AGI-30 sampler (q.v.) this is mimicked by the curvature of the impinger neck.) The pharynx divides to form the larynx and the oesophagus, with the epiglottis helping to prevent material entering the former during swallowing. The larynx with its vocal chords, offers air resistance and consequently the possibility for enhanced particle deposition. The larynx is supported by muscle, bone, cartilage and connective tissue and has a ciliated mucous membrane lining, with the mucus moving upwards.

Leading from the larynx is the trachea. This U-shaped thin-walled tube is supported by about 20 C or Y-shaped hyaline cartilages which encircle the whole length except at the flattened posterior. This portion comprises smooth muscle while the lining of the trachea has lymphocytes and ciliated epithelia well supplied with goblet cells secreting mucus and with serous glands. Both diameter and length of the trachea increase during inhalation. Some species such as the pig, goat and horse, have a bronchus which branches off the trachea to enter a lung lobe.

At its lower end, the trachea divides to form two major bronchi carrying air to the two lungs. The right bronchus divides into three branches while the left one forms two branches, thereby serving the lobes of the lung. Each branch subdivides about sixteen times (in humans) after which the alveoli appear. The bronchi are approximately circular in cross section and are supported by cartilaginous plates which become less evident as the bronchial diameter decreases with branching. These plates eventually disappear when the bronchioles attain a diameter of about 1 mm. The bronchioles themselves have an underlying layer of smooth muscle and a lining of ciliated columnar epithelia.

Mammalian tracheobronchial trees have two basic forms of branching, monopodial and regular dichotomous. The former is found in many animals and is characterized by having tapering airways with small lateral branches which make a 60° angle with the main tube. The latter, found in humans, differs in that when airways divide into two the original diameter is maintained. Phalen *et al.* (1973, 1978) have prepared and analysed silicone rubber replicas from human, dog, rat and hamster, while Phalen (1984) provides additional details and figures of them. Further division of the bronchioles gives rise to the respiratory bronchioles which in humans are about 0.5 mm diameter. They divide from two to five branches with an increasing number of alveoli opening to a common alveolar duct. The alveoli have thin walls with reticular and elastic fibres supporting myriad blood capillaries for gaseous exchange, hence respiratory bronchioles.

Between alveoli are ciliated epithelia which lie on a thin continuous membrane, while within the alveoli themselves there are no ciliated cells. Removal mechanisms from within the alveoli involve macrophages and lymph, but even so this region of the lung, perhaps of 200 m^2 surface area (von Hayek, 1960), seems especially susceptible to disease owing to the irreversible loss of these airways, accumulation of mucus and fibrosis with narrowing airways. Phalen (1984) indicates that these small airways are easily and irreversibly damaged, but because such abnormalities are not readily seen in radiographs, very specialized techniques are required for their detection (Bates, 1973; Ranga and Kleinerman, 1978).

Phalen (1984) describes three basic lung types differentiated by degree of lobe separation, characteristics of the plural membrane enclosing the lungs, presence/absence of terminal and respiratory bronchioles and the nature of the blood supply to the lung. Lung type I of cows, sheep, and pigs has a thick pleura, well-developed secondary lobulation and marked interlobular septa. Lung type II found in dogs, cats and rhesus monkeys, for example, has thin pleura, no secondary lobulation and ill-defined interlobular septa. Lung type III of the horse and humans, is intermediate with thick pleura, incompletely developed lobules and haphazard interlobular septa. Other differences between types include the extent to which the respiratory bronchioles have alveoli. That different lung types occur, should be taken into account in animal studies and in their relevance to other animals.

9.4.2 Cells and tissues

The ciliated mucosa found throughout virtually all the respiratory system comprises specialized cells having hairlike projections or cilia. There are also goblet cells and mucous and serous glands that secrete the components of mucus, primarily an acidic glycoprotein. Cilia being motile, beat several hundred times per minute in coordinated fashion thereby carrying mucus along the mucociliary 'escalator' (Gross *et al.*, 1966). Trapped particles, effete mucosal and other cells, likewise are transported at 15–18 mm/min to be swallowed. This action, dependent on the elastic properties of mucus, is responsible for maintaining the airways free of contagion and depends on both quantity and quality of mucus. Reduced cilia action and/or inadequate mucus caused by infections, medication, smoking, alcohol, etc., severely impede lung clearance with concomitantly increased risks to health.

The human alveolus is polyhedral in cross section and the enclosed airway is about 300 μm diameter. The alveolar epithelium itself has at least three types of cell:

1. alveolar lining cells (designated type I or A) representing about 90% of the alveolar surface and lying on a basement membrane;
2. granular pneumocytes (called type II or B) providing a surface-active secretion essential for preventing alveolar collapse (Pattle, 1965; Ryan and Vincent, 1967). They protrude into the air space of the alveolus and have microvilli. Also, they may detach from the alveolar walls to become phagocytic but are distinct from the alveolar macrophages as their origin is peribronchial lymph, tissue or blood monocytes (Pappagianis, 1969);
3. alveolar brush cells (type III) described by Wood (1978) as having pyramidal shape, like type II cells, protrude into the airspace and have microvilli. Their function is not known.

Pores 2–10 μm diameter occur in alveolar walls and are responsible for connecting juxtaposed alveoli; they are referred to as the pores of Kohn. According to Boatman and Martin (1963) and Boren (1962) they are smaller and unassociated with the larger pores observed in emphysema.

Essential for removal of foreign bodies are the alveolar macrophages which are mobile and change shape. Through phagocytosis solid materials are engulfed while pinocytosis leads to the inclusion of liquid materials. Macrophage behaviour is analogous to that of the *Amoeba* and is responsible for removing microorganisms and other particles. Phagocytosis requires particle recognition, formation of pseudopodia and particle inclusion, followed by antimicrobial action of catalase, lysozyme, hydrogen peroxide and superoxide. Whether such action is successful can depend on whether the microorganism is a spore, or has a capsule as discussed in Chapter 15. Another factor is whether cytotoxic dusts (e.g. particles of heavy metals, silica, asbestos and coal) also are present, while yet another is particle size, as Holma (1969) found when phagocytosis was limited to particles less than about 8 μm diameter, with 1.5 μm diameter being optimum.

Other cells and tissues involved are those of the nervous system and of muscle. Perhaps of especial importance are those involved in the cough reflex as this represents an essential lung clearance mechanism. Most walls of the airways have smooth muscle for the regulation of channel diameter.

9.5 INHALATION AND DEPOSITION

During normal breathing humans inhale about 0.5 l of air and of this volume about 0.35 l mixes with about 3 l of air already held in the lungs. On exhalation 0.35 l of alveolar air is removed; this is not the same as the inhaled air but it is derived from the mix of that air with the 3 l already held in the lungs. Consequently, it requires several breaths to exhale particles previously inhaled, given none deposited. The result is that particle residence times are several minutes, which is more than adequate time to ensure that water vapour condenses on all inhaled particles. Condensation leads to increased particle size which is very marked for hygroscopic materials. Deposition probability in lungs, therefore, is enhanced due to extended residence time as well as to increased particle size.

Whether inhalation is by nose or by mouth influences deposition patterns owing to the particle removal mechanism of the nose. However, only larger particles (> ca 6 μm) are affected, while smaller particles (ca < 6 μm) will behave similarly for both inhalation routes. Strictly, these diameters refer to particles with water condensed onto them. Even so, it should be noted that particle trapping in the nose, mouth, trachea, etc., is rarely 100% efficient. This means that each region of the respiratory tract does not have a sharp cut-off, so that the probability of finding particles of larger sizes decreases with increasing fineness of the respiratory tract (Table 9.1).

In practice, the nasopharyngeal region traps larger particles through their impaction onto the passage walls or onto the nasal hairs. Insoluble particles are removed through mucociliary action (to the throat) followed by swallowing and by 'blowing' of the nose. In the tracheobronchial region, particles impact at bifurcations and diffuse to the walls of the passages where they are removed by ciliary action aided by mucus secretion. Clearance of particles is to the throat where they are swallowed, their speed of removal being slowest in the finest airways. The pulmonary region i.e. the lung is invaded by the smallest particles which deposit by impaction or diffusion (depending on size). Removal of insoluble particles when it occurs, is through phagocytosis by macrophages and by transfer to the lymphatic system.

The efficiency of these depositions measured experimentally (van Wijk and Patterson, 1940) and calculated by Findeisen (1935) shows general agreement (Druett, 1967). Unfortunately, in the experimental work even though non-hygroscopic mineral dust was employed, no allowance was given for water vapour which condensed onto them and the concomitantly increased particle sizes. A better approach was that of Landahl and Herrman (1948) who compared samples taken with a six-stage collector of the exhaled aerosol with

Table 9.1. Retention in various regions of the respiratory tract. (Taken from Pappagianis, 1969) Reproduced by permission of Cambridge University Press

Region	300 cm^3/s 4-s cycle 450 cm^3 tidal air Particle diameter, μm					300 cm^3/s 8-s cycle 900 cm^3 tidal air Particle diameter, μm					300 cm^3/s 12-s cycle 1350 cm^3 tidal air Particle diameter, μm					1000 cm^3/s 4-s cycle 1500 cm^3 tidal air Particle diameter, μm				
	20	6	2	0.6	0.2	20	6	2	0.6	0.2	20	6	2	0.6	0.2	20	6	2	0.6	0.2
Mouth	15	0	0	0	0	14	1	0	0	0	14	1	0	0	0	18	1	0	0	0
Pharynx	8	0	0	0	0	8	1	0	0	0	8	1	0	0	0	10	1	0	0	0
Trachea	10	1	0	0	0	11	1	0	0	0	11	1	0	0	0	19	3	0	0	0
Primary bronchi	12	2	0	0	0	13	2	0	0	0	13	1	0	0	0	20	5	1	0	0
Secondary bronchi	19	4	1	0	0	17	4	1	0	0	18	5	1	0	0	21	12	2	0	0
Tertiary bronchi	17	9	2	0	0	20	9	2	0	1	21	10	2	0	0	9	20	5	0	0
Quaternary bronchi	6	7	2	1	1	8	7	1	1	1	8	7	1	0	1	1	10	3	1	1
Terminal bronchioles	6	19	6	4	6	6	24	7	4	6	6	24	8	4	6	1	9	3	2	4
Respiratory bronchi	0	11	5	3	4	0	10	7	6	6	0	12	11	3	5	0	3	2	2	4
Alveolar ducts	0	25	25	8	11	0	27	44	17	23	0	27	48	22	25	0	13	26	10	13
Alveolar sacs	0	5	0	0	0	0	5	4	2	3	0	5	11	9	10	0	18	17	6	7
Total	93	83	41	16	22	97	91	66	30	40	99	94	82	38	47	99	95	59	21	29

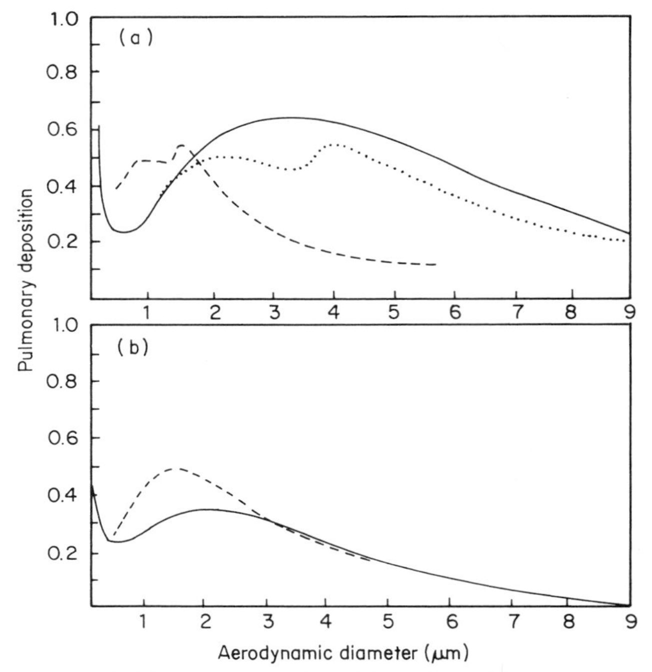

Figure 9.2 (a) Pulmonary deposition of particles entering trachea: ——— theoretical 15 resp/min, 700 cm^3 tidal volume, ---- diameters as on inhalation, ------- diameters equilibrated at 99.5% RH. (b) Pulmonary deposition for nasal breathing 15 resp/min, 750 cm^3 tidal volume: ——— theoretical, ------- Hatch's estimate. (From Druett (1967), *17th Symp. Soc. Gen. Microbiol.*) Reproduced by permission of Cambridge University Press

those of the incident aerosol. Unfortunately, while corresponding stages were compared, no allowance for hygroscopic growth was made. Later, Landahl *et al.* (1951, 1952) used monodisperse aerosols which greatly simplified sampling and analysis. They observed good agreement between practice and theory for larger particles whereas for smaller sizes less deposition than calculated was observed. More recently, Beeckmans (1965a, b) used a refined model and achieved very good agreement with experiment. The biggest uncertainty now is due to the difficulties of assessing the influence of hygroscopicity, particle shape and density inhomogeneity amongst various particles.

As a first approximation percent retention in various regions of the respiratory tract is likely to be as given in Table 9.1 and Figure 9.2. That a minimum in retention is at about 0.3 μm diameter is because particles smaller than 0.3 μm diffuse rapidly to surfaces where they adhere, whereas particles larger than 0.3 μm are trapped by inertial impaction. Alternatively, 0.3 μm particles are those least subject to diffusion and to inertial forces.

More specific details of the regional deposition of particles in the pulmonary region of man are provided by Walton (1971, 1982), Lippmann (1979) and Foord *et al.* (1978). Monodisperse polystyrene unit density aerosols in the range of 2.5 to 7.5 μm diameter, labelled with a radioactive isotope, were administered to human subjects under predetermined breathing patterns. Both the regional and total depositions were determined, the former by external NaI (Tl) radioactivity detectors, and found to be a function of a deposition parameter d^2F where d is the particle aerodynamic diameter and F is the average inspired airflow. Their results indicated also differences between subjects related to respiratory tract morphologies and precise breathing patterns. There were short-term effects caused by respiratory infection or allergic reaction, for example, and increased mucus flow.

Somewhat analogous measurements have been made by Berteau and Biermann (1977) on the respiratory deposition in mice and rats. For these experiments vegetable oil polydisperse aerosols (containing a radioactive tracer) of mass median diameter 2.1 μm and with 90% of particles between 0.4 and 6.3 μm were utilized. Counting of radioactivity in head, lung, trachea, oesophagus, stomach and duodenum gave the deposition and early distribution in these organs. For mice the values, respectively, were 9.1%, 3.8%, 0%, 4.7%, 59%, 7.0%, and for rats, 3.1%, 9.9%, 0%, 6.4%, 13%, 1.9%. With humans the amount entering the lungs is likely to be greater because of a much larger tracheal diameter (Pappagianis, 1969).

The importance of particle size in the degree of infectivity by the respiratory route is considered in more detail in Chapter 15. Suffice it to say here that for anthrax spores infectivity for guinea pigs and monkeys is highest for single spore aerosol particles, falling off as particle size increases (Druett *et al.*, 1953). But, particularly for biological and other hygroscopic particles, it is their hydrated size rather than that before inspiration that is important. For particles in the size range 0.4 to 6 μm Ferron (1977) provides a very detailed account of how to calculate hydrated sizes for hygroscopic particles. His results applied to the lung model of the Task Group on Lung Dynamics (1966) give lung depositions which differ considerably from their recommendations for hygroscopic particles. More specifically, for particles deposited in the nasopharyngeal region the calculated values are less by 20 to 30% in comparison to the recommended values. For the tracheobronchial region calculated depositions are greater than the recommended values, as they are in the pulmonary region where the difference is a factor of four, or so.

9.6 RESPIRATORY TRACT CLEARANCE

Clearance of particulate matter depends on the material as well as its deposition site (Walton, 1971). Soluble particles can dissolve, or absorb, to be carried away by the blood. Insoluble particles deposited in the nose may be removed by blowing, wiping, sneezing, coughing, etc., in addition to mucociliary action. This last action in healthy animals and humans clears the tracheobronchial tract by means of a moving layer of mucus called the mucociliary 'escalator'. It carries mucus along with entrained particles, cell debris and contagion to the oral pharynx where the mucus is swallowed. Some deposited particles require about 1 day for this process but whether there is complete removal within this time frame is questionable (Patrick and Stirling, 1977). On the other hand, Morrow *et al.* (1967) found that inhaled, deposited 10 μm diameter radioactive particles, had clearance times of 0.5, 2.5 and 6 h when deposited, respectively, in large airways, intermediate ones and small peripheral airways. These findings suggest a continuous velocity gradient for mucus which increases in intensity from the terminal bronchioles to the trachea. According to Druett (1967), the velocities are 0.15–0.30 cm/min in the bronchioles, 0.25–1.0 cm/min in the bronchi and 3–4 cm/min in the trachea. The upward flow, however, is non-uniform as it diverts round bronchiole junctions where particles frequently collect, a consequence of particular significance for pathogenesis.

Pavia *et al.* (1980) give a useful account of effects of various factors known to modify particle clearance rates, and this is summarized in Table 9.2. However, according to Phalen (1984) different species of animals are unlikely to have the same corresponding rates. The actual mechanisms involve, at least,

Table 9.2. Particle clearance from the human respiratory tract

Factor	Possible effects
Male/female	none
Increasing age	decrease clearance
Posture	none
Exercise	increase clearance
Sleep	decrease clearance
Smoking	decrease clearance
SO_2	variable
Hair spray	decrease clearance
Bronchitis	decrease clearance
Emphysema	normal or increased clearance
Bronchial asthma	decrease clearance
Influenza	decrease clearance lasting 2–3 months
Pneumonia	decrease clearance lasting 1 year
Bronchogenic carcinoma	no effect
Cystic fibrosis	decrease clearance
Asbestosis	probably none
Air pollution	decrease clearance
Barbiturates	decrease clearance
Hypoxia	decrease clearance
Alcohol	decrease clearance
Endotoxins	decrease clearance
Hypothermia	decrease clearance
Immunoglobulins	increase clearance
Interferon	increase clearance

phagocytosis (described in Section 9.4.2) as well as cilia action. Of particular significance is the high frequency with which airborne infections begin as infections of the ciliated epithelia (Tyrrell, 1973). Cilia themselves have nine fibres surrounded by a membrane and extending into the cell's cytoplasm. The cells are covered with apparently non-motile microvilli. Coordination between adjacent cells produces metachronal waves of beating cilia and one-way movement of overlying mucus.

Under these circumstances it would seem highly improbable that an infection could become initiated. That it can, may reflect a large infectious dose, or in the case of viruses a specific and high affinity for the cilia surface. That this effect of cilia is to comb out organisms rather than sweep them away is supported by electron microscopic evidence, at least for influenza and parainfluenza (Dourmashkin and Tyrrell, 1973). *Mycoplasma pneumoniae* and *Bordetella pertussis* may behave analogously (Tyrrell, 1973). Under these circumstances clearance rates may become negligible and the probability of infection greatly enhanced. At the same time, this mechanism related to tissue specificity may help to explain why a given disease is associated with a certain part of the respiratory tract, e.g. respiratory tract diseases and viruses. In contrast, *Haemophilus influenzae* (a bacterium responsible for meningitis, epiglottitis and arthritis) produces a ciliostatic substance but only after the initial implantation is established in the respiratory tract (Denny, 1973). *Mycoplasma pneumoniae* also reduces ciliary motility but probably through cell damage rather than direct action on the cilia (Taylor-Robinson, 1973).

Another mechanism invoked to reduce clearance is microbial resistance to phagocytosis. For example, virulent staphylococci will multiply in the cytoplasm of granulocytes, while anthrax spores may not be killed by phagocytosis and the capsule of *Francisella tularensis* protects it against intracellular bactericidal factors (Hood, 1977).

Druett (1967) provides the following useful summary of a comprehensive review of lung clearance (Task Group on Lung Dynamics, 1966):

1. particle deposits on ciliated epithelia have clearance half-times of minutes;
2. phagocytes plus ciliary mucus transport, clear the alveolar region with a half-time of 24 h;
3. after this initial phase, a slower phase occurs, the rate largely depending on the physicochemical properties of the particle;
4. elimination via the lymph tract and bloodstream accounts for about one-third, while the remaining two-thirds is via the gastrointestinal tract.

In addition, the action of pollutants (Bierstaker, 1979), meteorological factors (Hyslop, 1979), drugs, alcohol, hypothermia, etc., decrease clearance whereas immunoglobulins, interferon and exercise are likely to increase it.

9.7 CONCLUSIONS

Talking, singing, coughing, and sneezing can be effective dispersal processes as can the action of rain and wind in dislodging particles from surfaces. For respiratory sized particles, interception, impaction and diffusion are more important than is sedimentation in the landing of particles. Where they settle in the respiratory tract is a function also of its structure while the clearance of deposited insoluble particles is mainly by mucociliarly action together with phagocytosis. Some microorganisms can resist these clearance mechanisms and initiate disease through lesions of the ciliated epithelium.

REFERENCES

Bates, D. V. (1973). *J. occup. Med.*, **15**, 177–180.

Beeckmans, J. M. (1965a). *Can. J. Physiol. Pharmac.*, **43**, 157–173.

Beeckmans, J. M. (1965b). *Ann. Occ. Hyg.*, **8**, 221–233.

Berteau, P. E. and Biermann, A. G. (1977). *Toxicol. Appl. Pharmacol.*, **34**, 177–183.

Bierstaker, K. (1979). In *Biometeorological Survey*, vol. I, 1973–1978, Part A, *Human Biometeorology*, S. W. Tromp (Ed.), Heyden, London, Philadelphia, pp. 63–67.

Boatman, E. S. and Martin, H. B. (1963). *Amer. Rev. Respir. Disease*, **88**, 779–784.

Boren, H. (1962). *Amer. Rev. Respir. Disease*, **85**, 328.

Bovallius, Å. and Ånäs, P. (1980). In *First International Conference on Aerobiology*, Federal Environmental Agency (Ed.), Erich Schmidt Verlag, Berlin, pp. 222–231.

Carson, W. (1966). Personal communication to O. M. Lidwell (1967).

Chamberlain, A. C. (1967). *Symp. Soc. gen. Microbiol.*, **17**, 138–164.

Davies, C. B. and Peetz, C. V. (1956). *Proc. R. Soc. A*, **234**, 269–295.

Denny, F. W. (1973). In *Fourth International Symposium on Aerobiology*, J. F. Ph. Hers and K. C. Winkler (Eds), Oosthoek, Utrecht, The Netherlands, pp. 186–189.

Dourmashkin, R. R. and Tyrrell, D. A. J. (1973). In *Fourth International Symposium on Aerobiology*, J. F. Ph. Hers and K. C. Winkler (Eds), Oosthoek, Utrecht, The Netherlands, pp. 189–192.

Druett, H. A. (1967). *Symp. Soc. gen. Microbiol.*, **17**, 165–202.

Druett, H. A., Henderson, D. W., Packman, L. and Peacock, S. (1953). *J. Hyg. (Camb.).*, **51**, 359–371.

Duguid, J. F. (1945). *Edinburgh med. J.*, **52**, 385–401.

Edmonds, R. L. (Ed.) (1979). *Aerobiology. The Ecological Systems Approach.* Dowden, Hutchinson and Ross, Stroudsburg, PA.

Ferron, G. A. (1977). *J. Aerosol Sci.*, **8**, 251–267.

Findeisen, W. (1935). *Pflügers Arch. ges. Physiol.*, **236**, 367–379.

Foord, N., Black, A. and Walsh, M. (1978). *J. Aerosol Sci.*, **9**, 343–357.

Gross, P., Pfizer, E. and Hatch, T. (1966). *Amer. Rev. Respir. Disease*, **94**, 10–19.

von Hayek, H. (1960). *The Human Lung*, Hafner, New York.

Holma, B. (1969). *Arch. Environ. Hlth.*, **18**, 171.

Hood, A. M. (1977). *J. Hyg. (Camb.).*, **79**, 47–60.

Hyslop, N. St. G. (1979). In *Biometeorological Survey*, vol. I, 1973–1978. Part B. *Animal Biometeorology*, S. W. Tromp (Ed.), Heyden, London, Philadelphia, PA, pp. 169–178.

Ingold, C. T. (1967). *Symp. Soc. gen. Microbiol.*, **17**, 102–112.

Jones, I. S. and Pond, S. F. (1966). In *Surface Contamination*, B. R. Frisk (Ed.), Pergamon Press, Oxford.

Landahl, H. D. and Herrman, R. G. (1948). *J. ind. Hyg. Toxicol.*, **30**, 181–188.

Landahl, H. D., Tracewell, T. N. and Lassen, W. H. (1951). *Arch. ind. Hyg.*, **3**, 359–366.

Landahl, H. D., Tracewell, T. N. and Lassen, W. H. (1952). *Arch. ind. Hyg.*, **6**, 508–511.

Lidwell, O. M. (1967). *Symp. Soc. gen. Microbiol.*, **17**, 116–137.

Lippmann, M. (1979). In *Air Sampling Instruments*, 5th edn, American Conference of Government Industrial Hygienists, Cincinnati, Ohio, pp. G1–G23.
Morrow, P. E., Gibb, F. R. and Gaziogln, K. M. (1967). *Amer. Rev. Resp. Disease*, **96**, 1209–1222.
Pappagianis, D. (1969). In *An Introduction to Experimental Aerobiology*, R. L. Dimmick and Ann B. Akers (Eds), Wiley-Interscience, New York, London, pp. 390–406.
Patrick, G. and Stirling, C. (1977). *Proc. R. Soc. B*, **198**, 455.
Pattle, R. E. (1965). *Physiol. Rev.*, **45**, 48–79.
Pavia, D., Bateman, J. R. M. and Clarke, S. W. (1980). *Bull. Eur. Physiopathol. Respir.*, **16**, 335–366.
Phalen, R. F. (1984). *Inhalation Studies: Foundations and Techniques*, CRC Press, Florida.
Phalen, R. F., Yeh, H. C., Raabe, O. G. and Velasquez, D. J. (1973). *Anat. Rec.*, **177**, 255–264.
Phalen, R. F., Yeh, H. C., Schum, G. M. and Raabe, O. G. (1978). *Anat. Rec.*, **190**, 167–177.
Ranga, V. and Kleinerman, J. (1978). *Arch. Pathol. Lab. Med.*, **102**, 609–618.
Rao, C. (1982). In *Methods in Environmental Virology*, C. P. Gerba and S. M. Goyal (Eds), Marcel Dekker, New York, Basel, pp. 1–13.
Reid, L. (1973). In *Fourth International Symposium on Aerobiology*, J. F. Ph. Hers and K. C. Winkler (Eds), Oosthoek, Utrecht, The Netherlands, pp. 165–174.
Ryan, S. and Vincent, T. (1967). In *Ultrastructural Aspects of Disease*, D. W. King (Ed.), Haeber Medical, Harper and Row, New York.
Spendlove, J. C. and Fannin, K. F. (1982). In *Methods in Environmental Virology*, C. P. Gerba and S. M. Goyal (Eds), Marcel Dekker, New York, Basel, pp. 261–329.
Task Group on Lung Dynamics (1966). *Hlth. Phys.*, **12**, 173–208.
Taylor-Robinson, D. (1973). In *Fourth International Symposium on Aerobiology*, J. F. Ph. Hers and K. C. Winkler (Eds), Oosthoek, Utrecht, The Netherlands, pp. 196–200.
Tyrrell, D. A. J. (1965). *Common Colds and Related Diseases*, Williams and Wilkins, Baltimore.
Tyrrell, D. A. J. (1973). In *Fourth International Symposium on Aerobiology*, J. F. Ph. Hers and K. C. Winkler (Eds), Oosthoek, Utrecht, The Netherlands, pp. 183–185.
Walton, W. H. (Ed.) (1971). *Inhaled Particles III*, vol. I, Unwin Brothers, Surrey.
Walton, W. H. (Ed.) (1982). *Inhaled Particles V*, Unwin Brothers, Surrey.
van Wijk, A. M. and Patterson, H. S. (1940). *J. ind. Hyg. Toxicol.*, **22**, 31–35.
Wood, R. W. (1978). *Environ. Health Perspect.*, **26**, 69–76.

Chapter 10

Relative humidity and temperature

10.1 INTRODUCTION

On being exposed to the environment microorganisms lose water molecules that are replaced during host infection or transfer to some other aqueous region. Such movements of solvent water are relative humidity (RH) and temperature dependent and since they are common to all microbial aerosols they may be viewed as basic. Unfortunately, to date, only limited studies have been made of the effects of temperature and so at present we are restricted to considering the consequences of water movements mainly in terms of relative humidity only. Movements of solvent water are important not only for the dehydration–rehydration stress itself, but also because they modify the outcome of other applied stresses which are discussed in later chapters. The fundamental principles underlying the observed behaviour of dehydrated and rehydrated microorganisms provide the basis for this chapter.

10.2 PHAGES

The phages most widely studied are those belonging to the T series of coliphages (Cox and Baldwin, 1964, 1966; Cox *et al.*, 1974; Ehrlich *et al.*, 1964; Akers, 1969, 1973; Hatch and Warren, 1969; Warren *et al.*, 1969; Benbough, 1971; Trouwborst and de Jong, 1973a, b; Trouwborst *et al.*, 1972; Trouwborst *et al.*, 1974; Trouwborst and Kuyper, 1974). In addition, studies have been made of phages S-13 and MS2 (Warren *et al.*, 1969; Dubovi and Akers, 1970, 1973; Trouwborst and de Jong, 1973a, b), *Francisella pestis* bacteriophage (Hatch and Warren, 1969), and phage ϕX174 (Dubovi, 1971; Dubovi and Akers, 1973).

Unlike results obtained with many other microorganisms those with the T series are fairly consistent between different workers. Reasons for this appear to be the comparatively simple biochemical nature of coliphages, the limited conditions under which they may be grown and their lack of sensitivity to oxygen. Except for T1, T coliphages survive well when stored (in the absence of radiation) in purified air and nitrogen atmospheres at greater than about 75% RH, whereas below that RH ability to survive decreases progressively with RH (Figure 10.1). In contrast, rehumidification (Section 2.10) before sampling with an AGI-30 impinger enhances viability (Figure 10.1) in a way analogous to that for freeze-dried T3 and T7 (Cox *et al.*, 1974) (Figure 10.2). Similarly, T3

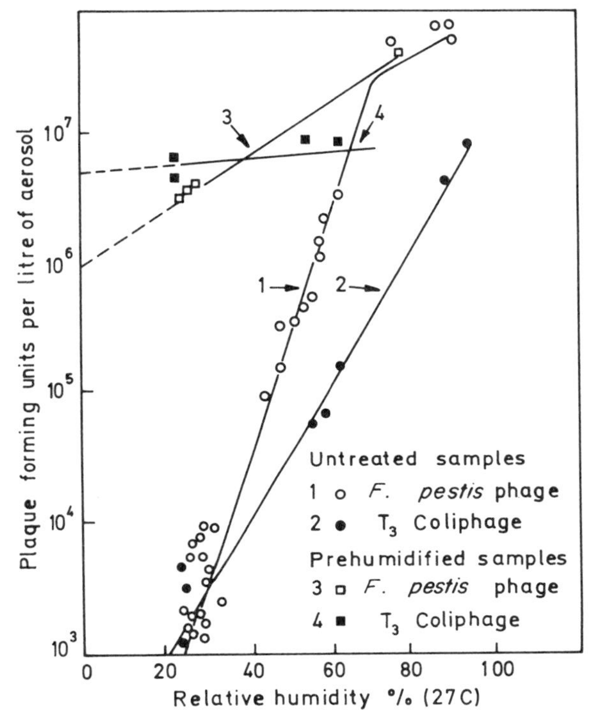

Figure 10.1 Effect of rehumidification on the survival of T3 coliphage and *F. pestis.* bacteriophage. (From Hatch and Warren (1969), *Appl. Microbiol.*, **17**, 685.) Reproduced by permission of American Society for Microbiology.

and T7 aerosol collection with gentle samplers (e.g. subsonic liquid impinger (May, 1966) or subsonic impactor) leads to recovered populations having higher viabilities compared to those collected by a sonic AGI-30 impinger (Benbough, 1971; Cox *et al.*, 1974). This implies that the sampling process *per se* can affect the viability of phages T3 and T7 recovered from aerosols.

The way in which different samplers can determine observed viability of recovered phages T3 and T7 is related to the intensity of shear forces applied through sampler operation. Samplers with high shear forces (e.g. AGI-30) give lower phage viabilities than subsonic samplers (e.g. impactor) which apply lower shear forces. In the case of these two phages, the bonds joining the phage head and tail are weak ones especially when desiccated and the phage tail is easily sheared from the phage head during aerosol sampling. Consequently, populations of phages T3 and T7 when gently recovered from aerosols demonstrate highest percent viability concomitantly with highest percent of phages having intact head-tail complexes (Cox *et al.*, 1974). In a related manner rehumidification of desiccated phages before recovery from aerosols with vigorous samplers (e.g. AGI-30) results in much higher percent viabilities

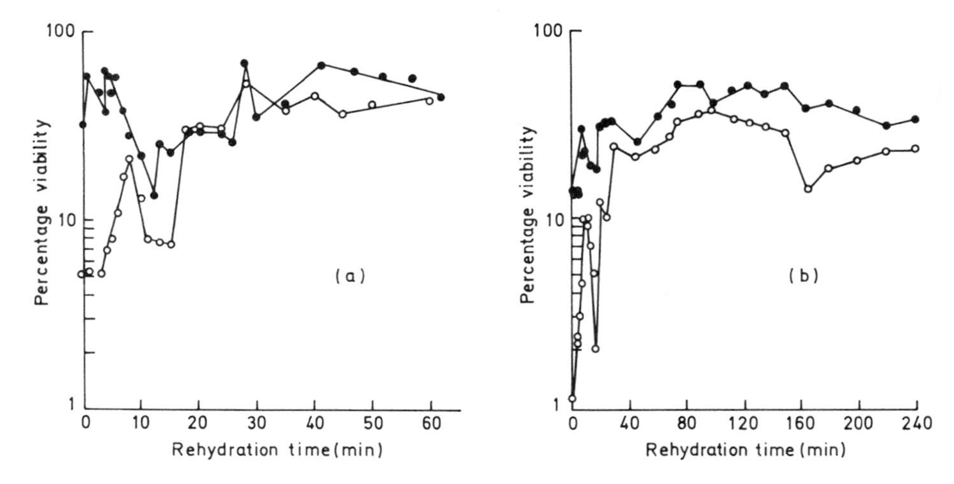

Figure 10.2 Effect of rehumidification on the survival of freeze-dried T3 coliphage. (a) Freeze-dried in 0.1 molal sucrose in nutrient broth, (b) freeze-dried in 0.2 molal sucrose in nutrient broth, ○: reconstituted with nutrient broth, ●: reconstituted with 2 molal sucrose in nutrient broth.

compared to when rehumidification is omitted. Likewise, there is a directly related increase in the percent of phages having intact head-tail complexes (Cox *et al.*, 1974).

Rehumidification of desiccated phages before sampling replaces in them water molecules lost through desiccation. Those replaced water molecules associated with the bonds joining phage head and tail either make the phage head-tail linkage more flexible (i.e. less brittle) or act as a glue. In their strengthened state phage head-tail complexes can withstand the rigours imposed by a high shear force sonic AGI-30 sampler and so the phages now survive.

Further support for this explanation for phages T2, T3 and T7 is that the aerosol survival of phage T7 when adsorbed onto *E. coli* B, increases as the pre-aerosolization time of contact between phage and *E. coli* increases (Cox and Baldwin, 1964, 1966). With longer pre-aerosolization contact time the percentage of absorbed phage whose DNA is transferred to the host bacteria increases. Once the phage DNA is transferred from the phage head via the tail into the host bacterium, separation of phage heads from phage head-tail-host complexes would not influence the subsequent development of phage progeny. Consequently, phage viability increases with pre-aerosolization contact time.

The events described so far for coliphages T2, T3 and T7 are those for comparatively short aerosol storage times of several minutes. When held for long periods, coliphages may die for additional reasons. Hatch and Warren (1969) found that when T3 phages are stored at low RH for 4 h (even with rehumidification) they do not completely survive. Similarly, phages T3 and T7 do not completely survive freeze-drying for 4 h, even when rehumidification is applied before reconstitution (Cox *et al.*, 1974). The cause of death under these circumstances is not through loss of phage tails or phage integrity and is not certain. However, it may have been due to loss of biological activity of phage DNA. The data of Cox and Baldwin (1964, 1966) indicate that coliphage DNA following transfer into the host bacterium *E. coli* B completely maintains its biological activity at short aerosol ages but when storage is prolonged phage DNA biological activity declines. While these data show that the molecular species DNA is not completely aerostable, it is not known whether a similar loss of DNA biological activity occurs when it is contained within an intact coliphage.

The results described above not only explain why phages T3 and T7 lose viability, but also suggest that at least two very different causes of loss of viability of coliphage may result from dehydration and rehydration. On a theoretical basis, this can be accounted for by combining the kinetic equations described later in this chapter with those of Chapters 11, 12 and 16, or by the approach of Bateman (1973).

Coliphage T1 aerosol survival is different to that described above for phages T2, T3 and T7. When aerosolized from clarified lysate survival at 1 h aerosol age is virtually complete over the range 20–95% RH (Benbough, 1971). However, deproteinization and desalting of this medium results in behaviour more like that for phages T2, T3 and T7 (Benbough, 1971). When sprayed from salt solutions aerosol stability is much reduced, with greatest loss of biological activity at about 75% RH (Trouwborst *et al.*, 1972). Unfortunately, it was not reported whether T1 aerosol survival was influenced by rehumidification or by sampling device.

Since coliphage T1 is much more aerostable than are phages T2, T3 and T7 (when sprayed as suspensions in culture lysates) the T1 head-tail complex must be much more stable to desiccation than are the head-tail complexes of phages T2, T3 and T7. On the other hand, in suspensions in salt solutions T1 is more readily surface-inactivated than are phages T3 and T5 (Trouwborst *et al.*, 1974). According to these workers, surface-inactivation of T1 in the aerosol droplet/air interface is the cause of loss of viability. This process then results in release of DNA from the T1 coliphage head (Trouwborst and de Jong, 1972). For their experiments, the DNA in phage T1 was labelled with P^{32}. After aerosolization and rapid rehydration in an AGI-30, phage viability exactly paralleled the degree of adsorption of P^{32} to the host bacterium (*Escherichia coli* B). Analyses of collected T1 aerosols on a Cscl gradient showed that the density of moieties corresponding to maximum radioactivity was 1.67 g/cm^3, while that for complete phage T1 was 1.52 g/cm^3. According to *The Handbook of Biochemistry* (Sober, 1968), the density of free phage T1 DNA is 1.705 g/cm^3 which is considerably greater than the value of 1.67 g/cm^3 for moieties containing P^{32}. Therefore, Trouwborst and de Jong (1972) may not have observed free T1 DNA, but rather T1 phage heads containing DNA but without tails. Therefore, T1 sprayed from suspensions in salt solutions may die for the same reasons as T2, T3 and T7 sprayed from suspensions in spent culture media, i.e. loss of tails. In which case the shear forces operating during rehydration must be greater than the forces operating in the aerosol droplet/air interface.

Further support for this mechanism of viability loss for T1 comes from the finding that phage T3 behaves like T1 when sprayed from suspensions in salt solutions (Trouwborst and Kuyper, 1974). Minimum stability occurs at about 70% RH, and rehumidification is without benefit (Section 2.10). The addition of 0.1% (w/v) peptone to the saline spray fluid (0.1% (w/v)) enhanced survival at high RH and caused the RH for minimum stability to decrease to about 60%. Rehumidification was still without effect, unlike the result observed when the spray fluid was 0.1% (w/v) peptone solution. Therefore, the action of NaCl possibly is to irreversibly weaken phage T3 head-tail bonding as previously described, or to cause other irreversible protein denaturation.

The toxicity of NaCl is both concentration and time dependent. Following aerosolization, as the storage RH is reduced, the NaCl concentration increases until it crystallizes at about 70% RH. Addition of peptone lowers the activity coefficient of the NaCl, so that a slightly lower RH is now required to cause NaCl crystallization. At RH values below that at which crystallization occurs the phages would be exposed to toxic salt concentrations for only the short periods of time required for water evaporation leading to NaCl crystallization. Since the evaporation rate increases with reduced humidity, the time of exposure of phages to these high concentrations of NaCl solution will decrease with decreasing RH. On the basis of this argument phage viability would be expected to decline with decreasing RH, until NaCl crystallization occurs at about 70% RH. Below this RH, viability would rise again as the humidity is further reduced (Figure 10.3).

Other phages have been studied to a lesser degree than the T coliphages. Pestis phage was found by Hatch and Warren (1969) to behave very much like coliphage T3 which implies that like T3 the Pestis phage head-tail complex may be stabilized by water molecules or the coat destabilized by salts. That such changes in the stability of biological structures is not confined to phage head-tail complexes is shown by the tail-less single stranded DNA phage S13 losing viability at mid and low RH (Dubrovi and Akers, 1970, 1973; Warren *et al.*, 1969). As for coliphages this loss of viability is much reduced by rehumidification. It seems most likely that S13 loss of viability is due to a denaturation of its coat since nucleic acid infectivity is more a function of primary rather than secondary or tertiary structure. Also, it seems much more reasonable that protein denaturation rather than damage to primary DNA structure can be reversed by rehumidification.

MS2, a single stranded RNA tail-less phage, also behaves like coliphage T1 in that when sprayed from suspension in tryptone survival is more or less complete and independent of storage RH (Dubovi and Akers, 1970). Likewise when sprayed from suspensions in NaCl solutions it becomes very sensitive to aerosolization (Dubovi and Akers, 1970, 1973; Trouwborst and de Jong,

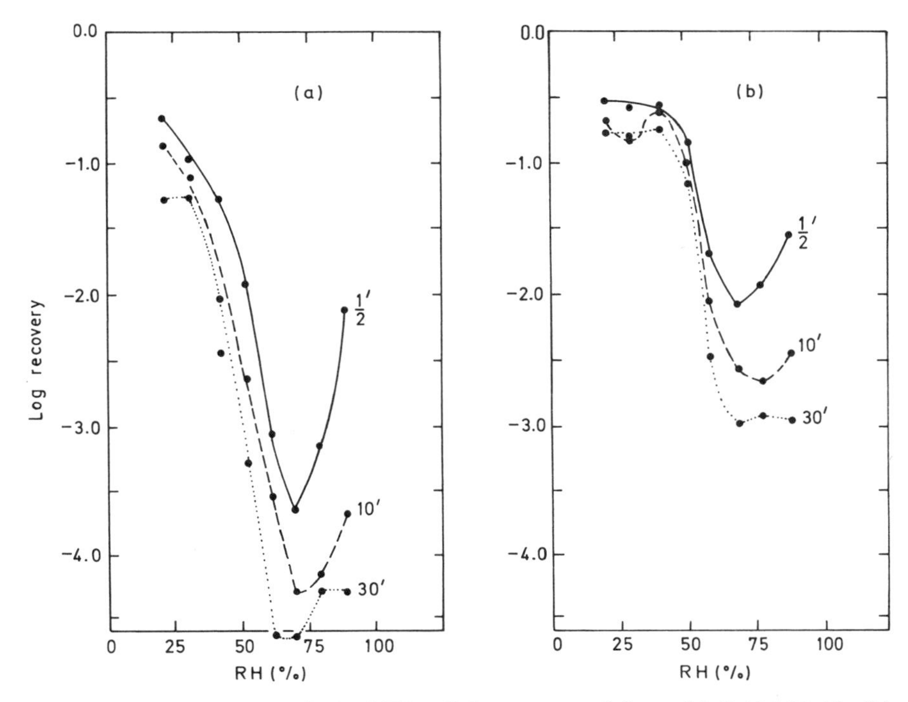

Figure 10.3 Aerosol survival of T1 coliphage sprayed from (a) 0.1 M NaCl; (b) 0.003 M NaCl at aerosol ages of ½, 10 and 30 min. (From Trouwborst *et al.* (1972), *J. Gen. Virol.*, **15**, 235.) Reproduced by permission of Cambridge University Press.

1973a, b). As for phage T1, phage MS2 is surface-inactivated in the presence of high concentrations of NaCl (Trouwborst and de Jong, 1973a, b; Trouwborst *et al.*, 1974). Such surface inactivation is prevented by materials which are adsorbed in liquid air interfaces (e.g. proteins and surface active agents). These materials also stabilize MS2 aerosolized from NaCl solutions (Trouwborst and de Jong, 1973a, b).

When MS2 is aerosol inactivated the process seems to be irreversible since rehumidification prior to sampling decreases rather than increases survival (Dubovi and Akers, 1970). Unfortunately survival studies using samplers applying low shear forces have not been reported. But as for ϕX174 (and other phages) the MS2 phage coat is the site for aerosol-induced damage since the nucleic acid when extracted from such aerosol inactivated phages retains its original infectivity (Dubovi, 1971; Dubovi and Akers, 1973).

10.3 VIRUSES

Many more viruses than phages have been studied from the standpoint of aerosol survival or infectivity (Akers, 1969, 1973). The emphasis was to determine biological activity patterns as a function of RH in relation to their airborne transmission. These papers are discussed in the review articles by Anderson and Cox (1967), Akers (1969), Donaldson (1978), Spendlove and Fannin (1982). Only those papers that are more directly concerned with causes of virus inactivation in aerosols will be referred to below.

Harper (1961) was one of the first investigators to study viral aerosols under carefully controlled experimental conditions while Harper (1963, 1965) and Webb *et al.* (1963) were among the first to show that viral aerosol inactivation could be greatly influenced by disseminating fluid composition. However, it was not until after the mid 1960s that the causes of inactivation of viruses in aerosols were understood. One of the best studied viruses in this context is poliovirus. Harper (1963) found that it survived best at high RH when sprayed from suspension in a variety of fluids, including water. These studies were extended by de Jong and Winkler (1968) and by Benbough (1971). In general, poliovirus behaves somewhat like coliphages T2, T3 and T7 in that aerosol stability is greatest above about 60% RH, and rehumidification of poliovirus aerosols before collection from low RH enhanced viability or infectivity by about 10-fold. Deproteinization and desalting similarly reduced survival while glucose has a protecting action when added to the spray fluid (Benbough, 1971). As with coliphages the most likely structures of poliovirus to be inactivated by dehydration and rehydration are those of the coat. This was shown by de Jong *et al.* (1973) to be so. In their experiments they found that under conditions when the whole virus quickly lost infectivity that of RNA subsequently extracted from aerosolized poliovirus was unaffected. Similarly, the infectivity of isolated RNA, when aerosolized, did not decline. These results contrast with those previously reported by de Jong and Winkler (1968) when they found no infective RNA in aerosol-inactivated poliovirus. This result was probably due to the presence of RNase activity in the fluids used for these earlier experiments (de Jong *et al.*, 1973).

Akers and Hatch (1968) observed a similar pattern for the picornavirus mengovirus 37A in that it is unstable in the aerosol state below 70% RH, whereas infectivity of isolated mengovirus 37A RNA is not affected by aerosolization (Figure 10.4). For this microorganism, in contrast to coliphages, rehumidification before sampling decreased the infectivity of the whole virus stored at low RH. These results show that aerosol-induced denaturation of microbial surface structures is not always reversed by vapour phase rehydration and that it may even be enhanced.

Another picornavirus encephalomyocarditis (EMC) virus similarly survives poorly at RH values below about 60% and its infectivity under these conditions also is not enhanced by rehumidification (de Jong *et al.*, 1974). The cause of this loss of infectivity as for other viruses in this group is because of a damaged viral protein coat. It is expressed as a loss of haemagglutination activity and of affinity for haemagglutination inhibiting antibodies, while the viral RNA extracted from recovered virus retains infectivity. As might be anticipated, therefore, phenol-extracted infectious EMC RNA on aerosolization does not lose infectivity (de Jong *et al.*, 1974). Sedimentation data indicate that

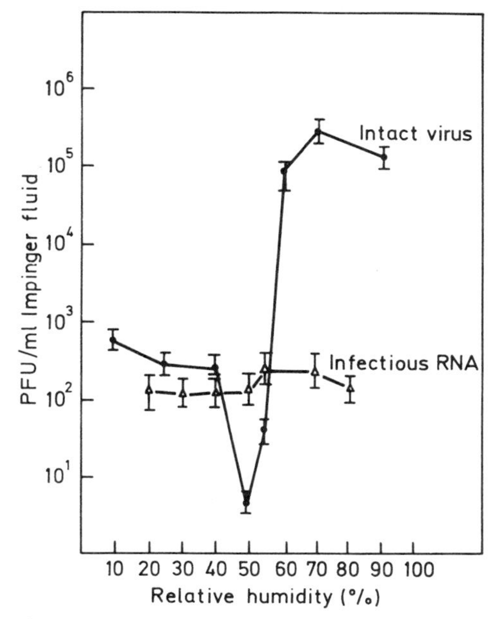

Figure 10.4 Effect of RH on the survival of airborne mengovirus 37A and isolated mengovirus 37A infectious RNA after aerosol storage for 2 h at 27 °C. (From Akers and Hatch (1968), *Appl. Microbiol.*, **16**, 1811.) Reproduced by permission of the American Society for Microbiology.

aerosol-induced damage to the coat protein did not cause any major change in the physical state of EMC virus. The surviving virus fraction was not sensitive to RNase whereas inactivated virions (while still containing their RNA) were sensitive. Therefore, the damage to the protein of EMC virus coat is such that the virus coat becomes permeable to the enzyme RNase, and probably to other small enzymes as well.

Whether the damage to the EMC coat which occurs following aerosolization took place in the aerosol state or during sampling is not absolutely certain. But for the above experiments de Jong *et al.* sampled aerosols with the bottom stage of the May three-stage all-glass impinger (May, 1966). This sampler imposes much lower degrees of shear than the AGI-30. It operates subsonically and introduces aerosol particles at a tangent to the collecting fluid rather than at right angles to it as in the AGI-30. Therefore, if the observed damage to the EMC coat were due to shear forces operating during sampling, this virus when desiccated must be very susceptible to these forces.

The presence of structural lipids in viral coats may affect the ability of such lipid-containing viruses to remain infective in the airborne state (Buckland and Tyrrell, 1962; Benbough, 1971; de Jong *et al.*, 1973). Poliovirus, which does not contain structural lipid in its coat, like the coliphages, is most stable at high

RH (Harper, 1963; Benbough, 1971), as is foot-and-mouth disease virus (Barlow, 1972; Donaldson, 1972). On the other hand, Langat and Semliki Forest virus (Benbough, 1971; Cox, 1976), vesicular stomatitis virus (Warren *et al.*, 1969), vaccinia, Venezuelan equine encephalomyelitis virus (Harper, 1961, 1963) and influenza virus (Harper, 1961; Schaffer *et al.*, 1976) are most stable at low RH. They contain structural lipids. Instability at mid RH at least for Langat and Semliki Forest virus can be greatly reduced by desalting and deproteinization (Benbough, 1969, 1971). The addition of bovine serum albumin to such suspensions greatly enhances survival at high RH (Benbough, 1969, 1971). In contrast such treatments for poliovirus (no structural lipid) are without marked effect (Benbough, 1971). Instability at mid and high RH for such lipid-containing viruses probably arises from components in the complex spray fluids destabilizing their lipoproteins and/or these lipoproteins are least stable in the presence of a few monolayers of water molecules. However, de Jong *et al.* (1976) report that purified SFV is less aerostable than crude preparations, although this could have been due to the spray fluid that they used. Interestingly, the cause of loss of infectivity of this lipid-containing virus is because of damage to virus coat as for the non-lipid-containing viruses (de Jong *et al.*, 1976). This suggests that lipoproteins denature most easily at mid to high RH, while proteins denature most easily at low RH. While the addition of polyhydroxy compounds, e.g. inositol, sorbitol, glucose, to complex spray fluids can mitigate against the viricidal action of desiccation and salts (Webb *et al.*, 1963; Benbough, 1971; Schaffer *et al.*, 1976) for non-lipid viruses, lipoproteins would seem to require more hydrophobic molecules like OED (de Jong *et al.*, 1976) and proteins (Benbough, 1969, 1971).

In contrast to RNA-containing viruses the DNA viruses pigeon pox virus (Webb *et al.*, 1963) and Simian virus 40 (Akers *et al.*, 1973) when aerosolized at ambient temperatures maintain their infectivity at very high levels, independently of RH. In view of the cause of inactivation of the RNA viruses it would seem that this is due to differences in viral coats rather than in the type of their nucleic acid.

Summarizing events for aerosolized phages and viruses, dehydration does not markedly denature their nucleic acids but inactivates surface structures directly or weakens them so that they are susceptible to the high shear forces imposed by some samplers. Vapour phase rehydration (or rehumidification) before sampling enhances the survival of phages and of viruses having no structural lipid, these microorganisms being most stable at high RH. Viruses with structural lipids which are least stable at higher RHs are not affected by rehumidification or are inactivated by it. Composition of disseminating materials can have a very profound effect upon the rate of denaturation of phage and viral structures and, therefore, their infectivity when airborne. Why solutes should demonstrate such an important role is further discussed in Section 10.4.

10.4 BACTERIA

Owing to the greater biochemical and structural complexity of bacteria, mechanisms for their aerosol inactivation can be expected to be more involved

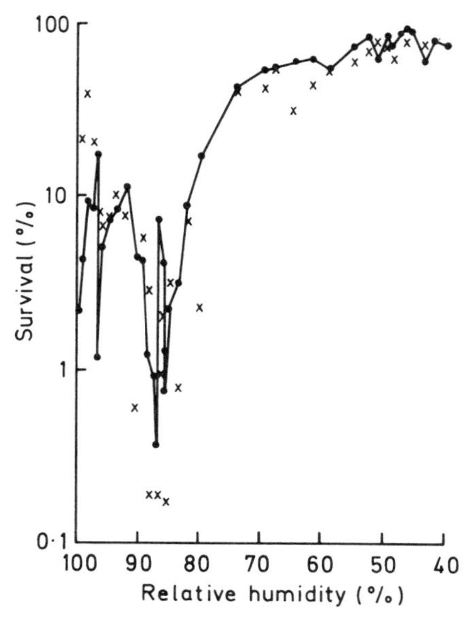

Figure 10.5 Aerosol survival of *E. coli* B sprayed from suspension in distilled water into nitrogen as a function of RH at an aerosol age of 30 min at 26.5 °C. ●: Collection by AGI-30 into phosphate buffer, X: collection by AGI-30 into phosphate buffer + M sucrose.

than those for phages and viruses (Winkler, 1973). Therefore, in order to resolve why aerosolized bacteria die, well-controlled experimental conditions are necessary (Heckly, 1973). Most published data for aerosolized bacteria are not suitable for detailed analyses of causes of death due to the simultaneous occurrence of many stresses. The review articles by Anderson and Cox (1967), Hatch and Wolochow (1969), Strange and Cox (1976) and Donaldson (1978) cover, in general terms, work with bacterial aerosols. Only those reports that are more directly concerned with death mechanisms will be considered.

When aerosolized from a suspension in distilled water into highly purified nitrogen, argon and helium atmospheres, *E. coli* B shows similar overall survival patterns as a function of RH (Figure 10.5, 10.6, 10.7). Under these conditions, survival is virtually complete at low RH but is critically dependent upon the RH at values above about 80% (Cox, 1968a). Such responses are similar to those for other stains of *E. coli* stored in highly purified nitrogen atmospheres (Cox, 1966a, 1966b, 1967, 1968b; Webb, 1969). However, a critical examination of data for the aerosol survival of *E. coli* B in nitrogen, argon and helium atmospheres suggests that these gases are not completely inert as minima do not exactly coincide (Cox, 1968a) (Figure 10.5, 10.6, 10.7). Therefore, either these gases slightly modify water structure through, for example, gas hydrate (clathrate) formation and water lattice modification (van

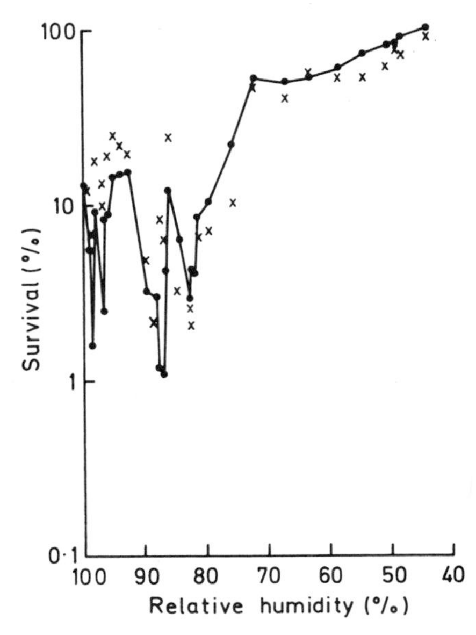

Figure 10.6 Aerosol survival of *E. coli.* B sprayed from suspension in distilled water into argon as a function of RH at an aerosol age of 30 min at 26.5 °C. ●: Collection by AGI-30 into phosphate buffer, X: collection by AGI-30 into phosphate buffer + M sucrose.

der Waals and Platteeuw, 1959) or affect the stability of biological structures. Another possibility is that the rate of evaporation of water influences *E. coli* aerosol survival, as suggested by Poon (1966). By comparing the evaporation rates of water droplets in the three atmospheres (Cox, 1968a) this possibility was shown to be of negligible effect. The actual causes for the slightly different survivals of *E. coli* B in nitrogen, argon and helium atmospheres have not been pursued further. However, since similar zones of instability are found with freeze-dried bacteria (Monk *et al.*, 1956; Monk and McCaffrey, 1957; Davis and Bateman, 1960; Bateman, 1968; Dewald *et al.*, 1967; Bateman *et al.*, 1961; Hess, unpublished data) they probably do not result from the shear forces operating during generation and collection. Such a conclusion is further supported by fact that the behaviour of *E. coli* B dry disseminated (low shear forces) is similar to that described above (Cox, 1970). In addition, *E. coli* aerosol survival is not enhanced by very gently sampling (Cox, 1976). Consequently *E. coli* mainly is inactivated in the aerosol state when stored at high RH in inert atmospheres.

Klebsiella pneumoniae aerosol stability has been studied by Maltman (1970), Maltman and Webb (1971) and Goldberg and Ford (1973). Unfortunately, the data of Maltman (1970) and Maltman and Webb (1971) were obtained only in

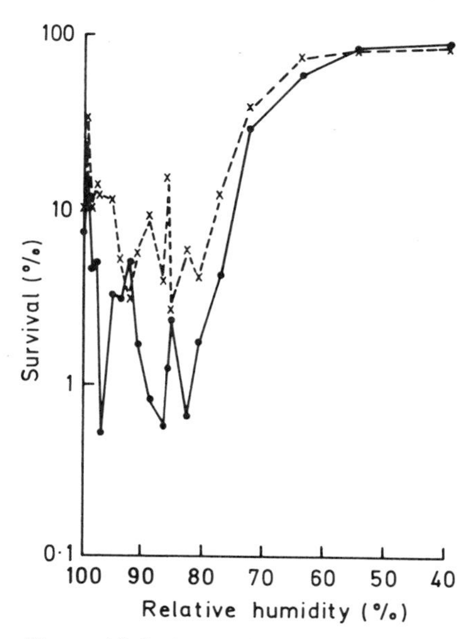

Figure 10.7 Aerosol survival of *E. coli* B sprayed from suspension in distilled water into helium as a function of RH at an aerosol age of 30 min at 26.5 °C. ●: Collection by AGI-30 into phosphate buffer, X: collection by AGI-30 into phosphate buffer + M sucrose.

air and are difficult to analyse due to this organism's oxygen sensitivity. Those data obtained by Goldberg and Ford (1973) were in nitrogen as well as in air (Figure 11.1). In nitrogen atmospheres, *K. pneumoniae* demonstrates maximum instability (when sprayed from suspension in distilled water) at about 55% RH. However, when vapour phase rehydration is used *K. pneumoniae* survives virtually independently of RH when sprayed into nitrogen. Mid RH sensitivity also is reduced by the spray fluid additives inositol and bovine serum albumin either singly or when combined. In this sense *K. pneumoniae* behaves somewhat like *E. coli*, since polyhydric compounds such as raffinose, dextran, glycerol, glutamate and inositol protect *E. coli* aerosols at high RH (Cox, 1966a, 1966b, 1967; Webb, 1969; Goldberg and Ford, 1973). For *E. coli* Jepp and K12 though the beneficial effects are much less, with glycerol becoming toxic at low RH (Cox, 1967, 1968b). Protecting action of such solutes depends as well upon the collection method (Cox, 1966a, 1966b, 1967, 1968b) (Figure 10.8). Raffinose, a trisaccharide, added to *E. coli* strains immediately before dissemination into nitrogen atmospheres at high RH only demonstrates a protecting action if the aerosol is collected into solutions of high solute concentration (e.g. 1 M/l sucrose) (e.g. Cox, 1966a) or is rehumidified before sampling (Cox, 1966b). The action of other more permeable solutes such as glycerol is similar (Cox, 1967) as is that of dextran (mol. wt. 115 000) (Cox, 1966a). Since small, large and very large solute molecules all can protect at high RH, surface structures of *E. coli* would seem to be those which are labile. This is suggested also by studies with the fibre technique (Cox, 1965) which showed how the distribution of added solute molecules within the bacteria/droplet system subsequently affected bacterial survival. Damage to surface structure was found also for phages and viruses as discussed in Sections 10.2 and 10.3. Furthermore, *E. coli* strains and *K. pneumoniae* (in the absence of oxygen, radiation, etc., induced death) as for viruses containing structural lipids are least aerostable at mid to high RH. The role of protecting spray fluid solutes is to stabilize dehydrated bacterial surface structures.

On aerosolization, events for bacteria such as *E. coli* and *K. pneumoniae* would seem to be as follows. When aerosolized into inert atmospheres at mid to high RH their surface structures become destabilized through loss of water molecules. Additives, such as polyhydroxy compounds which supersaturate, can stabilize these labile surface structures. According to Webb (1965) they replace *bound* water molecules, but this now seems unlikely. Instead these compounds by binding to sites on proteins, for example, cause conformational changes in them (Sokolowski *et al.*, 1969) and thereby stabilize proteins against denaturation (Henderson and Unwin, 1975; Unwin and Henderson, 1975; Cox, 1976).

In this context freeze-dried or vacuum-dried bacteria often turn brown during storage. This browning reaction possibly occurs through complicated Maillard reactions between reducing sugars and amino acids, involving carbon groups and possible release of CO_2. Such reactions are thought to be involved in the denaturation and unfolding of ovalbumin α-helix which occurs during drying, as well as the reaction of glucose with 70% of ovalbumin's available lysine molecules (Watanabe *et al.*, 1980). Similar reactions between other sugars, sorbitol and sodium glutamate have been reported for dried myofibrils which are stabilized by these solutes against denaturation induced by desiccation (Matsuda, 1979a, b, c, d, e). On the other hand, Kenehisa and Tsong (1979) present evidence that the unfolding of the globular protein, RNase, at low pH can involve proline isomerization. For collagen membranes sucrose seems to bind at two sites with different affinities producing independent conformational changes in the protein structure (Soddu and Vieth, 1980). There seems little doubt, therefore, that polyhydroxy compounds and amino acids can react together during desiccation, causing conformational changes in proteins. Perhaps then the role of sugar additives is to bind to coat proteins causing a conformational change in them. In the absence of these additives and free water molecules the coat proteins may react irreversibly through Maillard reactions with polyhydroxy coat moieties and cause loss of viability. However, in the presence of sugar additives, due to the conformational changes they cause in the coat proteins in this new configuration the latter do not react (or do so more slowly) with the polyhydroxy coat moieties. In addition, the sugar additives could compete with the polyhydroxy coat moieties for the

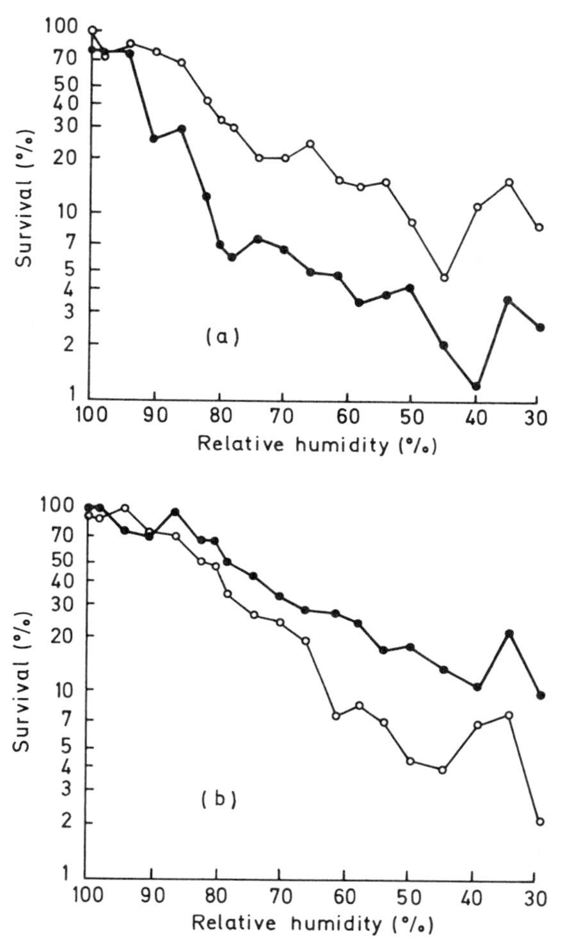

Figure 10.8 Aerosol survival of *E. coli* Jepp and collection method. (a) Sprayed from 0.3 M glycerol at an aerosol age of 25 minutes at the storage RH and 26.5 °C. ●: Collection into phosphate buffer, ○: collection into phosphate buffer + M sucrose. (b) Sprayed from 0.3 M glycerol at an aerosol age of 25 minutes at the storage RH and 26.5 °C followed by a change to 100% RH prior to collection. ●: Collection into phosphate buffer, ○: collection into phosphate buffer + M sucrose.

reaction site of the coat proteins or physically hinder that reaction's molecular collisions. The result in each case would then be more aerostable microorganisms. When in their more normal aqueous environment the Maillard reactions leading to this inactivation may not occur because the reaction sites are separated either by bulk water molecules or by water molecules bound at these sites. Removal of these water molecules by evaporation would then lead to the events proposed above. This mechanism not only provides a possible explanation as to why loss of water molecules can cause loss of microbial viability but provides an explanation for the protecting action of sugars and proteins, as well as effects of complex media spray fluids. In addition, the effects of other parameters (Chapter 2) seem to fall into place, for example, those of the actual microorganisms being tested, as well as the required growth conditions, since coat composition and structures will be dependent on them. Similarly, consideration should be given to the effect of incubation time of microorganism and additive before desiccation, since time may be required for the additive molecules to penetrate into the coat, to bind to coat-proteins and for the conformational changes in coat proteins. Likewise, effects of collecting fluids and vapour phase rehydration (Chapter 2) might be expected to occur in terms of the reversal or non-reversal of the changes induced through the addition of protecting compounds to microorganisms, or by desiccation.

There is much related work on the stability of isolated biological macromolecules in solution of high concentrations of a wide variety of solutes. Its relevance is extremely high since as shown above a major aspect of microbial aerosol survival is the stability of macromolecules in highly concentrated or crystallized solutions. In this context, dehydration is equivalent to greatly increasing solute concentration but as stability in solution is a large subject, only passing reference can be made to it.

Von Hippel *et al.* have made model studies on effects of neutral salts and non-electrolytes on the conformational stability of several isolated biological macromolecules (von Hippel and Wong, 1965; Schleich and von Hippel, 1969; von Hippel *et al.*, 1973; Hamabata and von Hippel, 1973; Hamabata *et al.*, 1973a, b; von Hippel, 1970; von Hippel and Hamabata, 1973), while Papahadajopoulos (1977) has studied the stability of phospholipid membranes. Similar studies have been made by Klotz (1965), Drost-Hansen (1965) and Hamaguchi and Geiduschek (1962). The basic tenets in such studies follow the classical Hofmeister series for effects of neutral salts, as well as changes in hydrogen and hydrophobic bonding, and phase transitions in water. In general, the observed responses for isolated macromolecules in concentrated solutions of solutes tend to parallel effects of solutes upon the aerosol stability of microorganisms, e.g., high concentrations of glycerol denature while glycerol is toxic to aerosols of *E. coli* at low RH (Cox, 1967). As another example Benbough (1971) measured the infectivity of Langat virus as a function of RH when sprayed from suspensions in different salt solutions (Figure 10.9). The stability order was no salt (most stable or high infectivity), 5% KCl, 5% NaCl, 5% LiCl (least stable or lowest infectivity). This is the same as the ranking given by von Hippel (1970) for the effectiveness of these salts in the denaturation of collagen-gelatin, ribonuclease and DNA (Figure 10.10). Similarly, for Semliki Forest virus stored at low RH (Benbough, 1969) the stabilizing order was sugars (most stable or highest infectivity), water, KCl,

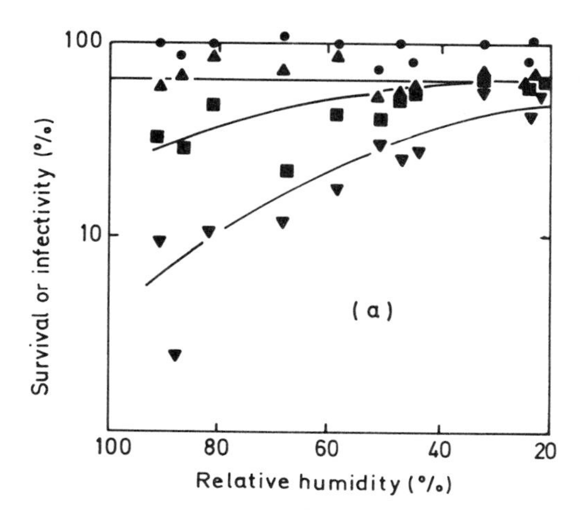

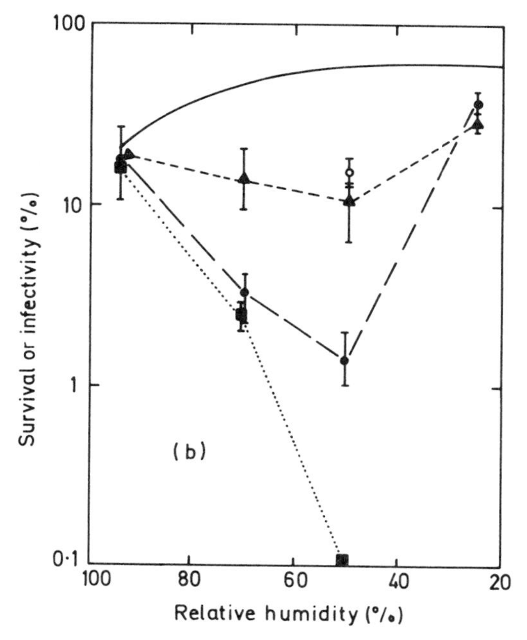

Figure 10.9 Survival of aerosolized Langat virus as a function of RH at aerosol ages of ●: 1 s; ▲: 5 min; ■: 1 h; ▼: 3 h. (a): sprayed from desalted suspensions. (b): sprayed from desalted suspensions with; —— no added salts; ○ added 0.1% (w/v) NaCl; ● added 5% (w/v) NaCl; ▲ added 5% (w/v) KCl; ■ added 5% (w/v) LiCl. (From Benbough (1971), *J. Gen. Virol.*, **10**, 209.) Reproduced by permission of the author.

	Helix ← Native Salting-out … Coil Denatured → Salting-in
Collagen-gelatin	$SO_4^{2-} < CH_3COO^- < Cl^- < Br^- < NO_3^- < ClO_4^- < I^- < CNS^-$ $(CH_3)_4N^+ < NH_4^+ < Rb^+: K^+: Na^+: Cs^+ < Li^+ < Mg^{2+} < Ca^{2+} < Ba^{2+}$ $(CH_3)_4N^+ < (C_2H_5)_4N^+ \ll (C_3H_7)_4N^+: (C_4H_9)_4N^+$
Ribonuclease	$SO_4^{2-} < CH_3COO^- < Cl^- < Br^- < ClO_4^- < CNS^-$ $(CH_3)_4N^+: NH_4^+. K^+: Na^+ < Li^+ < Ca^{2+}$ $(CH_3)_4N^+ < (C_2H_5)_4N^+ < (C_3H_7)_4N^+ < (C_4H_9)_4N^+$
DNA	$Cl^-: Br^- < CH_3COO^- < I^- < ClO_4^- < CNS^-$ $(CH_3)_4N^+ < K^+ < Na^+ < Li^+$
Polyvinylmethyl-oxazolidinone (Cloud point)	$SO_4^{2-} < CO_3^{2-} < F^- < Cl^- < Br^- < ClO_4^- < SCN^-$ $Na^+: K^+ < NH_4^+ < Li^+$

Figure 10.10 Relative effectiveness of various ions in stabilizing or destabilizing the native form of collagen, ribonuclease, DNA and polyvinylmethyl-oxazolidinone in aqueous solution. (From von Hippel (1970), *Colloques Internationeaux du CNRS*, no. 246.) Reproduced by permission of the author.

NaCl, glycerol (least stable or lowest infectivity). Again, this is the same ranking for their stabilizing action upon isolated biopolymers such as DNA, collagen, gelatin and myosin (Klotz, 1965; von Hippel and Wong, 1965; Lakshmi and Nandi, 1976). Likewise the rate of inactivation of the proteinaceous enzyme trypsin in aerosols depends upon the presence of solutes. NaBr enhances that rate to a greater extent than NaCl, although Na_2SO_4 does not have a stabilizing action (Trouwborst *et al.*, 1973). Possibly the lack of stabilizing action by Na_2SO_4 was because it crystallized in the aerosol. The ability of a solute to remain in solution seems to be an essential prerequisite for a stabilizing action in the aerosol (Cox, 1965; Silver, 1965). As shown by Silver (1965) the reason why mannitol can fail to protect is that it does not readily form supersaturated solutions. At low RH mannitol, when it crystallizes, loses its protecting ability. Sorbitol, differing from mannitol in the orientation of a single OH group is like glucose in that it forms a hydrate and will readily supersaturate. It therefore protects under conditions when mannitol will not.

Analogous studies also have been made with other isolated biological macromolecules when equilibrated with water vapour. The relevance is likewise extremely high, but again due to the vastness of the subject only brief reference can be made to it. Nucleic acids, proteins and polysaccharides all have water sorption isotherms which are similar to those for bacteria (e.g. Falk *et al.*, 1963a, b; Shiraishi *et al.*, 1977; Haly and Snarth, 1968; Clementi and Corongui, 1979; Grigera and Berendsen, 1979; Hagler and Moult, 1980; Bhaskara Rao and Bryan, 1978a, b; Nomura *et al.*, 1977; Pineri *et al.*, 1978;

Northcote, 1953; Baddiel *et al.*, 1972; Bull, 1951). Their conformations change with RH, and with temperature, due to concomitant changes in the numbers and states of hydrogen-bonded water molecules. These effects are dealt with more fully in Chapter 16.

Such phenomena as those described above help to explain also differences in observed responses when microorganisms are wet and dry disseminated. In general, when *E. coli* is dry disseminated into nitrogen atmospheres results are similar to those obtained for wet dissemination (Cox, 1970). Stability is greatest at low RH while at high RH the bacteria die more rapidly with the actual survival level depending fairly critically upon storage RH (Figure 10.11). On closer examination the RH values for minimum survival (when *E. coli* B are wet and dry disseminated) shift slightly (compare Figures 10.5 and 10.11). In contrast to this behaviour *Francisella tularensis* LVS demonstrates a big difference in the action of RH when wet and dry disseminated (Cox, 1971; Cox and Goldberg, 1972; Figure 10.12). In this example, the atmosphere is air but oxygen toxicity for aerosolized *F. tularensis* is not marked (Cox, 1971). Also shown in Figure 10.12 is what happens when the freeze-dried powder (used for dry dissemination) is reconstituted with distilled water and then disseminated. In this case the observed pattern is as if the freeze-drying step had been omitted. One possibility for the observed results is that due to hysteresis in water sorption isotherms exact numerical equivalence of the position of minimum survival for wet and dry dissemination would not occur. But these slight shifts observed with *E. coli* are in the wrong direction to be consistent with hysteresis (Figure 10.13) (Nothcote, 1953; Baddiel *et al.*, 1972; Falk *et al.*, 1963a,b; Bateman *et al.*, 1962). The much greater shift found for *F. tularensis* while occurring in a direction consistent with hysteresis is of too great a magnitude to be accounted for by this effect. Similarly, differences between wet and dry disseminated VEE virus (Ehrlich and Miller, 1971) are not consistent with hysteresis (see also Chapter 16). However, in terms of denaturation phenomena, such difference in survival patterns for wet and dry dissemination are possible. As discussed above vapour phase rehydration (or rehumidification) can increase, reverse or have no effect upon denaturation. Therefore, the greater instability of dry disseminated *F. tularensis* and VEE virus stored at high RH (i.e. when vapour phase rehydrated) is probably due to enhanced denaturation (see Chapter 16), as found for some phages and viruses (Sections 10.2 and 10.3). Likewise, effects caused by shifts in the storage RH reported by Hatch *et al.* (Anderson and Cox, 1967; Strange and Cox, 1976; Hatch and Wolochow, 1969) could be through changes in the degree of denaturation.

10.5 BIOCHEMICAL STUDIES WITH BACTERIA

Damage induced in bacteria by dehydration and rehydration is expressed not only as decreased viability but also as failure of various biochemical functions. Those directly involving the bacterial envelope are described in Chapter 14.

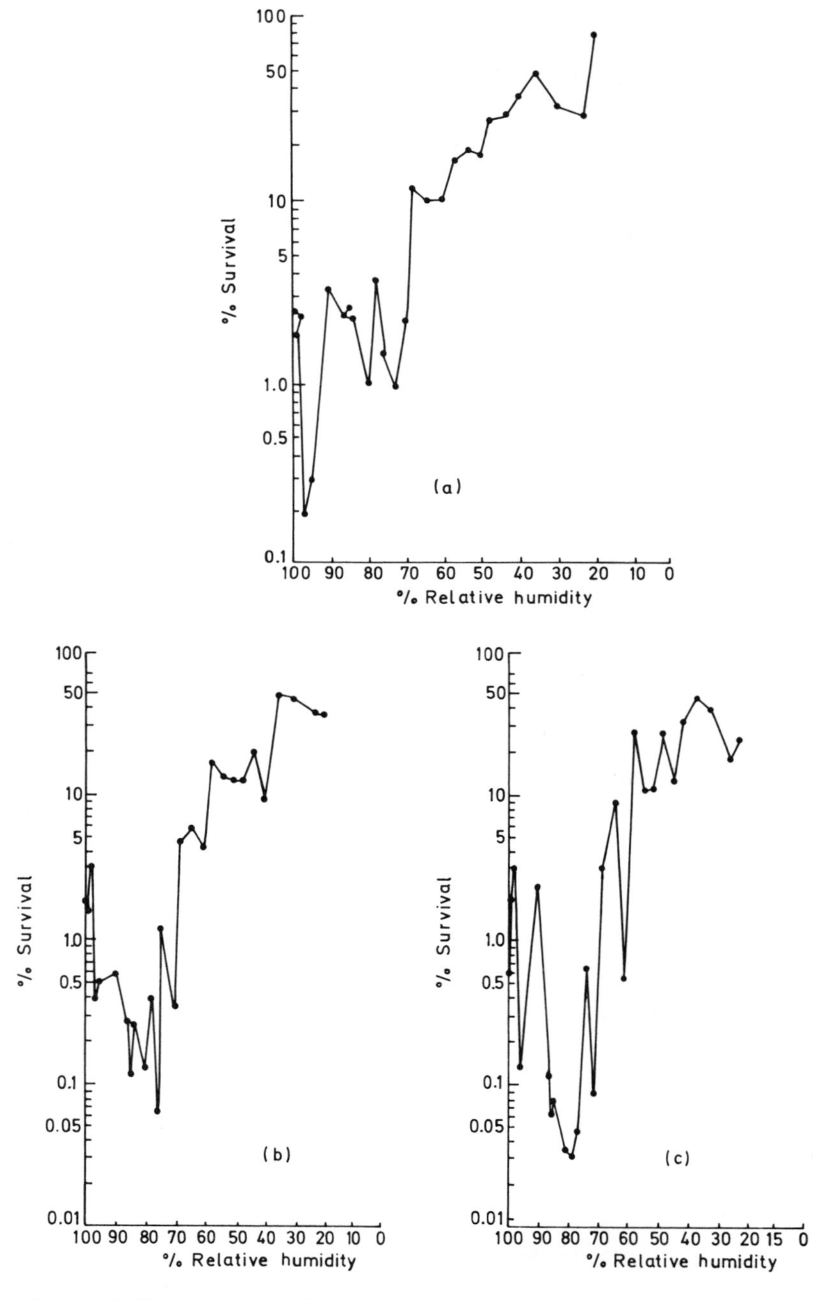

Figure 10.11 Aerosol survival of *E. coli* B powder dry disseminated into nitrogen at an aerosol age of (a) 2 min, (b) 15 min, (c) 30 min.

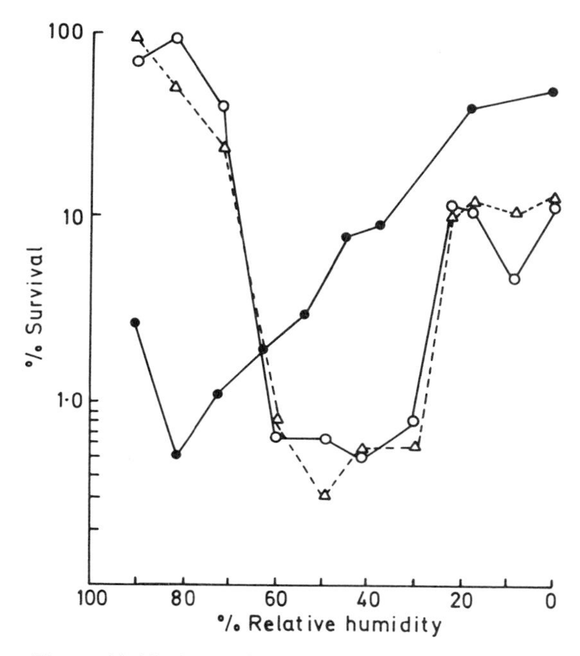

Figure 10.12 Aerosol survival in air of *F. tularensis* LVS when wet and dry disseminated. Aerosol age 15 min; temperature 26.5 °C. ○: Disseminated from the wet state, ●: disseminated from the dry state, △: disseminated from the wet state: suspension prepared by freeze-drying the bacteria and then reconstituting with distilled water.

The ability of *E. coli* to synthesize β-galactosidase in the presence of a specific inducer is considerably impaired in the period immediately following collection from the aerosol. But the loss of this ability precedes loss of ability to form colonies (Anderson, 1966; Webb and Walker, 1968; Webb *et al.*, 1965). However, more recent studies (Benbough *et al.*, 1972) indicate that it is the active transport mechanism for the uptake of inducer that is impaired by aerosolization rather than an impaired ability to synthesize β-galactosidase. Loss of potassium also occurs, but is not in itself a lethal event (Anderson and Dark, 1967). On the other hand, greatest breakdown of RNA in *E. coli* K12 HfrC (collected from aerosols stored in nitrogen) is associated with lowest survival (Cox, 1969). This breakdown possibly results from loss of moieties necessary for RNA stability as could the reduced ability to synthesize RNA and protein which parallels loss of viability (Cox, 1969). As well as these defects in metabolism, DNA synthesis is slightly reduced (Cox *et al.*, 1971) while the ability of *E. coli* strains to utilize oxygen is markedly affected and follows ability to form colonies (Cox, 1969; Cox *et al.*, 1971). Because of this wide range of biochemical lesions it is not too surprising that the ability of *E. coli* B

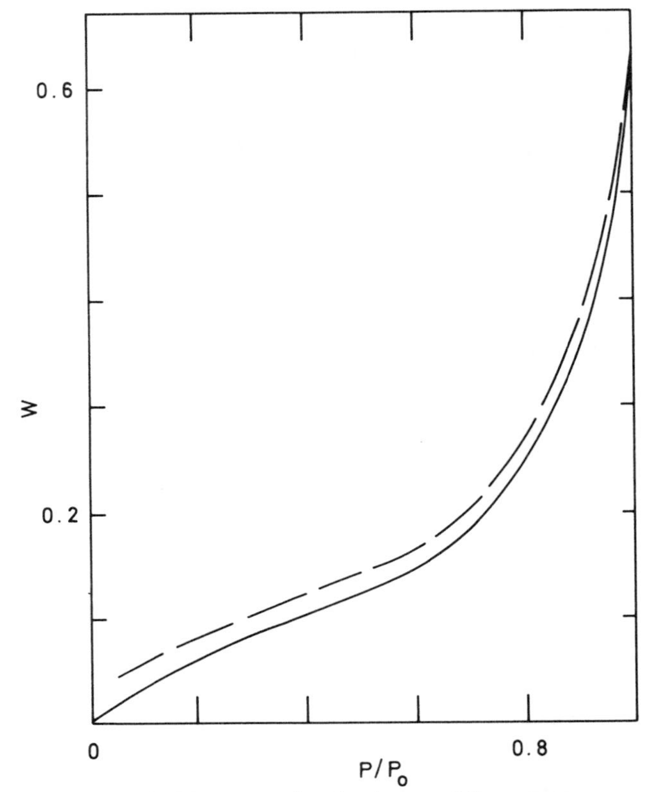

Figure 10.13 Water sorption isotherm of *Serratia marcescens* as a function of water activity (i.e. % RH as a fraction). Weight of water (w) per gram of dry *Serratia marcescens* as a function of water activity (P/P) ——— adsorption, —— desorption.

(recovered from aerosols in nitrogen) to support replication of phage T7 parallels ability to form colonies (Cox and Baldwin, 1966).

To summarize the findings with *E. coli* strains following aerosolization into inert atmospheres, they suffer damage to their surface structures. This damage results in leakage of ions, reduced DNA, RNA and protein synthesis, impaired active transport and greatly decreased oxygen consumption. Surface damage arising through denaturation, possibly of proteins or lipoproteins occurs to the greatest extent at high RH. Of the biochemical failures loss of ability to utilize oxygen seems to be the most marked and the most fundamental because cells which cannot produce energy cannot possibly divide to form colonies (see also Chapter 14).

A different type of biochemical study has been conducted by Dimmick and co-workers (Dimmick *et al.*, 1975, 1977, 1979; Straat *et al.*, 1977). They found that under certain conditions *Serratia marcescens* can metabolize and divide in aerosol droplets. When aerosolized (for example from suspensions in growth media and glycerol) into air at 95% RH and 30 °C these bacteria produce $^{14}CO_2$ from ^{14}C-labelled glucose, incorporate ^{3}H-thymidine into DNA and divide for at least two generations. Such behaviour is likely also for other bacteria when aerosolized under conditions conducive for their survival and cell division, e.g. being sprayed from suspensions of protecting additives and stored at very high relative humidity. These results, and those that demonstrate very significant enzyme activity in solutions of high concentration of solutes which stabilize proteins (Cox, 1961), indicate that airborne bacteria should not be considered inactive metabolically. Furthermore, such results may have implications for the aerosol dispersal of microorganisms in nature. These are discussed in Chapter 13.

10.6 OTHER MICROORGANISMS

Little is known about the causes of inactivation of psittacosis, chlamydia, rickettsia and fungi when airborne (Anderson and Cox, 1967; Akers, 1969; Donaldson, 1978). Mycoplasmas, bacterial L-forms, and algae have received only slight attention.

Mycoplasmas (about 125 mμ diameter) are one of the smallest lifeforms that can replicate independently of other cells. Even though they have no cell walls, when suspended in complex media they are not particularly sensitive to shear stresses imposed in a Wells atomizer. When suspended in 0.85% NaCl, though, they are sensitive to these stresses (Wright *et al.*, 1968b).

Mycoplasma pneumoniae and *M. gallisepticum* appear to be most rapidly inactivated in aerosols in air at 40–60% RH (Wright *et al.*, 1968a, b; Wright *et al.*, 1969). Unfortunately, a tracer was not used in this work and recoveries were normalized to the number of mycoplasmas surviving 5 min in the aerosol.

These data on effects of RH, therefore, need to be treated with caution, since the initial decay rate may have been high and RH dependent (see Section 2.11). As well as showing mid-RH instability, *M. pneumoniae* is unstable, when airborne, in a narrow zone at about 80% RH. Such behaviour is similar to that for *E. coli* strains (Section 10.4). A more detailed analysis of the causes of inactivation of mycoplasma aerosols is not possible as there are no reports of whether or not these organisms are killed by oxygen. This is also the case for the work with airborne streptococcal L-forms which demonstrate maximum inactivation in aerosols at about 40% RH and minimum inactivation at about 20% RH (Stewart and Wright, 1970).

Two algal species, *Nannochloris atomus* and *Synechococcus* sp. (R-3), have been examined for aerosol stability (Ehresmann and Hatch, 1975). While they do not appear to be inactivated by the shear forces operating during atomization, the prokaryot *Synechococcus* sp. is most rapidly inactivated in aerosols at 80% RH, and is most stable at 94% RH. In contrast, the aerosol stability of the eukaryotic alga, *N. atomus*, was different in that no viable airborne cells could be recovered below 92% RH. When stored at that RH, *N. atomus* was more aerostable than *Synechococcus*, with both algal species remaining viable for a considerable period of time.

Spores of fungi and actinomycetes like bacterial spores probably are resistant to desiccation (Donaldson, 1978) but as pointed out by Gregory (1966) little is known about their survival in aerosols.

10.7 KINETIC MODEL

Although not directly mentioned so far, inactivation of microbial aerosols is time dependent. Many workers quote values for decay rates based on simple exponential decay, viz.

$$\ln V_t = \ln V_0 - kt \qquad (10.1)$$

where, V_0 = % viability at time = $t = 0$ (normally 100%)

V_t = % viability at time, t

k = decay constant

Such a procedure is far from satisfactory as loss of viability (or infectivity) rarely follows exponential decay. When death is due to dehydration and rehydration, the rate of loss of viability usually is initially high and then decreases with time (e.g. Figure 10.14). Such an effect of time upon loss of viability cannot be accounted for by simple exponential decay (equation 10.1). One kinetic model which is much more applicable (Cox, 1976) involves a mechanism which supposes that microorganisms contain a molecular species $B(n)H_2O$, the biological activity of which is essential for a microbial cell to replicate, i.e. to be viable or infective. $B(n)H_2O$, when exposed to an environment of lowered water activity (or RH), forms a series of hydrates in a manner similar to that for most biological macromolecules (Section 10.4). It is further supposed that some of these hydrates are unstable and spontaneously denature through a first order process, i.e.

$$\frac{-dx}{dt} = kx \qquad (10.2)$$

where x is the concentration of the species which denatures. Hence the model is,

$$\begin{array}{ccccc}
B(n)H_2O \rightleftharpoons & B(n-x)H_2O + xH_2O & \underset{k_-}{\overset{k_+}{\rightleftharpoons}} & B(n-x-y)H_2O + yH_2O \rightleftharpoons & \\
& \downarrow k_x & & \downarrow k_y & B + iH_2O) \\
& \text{denatured form} & & \text{denatured form} & (10.3)
\end{array}$$

For a population of microorganisms of size N which is homogeneous with respect to the nature of species $B(n)H_2O$, but heterogeneous with respect to its concentration, the average concentration of $B(n)H_2O$ is given by,

$$[B(n)H_2O] = \sum_{1=1}^{1=N}\left[\frac{B_1\ (n)H_2O}{N}\right] \tag{10.4}$$

If all members of the population were to have the same concentration of $B(n)H_2O$ they would all die at the same instant of time. Such behaviour, though very rarely found, was observed with synchronously growing bacterial cultures by Dimmick (personal communication).

Given that the average concentration within the population being tested is $[B(n)H_2O]_0$ and that following aerosolization its value (because of denaturation) decreases with time, the level of survival at time t must be related to that average concentration at time t. The form of the relationship between % viability (or infectivity) and $[B(n)H_2O]_t$ can be derived in various ways. One possible form of the relationship is given by Cox *et al.* (1973) and is shown below. Another form of this relationship is discussed in Chapter 16.

Given that the probability of death P_d is

$$P_d \cdot N_l = -\delta N_l \tag{10.5}$$

and that the probability of death, P_d, is directly proportional to the corresponding negative change in the concentration of species B, i.e.

$$P_d = -K_l\ \delta[B] \tag{10.6}$$

where, N_l = number of live microorganisms at time t,
$[B]$ = average concentration of species B at time t

then,

$$\frac{\delta N_l}{\delta[B]} = K_1 N_l \tag{10.7}$$

i.e.

$$\frac{\delta V}{\delta[B]} = K_1 V \tag{10.8}$$

where, $V = \%\ \text{viability} = \dfrac{100 \times N_l}{N_l + N_d}$

N_d = number of dead microorganisms at time t.

Integration of equation 10.8 gives:

$$\ln V = K_1\ [B] + K_2 \tag{10.9}$$

where K_2 = the integration constant.

The boundary condition, $t = 0$, $V = 100\%$, $[B] = [B]_0$, gives:

$$\ln V = K_1[B] - K_1[B]_0 + \ln 100 \tag{10.10}$$

For $0 < a_w < 1$, the differential equations representing equation 10.3, when integrated and combined with equation 10.10 gives:

$$\ln V = K_1[B(n - x)H_2O]_0([B_x] + [B_y] - 1) + \ln 100 \tag{10.11}$$

It is assumed that the rate of liquid water evaporation is very high so that species $B(n - x)H_2O$ effectively is present at the instant of aerosol generation (i.e. at $t = 0$). The concentration terms $[B_x]$ and $[B_y]$ are complex functions of the parameters, k_+, k_-, k_x, k_y, RH, and time (Cox, 1976). Provided that $k_+ \simeq k_- \gg k_x \simeq k_y$, equation 10.11 simplifies to:

$$\ln V = K_1[B(n - x)H_2O]_0(e^{-kt} - 1) + \ln 100 \tag{10.12}$$

where k is a first order denaturation constant (Cox, 1976).

Several hundred viability-time curves (obtained under conditions where denaturation was the likely cause of loss of viability or infectivity) have been analysed using this equation (Cox, 1976). In most cases the equation agreed extremely well with experimental data. Examples are given in Figures 10.14, 10.15 and 10.16, where the points are experimental data, and the curves were calculated with equation 10.12.

Data for the survival of *E. coli* strains in nitrogen atmospheres as a function of storage RH (Cox, 1966a, b, 1970) were analysed with equation 10.12 and derived values for $K_1[B(n - x)H_2O]_0$ and k are shown in Figures 10.17, 10.18 and 10.19. The survival of *E. coli* B sprayed from suspensions in distilled water into nitrogen is characterized by poor survival at 100, 97, 85, 50 and 20% RH (Cox, 1966a). These RH values correspond exactly to maximum derived values for k, except at 97% RH, whereas the derived values of $K_1[B(n - x)H_2O]_0$ are independent of RH except at 97% and 85% (Figure 10.17). These discrepancies, however, coincide with the highest rates of loss of viability, and presumably the inequality $k_+ \simeq k_- \gg k_x \simeq k_y$ used to derive equation 10.12 was not fulfilled at 97% and 85% RH, i.e. the equation fails (strictly) precisely when the approximation might be expected to fail. Equation 10.12 is more strictly applicable when the biological decay rate is much less than the rate at which aerosolized microorganisms equilibrate (i.e. $k_+ \simeq k_- \gg k_x \simeq k_y$). This situation is shown in Figure 10.18 for *E. coli* B protected with raffinose, and in Figure 10.19 for the more robust *E. coli* commune strain sprayed from distilled water. In these cases, $K_1[B(n - x)H_2O]_0$ is independent of RH in the range 20–100%, and k attains a single maximum.

That $K_1[B(n - x)H_2O]_0$ is independent of RH could mean that species B must reside where its concentration does not change with RH. If this were so, a location other than on the cell surface is difficult to envisage (Cox, 1976), since only there can water evaporate at 100% RH.

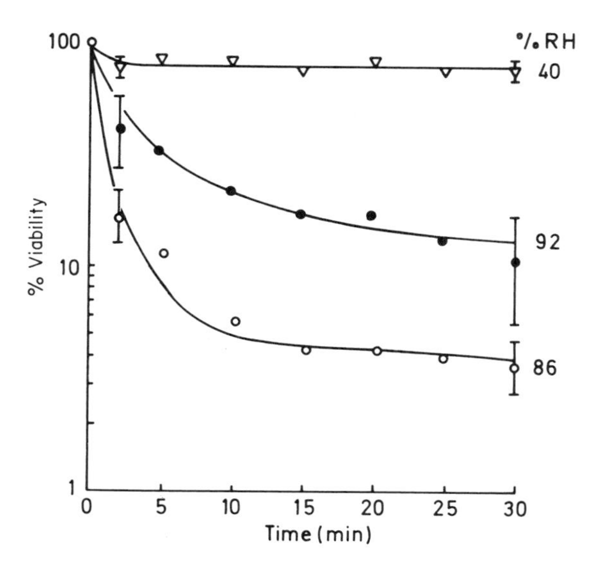

Figure 10.14 Survival of *E. coli* Jepp wet disseminated into nitrogen; points experimental data, lines calculated by equation 10.12.

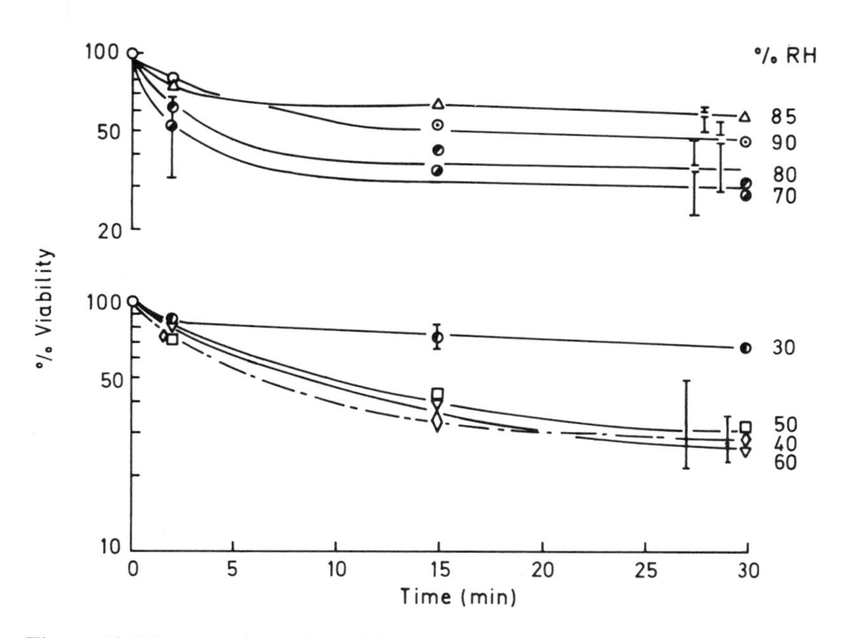

Figure 10.15 Aerosol survival of Semliki Forest virus suspended in medium 199 + 10% (vol/vol) calf serum, sprayed into air; points experimental data, lines calculated by equation 10.12.

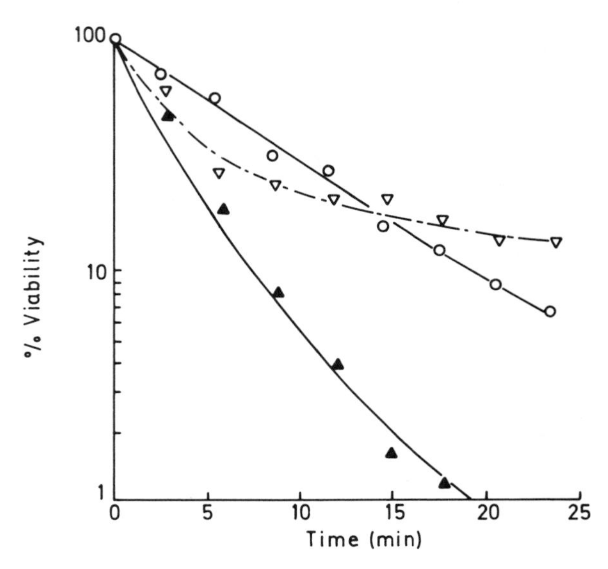

Figure 10.16 Survival of Semliki Forest virus in liquid suspension at 50 °C. Points experimental data, lines calculated by equation 10.12. ○: After one L-cell passage, △: original mouse brain material, ▽: after one chicken embryo cell passage. (Experimental data of Fleming (1973), *J. Gen. Virol.*, **19**, 353–367.)

Further examples of first order denaturation kinetics are given by Cox (1976) and in Chapter 16. In the latter it is also shown that while some bacterial and viral species appear to die through a first order denaturation process others appear to die through a second order denaturation process.

10.8 TEMPERATURE

Aerosol storage temperature is another environmental parameter known to affect aerosol survival and infectivity (Anderson and Cox, 1967; Akers, 1969, 1973; Strange and Cox, 1976; Spendlove and Fannin, 1982; Donaldson, 1978; Hatch and Warren, 1969). Possibly, temperature effects arise because it is the absolute humidity (a function of both relative humidity and temperature) rather than the relative humidity which controls aerosol survival. Alternatively temperature may affect the inactivation of microbial aerosols through reaction rate as it does most chemical and biochemical processes. Other possibilities exist, which include temperature-dependent changes in the conformational structures of biopolymers, e.g. the temperature-dependent melting of nucleic acids or the α-helix $\leftrightarrows$ β-pleated sheet temperature-dependent transition of keratin. In such cases one form of the biopolymer may be less stable than others.

To date the particular biomolecular moieties of phages, viruses and bacteria

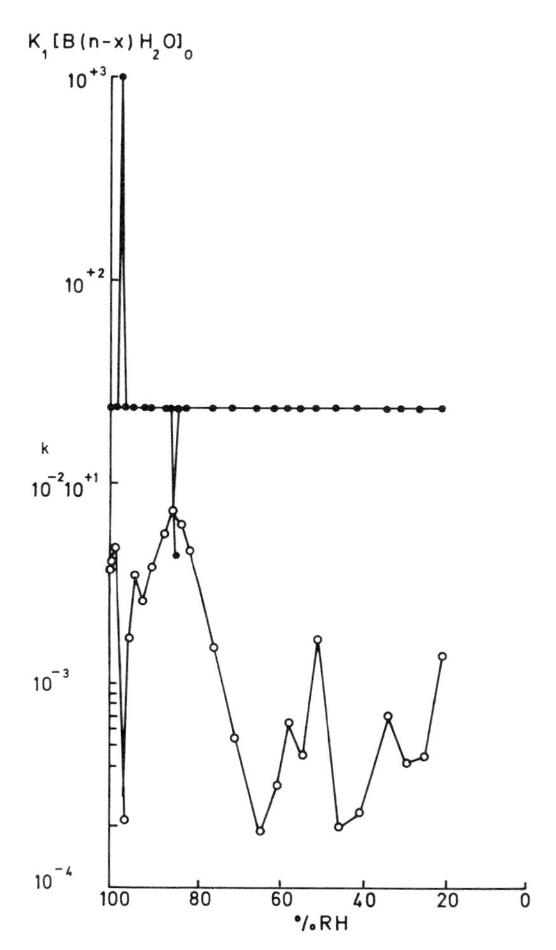

Figure 10.17 Values of derived constants for *E. coli* B sprayed from suspension in distilled water into nitrogen. Values derived by fitting equation 10.12 to experimental viability-time curves.

which denature (as a consequence of aerosolization) have not been isolated. As a result, the only approach presently available for understanding temperature effects is that of kinetic analyses. Due to this necessity, further discussion of effects of temperature will be found in Chapter 16 where catastrophe theory is applied and where equations allowing for second order denaturation also are derived.

10.9 CONCLUSIONS

On being exposed to the atmosphere, microorganisms lose water that is

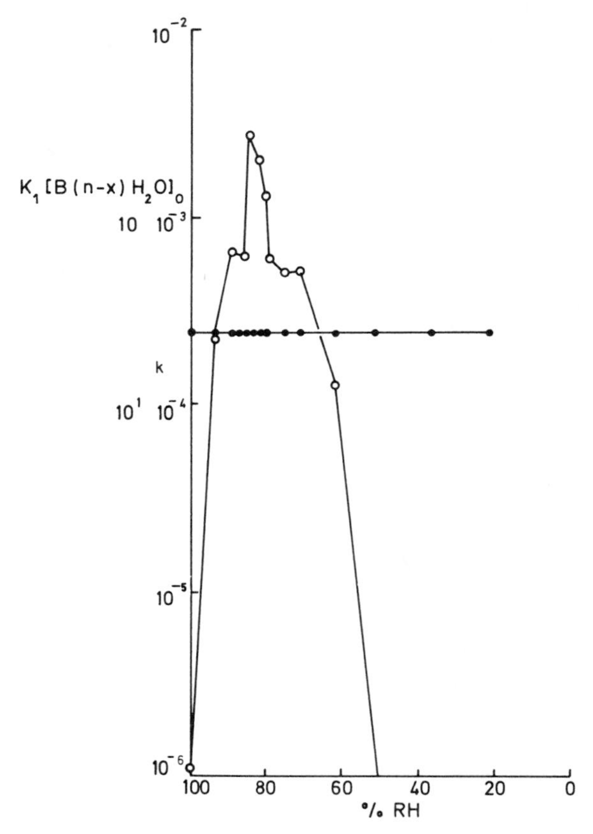

Figure 10.18 As for Figure 10.17 except sprayed from suspension in 0.13 M raffinose.

replaced during host infection, or transfer to another aqueous environment. These movements of water molecules are relative humidity (RH) and temperature dependent and common to all microbial aerosols. Their effects range from weakening of the head-tail bonding of phages to irreversible changes in the protein– and lipoprotein–carbohydrate structures of the coats of viruses and bacteria. Kinetic analyses of viability–time curves leads to similar conclusions, while biochemical analyses indicate that, at least for *E. coli*, the denaturation causes loss of ability to utilize oxygen. Without oxidative metabolism repair mechanisms (q.v.) are inoperative and viability loss inevitable. For viruses denaturation of surface structures prevents cell invasion, thereby causing infectivity loss.

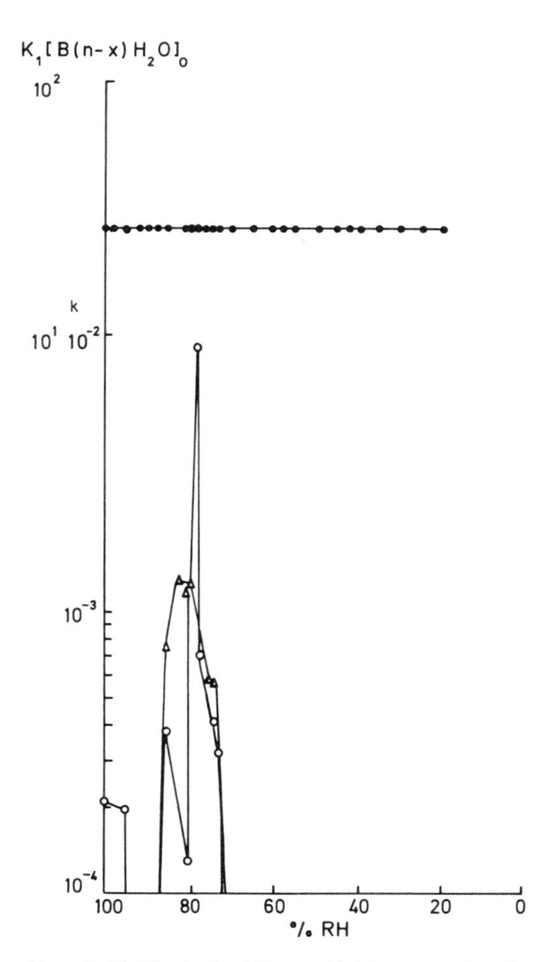

Figurė 10.19 As for Figure 10.17 except for *E. coli* commune sprayed from suspension in 0.13 M raffinose (○) and 0.3 M raffinose (△).

REFERENCES

Akers, T. G. (1969).In *An Introduction to Experimental Aerobiology*, R. L. Dimmick and Ann B. Akers (Eds), Wiley-Interscience, New York, London, pp. 296–339.

Akers, T. G. (1973). In *Fourth International Symposium on Aerobiology*, J. F. Ph. Hers and K. C. Winkler (Eds), Oosthoek, Utrecht, The Netherlands, pp. 73–81.

Akers, T. G. and Hatch, M. T. (1968). *Appl. Microbiol.*, **16**, 1811–1813.

Akers, T. G., Prato, C. M. and Dubovi, E. (1973). *Appl. Microbiol.*, **26**, 146–148.

Anderson, J. D. (1966). *J. Gen. Microbiol.*, **45**, 303–313.

Anderson, J. D. and Cox, C. S. (1967). *Symp. Soc. Gen. Microbiol.*, **17**, 203–226.

Anderson, J. D. and Dark, F. A. (1967). *J. Gen. Microbiol.*, **46**, 95–105.

Baddiel, C. B., Breuer, M. M. and Stephens, R. (1972). *J. Coll. Int. Sci.*, **40**, 429–436.

Barlow, D. F. (1972). *J. Gen. Virol.*, **15**, 17–24.
Bateman, J. B. (1968). *Amer. J. Epidemiol.*, **87**, 349–366.
Bateman, J. B. (1973). In *Fourth International Symposium on Aerobiology*, J. F. Ph. Hers and K. C. Winkler (Eds), Oosthoek, Utrecht, The Netherlands, pp. 117–124.
Bateman, J. B., McCaffrey, P. A., O'Connor, R. J. and Monk, G. W. (1961). *Appl. Microbiol.*, **9**, 567–571.
Bateman, J. B., Stevens, C. L., Mercer, W. B. and Carstensen, E. L. (1962). *J. Gen. Microbiol.*, **29**, 207–219.
Benbough, J. E. (1969). *J. Gen. Virol.*, **4**, 473–477.
Benbough, J. E. (1971). *J. Gen. Virol.*, **10**, 209–220.
Benbough, J. E., Hambleton, P., Martin, K. L. and Strange, R. E. (1972). *J. Gen. Microbiol.*, **72**, 511–520.
Bhaskara Rao, P. and Bryan, W. P. (1978a). *Biopolymers.*, **17**, 291–314.
Bhaskara Rao, P. and Bryan, W. P. (1978b). *Biopolymers.*, **17**, 1957–1972.
Buckland, F. E. and Tyrrell, D. A. J. (1962). *Nature (Lond.)*, **195**, 1063–1064.
Bull, H. B. (1951). *Physical Biochemistry*, Wiley, New York.
Clementi, E. and Corongui, G. (1979). *Biopolymers.*, **18**, 2431–2450.
Cox, C. S. (1961). *Enzyme Activity in Modified Solvent*, Ph. D. Thesis, University of Bristol.
Cox, C. S. (1965). In *First International Symposium on Aerobiology*, R. L. Dimmick (Ed.), Naval Biological Laboratory, Naval Supply Center, Oakland, California, pp. 345–368.
Cox, C. S. (1966a). *J. Gen. Microbiol.*, **43**, 383–399.
Cox, C. S. (1966b). *J. Gen. Microbiol.*, **45**, 283–288.
Cox, C. S. (1967). *J. Gen. Microbiol.*, **49**, 109–114.
Cox, C. S. (1968a). *J. Gen. Microbiol.*, **50**, 139–147.
Cox, C. S. (1968b). *J. Gen. Microbiol.*, **54**, 169–175.
Cox, C. S. (1969). *J. Gen. Microbiol.*, **57**, 77–80.
Cox, C. S. (1970). *Appl. Microbiol.*, **19**, 604–607.
Cox, C. S. (1971). *Appl. Microbiol.*, **21**, 482–486.
Cox, C. S. (1976). *Appl. Environ. Microbiol.*, **31**, 836–846.
Cox, C. S. and Baldwin, F. (1964). *Nature (Lond.)*, **202**, 1135.
Cox, C. S. and Baldwin, F. (1966). *J. Gen. Microbiol.*, **44**, 15–22.
Cox, C. S. and Goldberg, L. J. (1972). *Appl. Microbiol.*, **23**, 1–3.
Cox, C. S., Baxter, J. and Maidment, B. J. (1973). *J. Gen. Microbiol.*, **75**, 179–185.
Cox, C. S., Bondurant, M. C. and Hatch, M. T. (1971). *J. Hyg. (Camb.)*, **69**, 661–672.
Cox, C. S., Harris, W. J. and Lee, J. (1974). *J. Gen. Microbiol.*, **81**, 207–215.
Davis, M. S. and Bateman, J. B. (1960). *J. Bact.*, **80**, 580–584.
Dewald, R. R., Browall, K. W., Schaefer, L. D. and Messer, A. (1967). *Appl. Microbiol.*, **15**, 1299–1302.
Dimmick, R. L., Straat, P. A., Wolochow, H., Levine, G. V., Chatigny, M. A. and Schrot, J. R. (1975). *J. Aerosol. Sci.*, **6**, 387–393.
Dimmick, R. L., Chatigny, M. A., Wolochow, H. and Straat, P. (1977). *Cospar Life Sci. Space Res.*, **15**, 41–45.
Dimmick, R. L., Wolochow, H. and Chatigny, M. A. (1979). *Appl. Environ. Microbiol.*, **38**, 642–643.
Donaldson, A. I. (1972). *J. Gen. Virol.*, **25**, 25–33.
Donaldson, A. I. (1978). *Vet. Bull.*, **48**, 83–94.
Drost-Hansen, W. (1965). *Ann. N. Y. Acad. Sci.*, **125**, 471–501.
Dubovi, E. J. (1971). *Appl. Microbiol.*, **21**, 761–762.
Dubovi, E. J. and Akers, T. G. (1970). *Appl. Microbiol.*, **19**, 624–628.
Dubovi, E. J. and Akers, T. G. (1973). In *Fourth International Symposium on Aerobiology*, J. F. Ph. Hers and K. C. Winkler (Eds), Oosthoek, Utrecht, The Netherlands, pp. 130–131.

Ehresmann, D. W. and Hatch, M. T. (1975). *Appl. Microbiol.*, **29**, 352–357.
Ehrlich, R. and Miller, S. (1971). *Appl. Microbiol.*, **22**, 194–199.
Ehrlich, R., Miller, S. and Idoine, L. S. (1964). *Appl. Microbiol.*, **12**, 479–482.
Falk, M., Hartman, Jr., K. A. and Lord, R. C. (1963a). *J. Amer. Chem. Soc.*, **85**, 387–391.
Falk, M., Hartman, Jr., K. A. and Lord, R. C. (1963b). *J. Amer. Chem. Soc.*, **85**, 391–394.
Goldberg, L. J. and Ford, I. (1973). In *Fourth International Symposium on Aerobiology*, J. F. Ph. Hers and K. C. Winkler (Eds), Oosthoek, Utrecht, The Netherlands, pp. 86–89.
Gregory, P. H. (1966). In *The Fungus Spore*, M. F. Madelin (Ed.), Butterworth, London, pp. 1–13.
Grigera, J. A. and Berendsen, H. J. C. (1979). *Biopolymers*, **18**, 47–57.
Hagler, A. J. and Moult, J. (1980). *Biopolymers*, **19**, 395–418.
Haly, A. R. and Snarth, J. W. (1968). *Biopolymers*, **6**, 1355–1377.
Hamabata, A. and von Hippel, P. H. (1973). *Biochemistry*, **12**, 1264–1271.
Hamabata, A., Chang, S. and von Hippel, P. H. (1973a). *Biochemistry*, **12**, 1271–1278.
Hamabata, A., Chang, S. and von Hippel, P. H. (1973b). *Biochemistry*, **12**, 1278–1282.
Hamaguchi, K. and Geiduschek, E. D. (1962). *J. Amer. Chem. Soc.*, **84**, 1329–1338.
Harper, G. J. (1961). *J. Hyg. (Camb.)*, **59**, 479–486.
Harper, G. J. (1963). *Arch. ges. Virusforsch.*, **13**, 64–71.
Harper, G. J. (1965). In *First International Symposium on Aerobiology*, R. L. Dimmick (Ed.), Naval Biological Laboratory, Naval Supply Center, Oakland, California, pp. 335–343.
Hatch, M. T. and Warren, J. C. (1969). *Appl. Microbiol.*, **17**, 685–689.
Hatch, M. T. and Wolochow, H. (1969). In *An Introduction to Experimental Aerobiology*, R. L. Dimmick and Ann B. Akers (Eds), Wiley Interscience, New York, London, pp. 267–295.
Heckley, R. J. (1973). In *Fourth International Symposium on Aerobiology*, J. F. Ph. Hers and K. C. Winkler (Eds), Oosthoek, Utrecht, The Netherlands, pp. 105–107.
Henderson, R. and Unwin, P. N. T. (1975). *Nature (Lond.)*, **257**, 28–32.
von Hippel, P. H. (1970). *Colloques internationaux du C. N. R. S.*, No. 246, pp. 19–26.
von Hippel, P. H. and Hamabata, A. (1973). *J. Mechanochem. Cell Mobility*, **2**, 127–138.
von Hippel, P. H. and Wong, K. Y. (1965). *J. Biol. Chem.*, **240**, 3909–3923.
von Hippel, P. H., Peticolas, V., Schack, L. and Karlson, L. (1973). *Biochemistry*, **12**, 1256–1263.
de Jong, J. C. and Winkler, K. C. (1968). *J. Hyg. (Camb.)*, **66**, 557–565.
de Jong, J. C., Harmsen, M. and Trouwborst, T. (1973). *J. Gen. Virol.*, **18**, 83–86.
de Jong, J. C., Harmsen, M., Trouwborst, T. and Winkler, K. C. (1974). *Appl. Microbiol.*, **27**, 59–65.
de Jong, J. C., Harmsen, M., Plantinga, A. D. and Trouwborst, T. (1976). *Appl. Environ. Microbiol.*, **32**, 315–319.
Kenehisa, M. I. and Tsong, T. Y. (1979). *J. Mol. Biol.*, **133**, 279–284.
Klotz, I. M. (1965). *Fed. Proc.*, **24**, S-24 to S-33.
Lakshmi, T. S. and Nandi, P. K. (1976). *J. Phys. Chem.*, **80**, 249–252.
Maltman, J. R. (1970). *La Revue d'Hygiene et Medecine Sociale*, **18**, 395–407.
Maltman, J. R. and Webb, S. J. (1971). *Can. J. Microbiol.*, **17**, 1443–1450.
Matsuda, Y. (1979a). *Bull. Jpn. Soc. Sci. Fish.*, **45**, 573–580.
Matsuda, Y. (1979b). *Bull. Jpn. Soc. Sci. Fish.*, **45**, 581–584.
Matsuda, Y. (1979c). *Bull. Jpn. Soc. Sci. Fish.*, **45**, 841–844.

Matsuda, Y. (1979d). *Bull. Jpn. Soc. Sci. Fish.*, **45**, 737–744.
Matsuda, Y. (1979e). *Bull. Jpn. Soc. Sci. Fish.*, **45**, 733–736.
May, K. R. (1966). *Bact. Rev.*, **30**, 559–570.
Monk, G. W. and McCaffrey, P. A. (1957). *J. Bact.*, **73**, 85–88.
Monk, G. W., Elbert, M. L., Stevens, C. L. and McCaffrey, P. A. (1956). *J. Bact.*, **72**, 368–372.
Nomura, S., Hiltner, A., Lando, J. B. and Baer, E. (1977). *Biopolymers*, **11**, 231–246.
Northcote, D. H. (1953). *Biochim. Biophys. Acta*, **11**, 471–479.
Papahadajopoulos, D. (1977). *J. Coll. Int. Sci.*, **58**, 459–470.
Pineri, M. H., Escoubes, M. and Roche, G. (1978). *Biopolymers*, **17**, 2799–2815.
Poon, C. P. C. (1966). *Am. J. Epidem.*, **84**, 1–39.
Schaffer, F. C., Soergel, M. E. and Straube, D. C. (1976). *Arch. Virol.*, **51**, 263–273.
Schleich, T. and von Hippel, P. H. (1969). *Biopolymers*, **7**, 861–877.
Shiraishi, H., Hiltner, A. and Baer, E. (1977). *Biopolymers*, **16**, 2801–2806.
Silver, I. H. (1965). In *First International Symposium on Aerobiology*, R. L. Dimmick (Ed.), Naval Biological Laboratory, Naval Supply Center, Oakland, California, pp. 319–333.
Sober, H. A. (Ed.) (1968). *Handbook of Biochemistry*, Chemical Rubber Co., Cleveland, Ohio, Section H.
Soddu, A. and Vieth, W. R. (1980). *J. Mol. Catal.***7**, 491–500.
Sokolowski, M. B., Weneck, E. J., Trkula, D. and Bateman, J. B. (1969). *Biophys. J.*, **9**, 950–953.
Spendlove, J. C. and Fannin, K. F. (1982). In *Methods in Environmental Virology*, C. P. Gerba and S. M. Goyal (Eds), Marcel Dekker, New York, Basel, pp. 261–329.
Stewart, R. H. and Wright, D. N. (1970). *Appl. Microbiol.*, **19**, 865–866.
Straat, P. A., Wolochow, H., Dimmick, R. L. and Chatigny, M. A. (1977). *Appl. Environ. Microbiol.*, **34**, 292–296.
Strange, R. E. and Cox, C. S. (1976). *Symp. Soc. Gen. Microbiol.*, **26**, 111–154.
Trouwborst, T. and de Jong, J. C. (1972). *Appl. Microbiol.*, **23**, 938–941.
Trouwborst, T. and de Jong, J. C. (1973a). In *Fourth International Symposium on Aerobiology*, J. F. Ph. Hers and K. C. Winkler (Eds), Oosthoek, Utrecht, The Netherlands, pp. 137–140.
Trouwborst, T. and de Jong, J. C. (1973b). *Appl. Microbiol.*, **26**, 252–257.
Trouwborst, T. and Kuyper, S. (1974). *Appl. Microbiol.*, **27**, 834–837.
Trouwborst, T., de Jong, J. C. and Winkler, K. C. (1972). *J. Gen. Virol.*, **15**, 235–242.
Trouwborst, T., de Jong, J. C. and Winkler, K. C. (1973). *J. Coll. Int. Sci.* **45**, 198–208.
Trouwborst, T., Kuyper, S., de Jong, J. C. and Plantinga, A. D. (1974). *J. Gen. Virol.*, **24**, 155–165.
Unwin, P. N. T. and Henderson, R. (1975). *J. Mol. Biol.*, **94**, 425–440.
van der Waals, J. H. and Platteeuw, J. C. (1959). *Advanc. Chem. Phys.*, **2**, 1.
Warren, J. C., Akers, T. G. and Dubovi, E. J. (1969). *Appl. Microbiol.*, **18**, 893–896.
Watanabe, K., Sato, Y. and Kato, Y. (1980). *J. Food Process. Preserv.*, **4**, 263–274.
Webb, S. J. (1965). *Bound Water in Biological Integrity*, Charles C. Thomas, Springfield, Illinois.
Webb, S. J. (1969). *J. Gen. Microbiol.*, **58**, 317–326.
Webb, S. J. and Walker, J. L. (1968). *Can. J. Microbiol.*, **14**, 565–572.
Webb, S. J., Bather, R. and Hodges, R. W. (1963). *Can. J. Microbiol.*, **9**, 87–92.
Webb, S. J., Dumasia, M. D. and Singh Bhorjee, J. (1965). *Can. J. Microbiol.*, **11**, 141–149.

Winkler, K. C. (1973). In *Fourth International Symposium on Aerobiology*, J. F. Ph. Hers and K. C. Winkler (Eds), Oosthoek, Utrecht, The Netherlands, pp. 1–11.
Wright, D. N., Bailey, G. D. and Goldberg, L. J. (1969). *J. Bact.*, **99**, 491–495.
Wright, D. N., Bailey, G. D. and Hatch, M. T. (1968a). *J. Bact.*, **95**, 251–252.
Wright, D. N., Bailey, G. D. and Hatch, M. T. (1968b). *J. Bact.*, **96**, 970–974.

Chapter 11

Oxygen

11.1 INTRODUCTION

Oxygen inactivates only some species of vegetative bacteria and algae; other microorganisms, such as spores, phages and viruses, survive equally in air and in pure nitrogen. Oxygen susceptibility usually increases with degree of desiccation, increasing oxygen concentration, and time, whether dehydration is through aerosolization, freeze-drying, or through drying on surfaces. Oxygen does not appear to be toxic for frozen bacteria such as *Escherichia coli* B and *Serratia marcescens* 8UK (Cox and Heckley, 1973). These bacteria and *Micrococcus candidus*, *Klebsiella pneumoniae* and *Francisella tularensis* have been studied most and show greatest sensitivity to oxygen when in the log phase of growth and when the dehydrated product is derived from suspensions of them in distilled water (Strange and Cox, 1976). Spent culture fluids, anti-oxidants, polyhydric alcohols and sugars often have a protecting action. Their beneficial effects, however, may not be apparent unless high osmotic pressure collecting fluids and/or rehumidification techniques (Chapter 2) also are employed. These sometimes are required because dehydration–rehydration stress (Chapter 10) and oxygen-induced damage occur concomitantly, i.e. when aerosolized into air, bacteria may suffer simultaneous dehydration–rehydration stress and oxygen-induced damage. To separate these two effects results need to be obtained in atmospheres containing oxygen and in inert atmospheres under otherwise identical conditions. For such comparisons, since air may contain trace quantities of other toxic gases and vapours, mixtures of pure nitrogen and pure oxygen are preferable to air, unless purified. Difference between aerosol survival or infectivity in such atmospheres and in pure nitrogen then are attributable to oxygen.

As mentioned above oxygen toxicity usually increases with degree of dehydration, with increased oxygen concentration and with time. Above an RH of about 70%, survival in air, nitrogen + 20% (v/v) oxygen and nitrogen is about the same for wet disseminated *E. coli* B (Cox and Baldwin, 1967), i.e. these bacteria and other strains have to lose most of the water normally associated with them before oxygen toxicity is apparent. In contrast, dry disseminated *E. coli* B while demonstrating oxygen toxicity at low RH survives better in air than in nitrogen atmospheres at high RH (Cox, 1970). This beneficial action of oxygen is not understood but could be due to oxygen decreasing the rate of denaturation of *E. coli* surface structures caused by vapour phase rehydration (Chapter 10). *Francisella tularensis* demonstrates an even more complex response to oxygen. When sprayed from suspensions in distilled water (which is possible even though this organism is osmotically sensitive) survival is decreased at low RH by oxygen. This is similarly so when peptone (or cysteine) broth replaces water as the spray fluid (Cox, 1971). However, when the spray fluid is spent culture medium survival in air is greater than that in nitrogen over the range 20–95% RH. On the other hand, when dry disseminated as a powder produced by freeze-drying in spent culture medium survival in air and nitrogen is the same (Cox, 1971). This complicated pattern is partly due to the presence of a toxic component in the spent culture medium which loses toxicity by reaction with oxygen (Cox, 1971).

Wet disseminated *E. coli* and *S. marcescens* respond as previously indicated (Hess, 1965; Cox and Baldwin, 1967; Cox *et al.*, 1973; Cox *et al.*, 1974). But the response of *K. pneumoniae* appears at first sight to be more involved since, like *F. tularensis*, survival in air is lowest at mid-RH. However, as shown by Goldberg and Ford (1973) this mid-range instability is due to desiccation and not oxygen because when sprayed into nitrogen minimum survival also was at mid-RH. When aerosols of *K. pneumoniae* in air are rehumidified (a process which can reverse desiccation damage) before sampling the pattern shown for *E. coli* and *S. marcescens* is observed (Figure 11.1). Likewise data obtained in air and corrected for viability loss in nitrogen follow this same pattern.

11.2 BIOCHEMICAL STUDIES

Determination of lesions caused by oxygen toxicity is complicated by the simultaneous occurrence of desiccation damage and oxygen damage. Unless care is taken effects attributed to one may actually be due to the other. One way to limit this difficulty is to choose a test organism which is stable in nitrogen at low RH. Then, samples collected from such aerosols can be used as good controls for comparisons with samples collected under identical conditions except that oxygen also is present. By means of this approach it was found that the ability of *E. coli* B (recovered from aerosols in air at low RH) to form colonies was very limited, whereas the ability to support replication of phage T7 was very much greater. The extent of phage replication approaches that obtained in the absence of oxygen (Cox and Baldwin, 1964, 1966). Somewhat similar results were obtained by Webb *et al.* (1965). Therefore, the damage in *E. coli* due to oxygen must be such that processes involved in phage T7 replication are largely unimpaired, i.e. those metabolic pathways involved in DNA synthesis, RNA synthesis, protein synthesis and production of energy. Thus the most obvious sites for oxygen-induced damage in *E. coli* B are DNA synthesis directed by *E. coli* B DNA, cell wall synthesis and cell division. Benbough (1967) and Cox *et al.* (1971a, b) observed that DNA synthesis

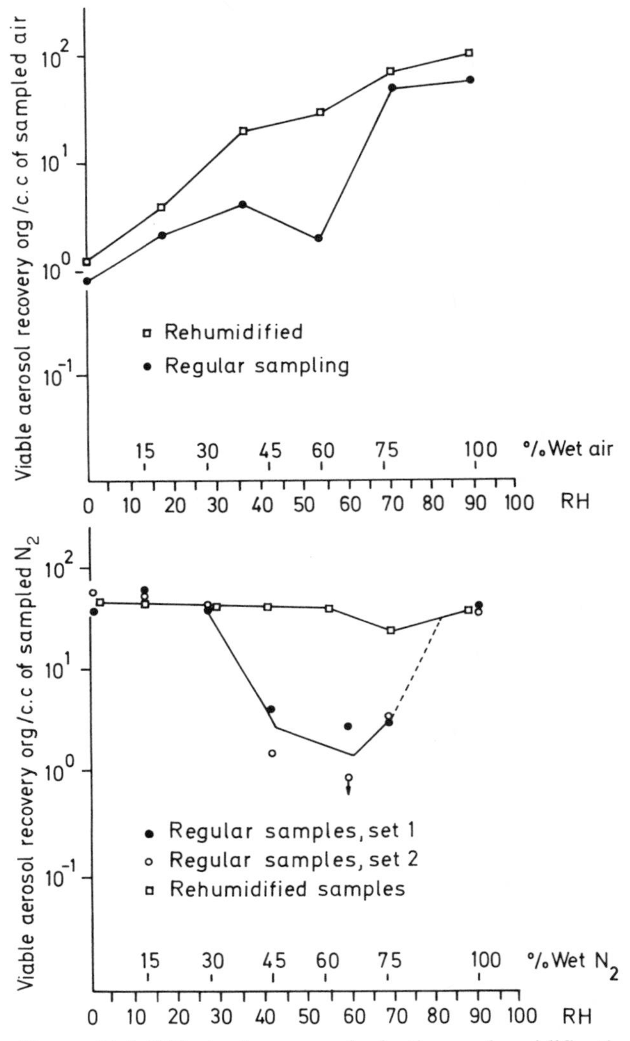

Figure 11.1 Effect of oxygen, hydration, rehumidification and protecting additives on the survival of aerosolized *Klebsiella pneumoniae*. (From Goldberg and Ford (1973), in *Fourth International Symposium on Aerobiology*, Oosthoek, Utrecht, The Netherlands.) Reproduced by permission of the Oosthoek Publ. Co.

directed by *E. coli* DNA was only slightly diminished when these bacteria were killed by oxygen. Oxygen uptake by oxygen-killed *E. coli* B/r and *E. coli* B_{S-1} similarly was unaffected. In addition the initial loss of ^{43}K by populations of *E. coli* B recovered from aerosols was not altered by the presence of oxygen (Anderson and Dark, 1967). Consequently, aerosolized *E. coli* strains are affected by oxygen-induced damage to cell wall synthesis or cell division, or

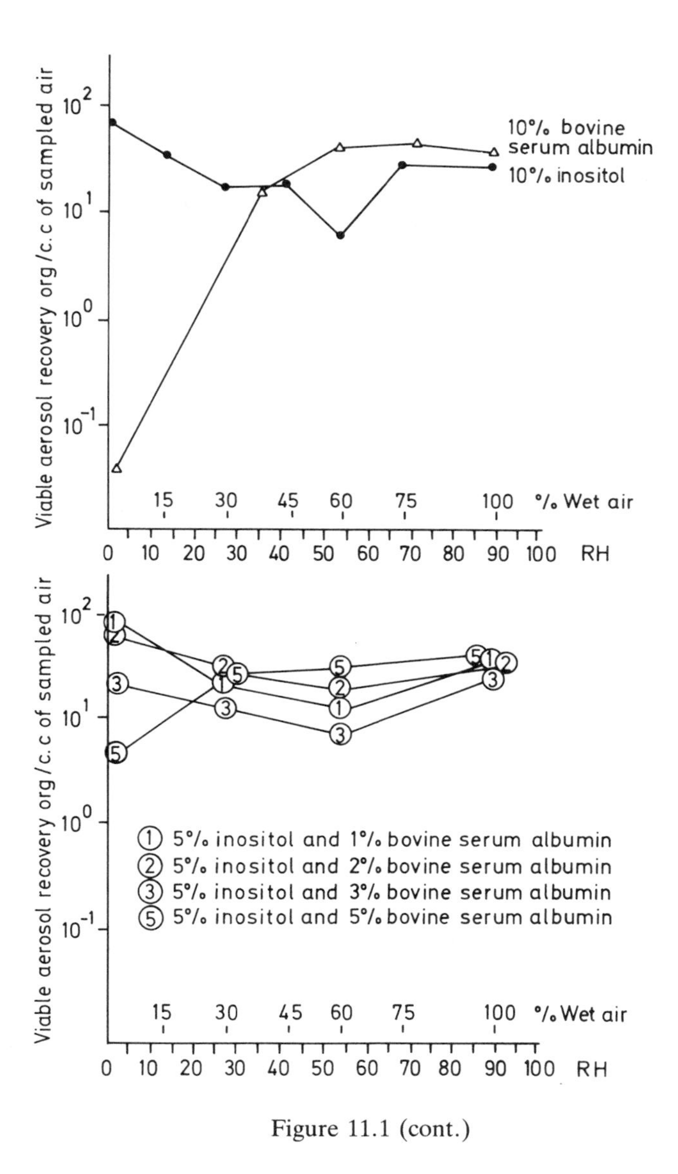

Figure 11.1 (cont.)

both. That *E. coli* B collected from aerosols after exposure at low RH to oxygen grow into extremely long filaments (e.g. 200 μm long) indicates it is the process of cell division which is involved rather than cell wall synthesis (Cox and Baldwin, 1966). While this occurs also for other *E. coli* strains such as commune, it is not known whether it is a general phenomenon. In this context, it may be significant that oxygen is not toxic for dried bacterial spores, bacteriophages and viruses (see Chapter 10).

Failure of cell division in *E. coli* seems the major result of oxygen-induced damage but other mechanisms may take place concurrently. For example, the extent of phage replication in oxygen-killed *E. coli* B, while being very much greater than the extent of colony formation, is slightly less than that for controls stored in nitrogen (Cox and Baldwin, 1966). Likewise, DNA synthesis is slightly lower (Cox *et al.*, 1971a, b) as observed also by Israeli (1973), who reported an irreversible blockage of the β-galactosidase induction system as well. The validity, though, of this latter result is in question with respect to oxygen-induced death. As discussed in Chapter 10, Benbough *et al.* (1972) showed that it is the active transport mechanism for the uptake of inducer that is impaired. Unfortunately these workers did not report whether this impairment occurred in nitrogen atmospheres as well as in air. But since impairment of active transport was detected within 1 s of aerosol generation (i.e. in a time comparable to that required for water evaporation, but not bacterial equilibration) it seems most likely to be through dehydration rather than reaction with oxygen. Possibly then in addition to its primary action in inhibiting cell division of *E. coli* B oxygen also may have a second order inhibitory action upon DNA synthesis when it is directed by either *E. coli* DNA or phage T7 DNA.

11.3 KINETIC MODELS

Kinetic models have been developed which account extremely well for oxygen toxicity (Cox, 1973a; Cox *et al.*, 1973; Cox *et al.*, 1974). In an analogous manner to that for desiccation damage the proposed mechanism is that bacteria contain a hypothetical crucial moiety which forms a series of hydrates, i.e.

$$A(n)H_2O \underset{+xH_2O}{\rightleftharpoons} A(n-x)H_2O \underset{k_-}{\overset{k_+}{\rightleftharpoons}} A(n-y)H_2O \underset{+yH_2O}{\rightleftharpoons} A + iH_2O \tag{11.1}$$

Furthermore, a carrier X is postulated to combine reversibly with oxygen, i.e.

$$X + O_2 \rightleftharpoons XO_2 \tag{11.2}$$

and that this complex reacts with each hydrate. This reaction produces free carrier X and an oxidized hydrate which biologically is inactive. Reaction between XO_2 and species A occurs without the formation of the ternary complex $A \cdot XO_2$. Integration of the differential equations representing this mechanism and use of the form of the relationship given in equation 10.10 between viability and concentration of the critical molecular species gives the expression,

$$\ln V = K_1[A(n-x)H_2O]_0([Ax] + [Ay]) - K_1[A(n-x)H_2O]_0 + \ln 100 \tag{11.3}$$

The terms $[Ax]$ and $[Ay]$ are complex functions of the velocity constants, the carrier concentration $[X]_0$, $[O_2]$ the oxygen concentration, RH and time (Cox *et al.*, 1974).

A simple form of this relationship, which does not take account of RH, is given (Cox, 1973a, Cox *et al.*, 1973) as:

$$\ln V = K_1[A]_0\left\{exp\left(-\frac{k[X]_0[O_2]t}{K_x + [O_2]}\right) - 1\right\} + \ln 100 \qquad (11.4)$$

Cox *et al.* (1973) found that equation 11.4 could be accurately fitted to the experimental data (for *S. marcescens* 8UK) of Hess (1965) and of Cox and Heckly (1973) whereas it did not fit those of Dewald (1966) quite as well. Since the data set of Hess was obtained with aerosols while the other data sets were for freeze-dried *S. marcescens* 8UK the response to oxygen concentration and time are similar for aerosolized and freeze-dried *S. marcescens* 8UK. Consequently, the toxic action of oxygen on aerosolized bacteria does not arise through the processes of aerosol generation and collection but must take place during aerosol storage.

The above data sets were not obtained as a function of RH unlike the set obtained by Cox *et al.* (1974). Figures 11.2 and 11.3 show examples of their results and the degree to which the calculated curves agree with their experimental data. Fitting equation 11.3 to these data provided estimates for the values of the six constants involved in that equation as a function of RH (Figure 11.4). Of these constants, K_x the equilibrium constant for the reaction between carrier X and O_2, was independent of RH which is consistent with the postulated reaction not involving water molecules. The constant $K_1[A(n - x)H_2O]_0$ (i.e. a normalization constant times the concentration of species A at time zero) decreased with increasing RH and approached zero. An analysis of the value of $K_1[A(n - x)H_2O]_0$ as a function of RH indicated that the species A would have to reside in a region where most of the bacterial water was lost, between 100 and 60% RH (Cox *et al.*, 1974). This contrasts with desiccation damage when the value of the corresponding constant $K_1[B(n - x)H_2O]_0$ was independent of RH indicating that species B could be on the cell surface (Chapter 10). Therefore, species A, if it exists, would do so in a region of the bacterium where there is weakly bound water. The interspace between the cell wall and cytoplasmic membrane is such an environment which is consistent with oxygen toxicity inhibiting cell division (Section 11.2).

Lack of oxygen toxicity at RH values above about 65% can be explained in terms of dehydration causing an increase in the concentration of species A. As the RH is decreased water evaporation causes the concentration of species A to increase. Since reaction rates increase with reactant concentrations (mass per unit volume) oxygen toxicity correspondingly increases with decreased RH. It is not until the RH is less than about 65% that reaction between species A and XO_2 attains a sufficiently high level for the rate of their reaction to be observable. Alternatively, as the RH increases the value of the constant $k_-[H_2O]^y$ increases greatly whereas the constant $k_+[H_2O]^x$ decreases. Therefore, in terms of equation 11.1, the equilibrium between the hydrates $A(n - x)H_2O$ and $A(n - y)H_2O$ favours the existence of the higher

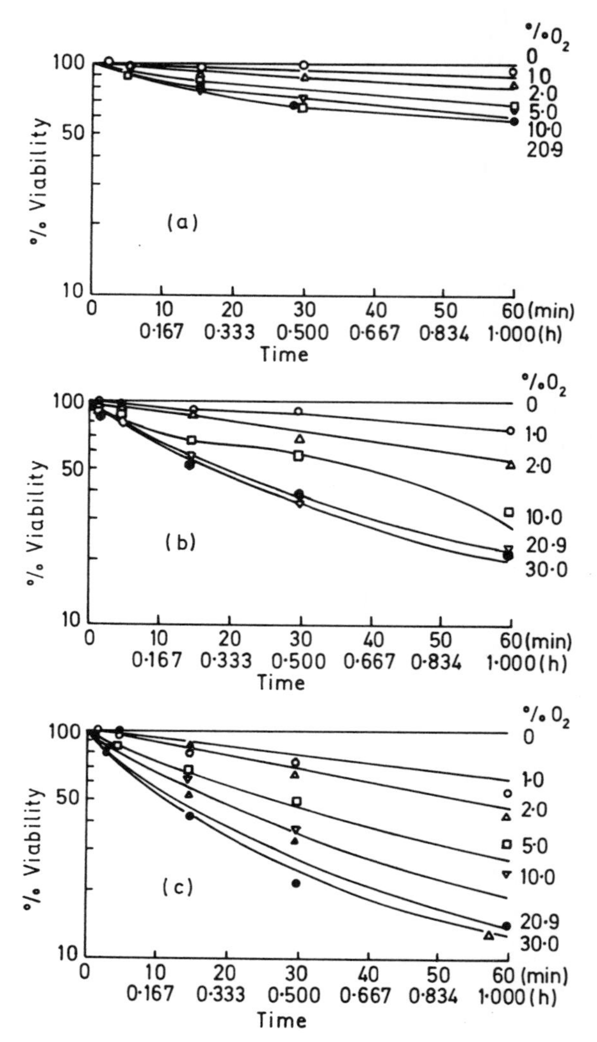

Figure 11.2 (a)–(c) Aerosol survival of *Serratia marcescens* 8 UK sprayed from suspension in distilled water into different nitrogen plus oxygen mixtures at 25 °C. Points are experimental data; lines are calculated by equation 11.3 and values of the constants are given in Figure 11.4. (a) 60% RH, (b) 50% RH, (c) 20% RH.

order hydrate, $A(n - x)H_2O$. As discussed in Section 10.4 and Chapter 16, higher order hydrates due to conformational differences may be less reactive than lower order hydrates. If so, oxygen toxicity would decrease with increasing RH. In this context additives which decrease oxygen toxicity (e.g. spent culture fluids, polyhydric compounds, etc.) may do so for reasons

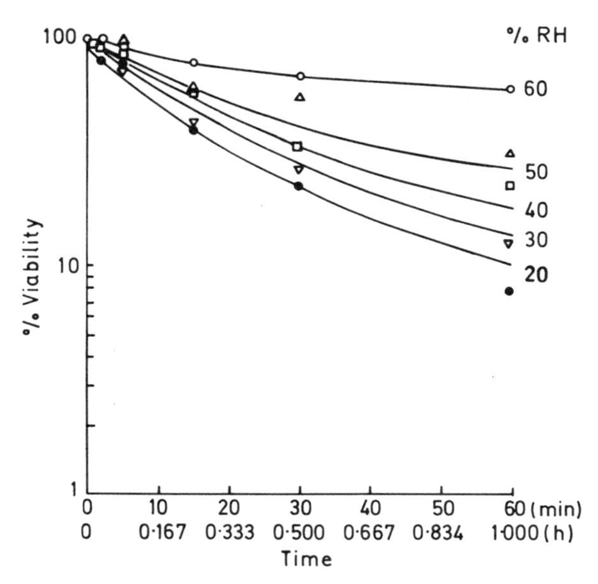

Figure 11.3 As for Figure 11.2 but at 10% oxygen and different relative humidities.

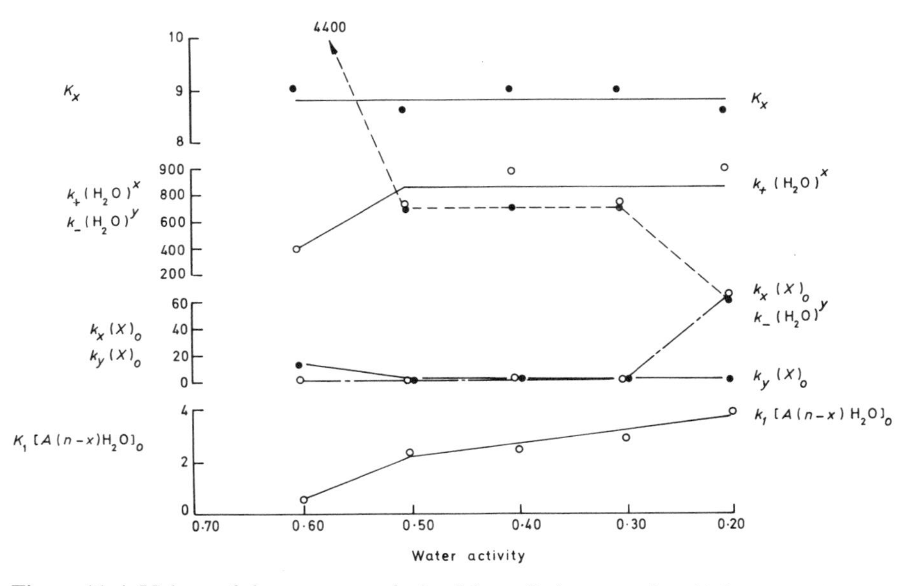

Figure 11.4 Values of the constants derived from fitting equation 11.3 to experimental viability-time curves.

analogous to those given in Section 10.4 for reducing desiccation damage. In any event, for *S. marcescens* 8UK sprayed from distilled water, loss of viability due to oxygen toxicity at RH values above about 70% RH is only 1/300th of that at 20% RH. Consequently, oxygen-induced loss of viability at high RH would be virtually unobservable (Cox *et al.*, 1974).

An alternative form of the kinetic model is given in Chapter 16 where catastrophe theory rather than probability theory is used for deriving the relationship between viability and the concentration of species A. But, as there are no other suitable experimental data sets against which the equations may be compared it is not known how generally they may apply.

11.4 FREE RADICALS

The kinetics of oxygen toxicity seem to be the same for aerosolized and freeze-dried *S. marcescens* 8UK and possibly for other bacteria as well (Section 11.3). It has been suggested that such oxygen-induced loss of viability is associated with the formation of free radicals (Cox and Baldwin, 1967; Dimmick *et al.*, 1961; Heckly and Dimmick, 1967, 1968; Heckly *et al.*, 1963; Lion *et al.*, 1961; Schwartz, 1971). However, in these investigations parallel studies of the kinetics of viability loss and of free radical formation were lacking. In 1973, Cox and Heckly reported such parallel measurements for *S. marcescens* 8UK. As shown in Figure 11.5 there is a lag in the appearance of oxygen-induced free radicals compared to the onset of oxygen-induced viability loss. At 4 h storage time the viability of the freeze-dried *S. marcescens* 8UK had declined to about 0.1% without any free radicals being detected. After 24 h storage viability was 0.000016% whereas oxygen-induced free radical formation was not complete. This negative correlation between oxygen-induced loss of viability and free radical formation is opposite to that reported by Heckly *et al.* (1963) and by Heckly and Dimmick (1968). In their work viability and free radical content was not measured until considerable reaction time had elapsed and therefore the lag in the formation of oxygen-induced free radicals was not seen. In many other respects the phenomena due to oxygen of viability loss and free radical formation show many similarities. For example, drying fluid additives which enhance survival concomitantly decrease free radical content. Similarly, increasing the oxygen concentration lowers survival and increases free radical content. A possible explanation for these results has been advanced (Cox, 1973b). For loss of viability induced by oxygen the mechanism appears to be (Section 11.3):

$$X + O_2 \underset{k_x}{\rightleftharpoons} XO_2 \tag{11.5}$$

and,

$$A + XO_2 \xrightarrow{k} AO_2 + X \tag{11.6}$$

In this case the equilibrium between X, O_2 and XO_2 must be rapidly established otherwise a lag would result before viability declines with time (see Chapter 12). On the other hand, if O_2 also reacted with another substance Y, i.e.

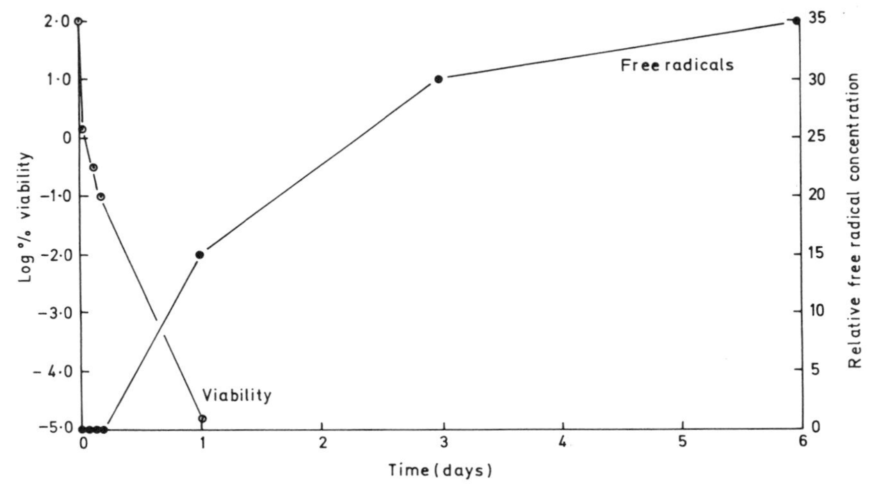

Figures 11.5 Free radical formation and loss of viability as a function of time in freeze-dried *Serratia marcescens* 8 UK/2 under 10% oxygen at zero per cent relative humidity.

$$O_2 + Y \underset{k_-}{\overset{k_+}{\rightleftharpoons}} YO_2 \qquad (11.7)$$

but with the equilibrium being slowly established then the concentration of YO_2 would show a lag as a function of time. Integration of the differential equations representing equation 11.7 gives,

$$\frac{[YO_2]}{[Y]_0} = \left(1 - \frac{k_-}{k_+[O_2] + k_-}\right) - \left\{\frac{k_+[O_2]}{k_+[O_2] + k_-} e^{-(k_+[O_2]+k_-)}\right\} \qquad (11.8)$$

where, $[Y]_0$ = total concentration of species Y,

$[YO_2]$ = concentration of species YO_2 at time = t,

k_+ = forward reaction velocity constant,

k_- = backward reaction velocity constant,

$[O_2]$ = oxygen concentration.

Values of $10 \times ([YO_2]/[Y]_0)$ calculated from equation 11.8 for an oxygen concentration of 10% and with $k_+ = 8.0 \times 10^{-4} h^{-1}$ and $k_- = 0.014 h^{-1}$ agree extremely well with the experimental data for free radical formation (see Figure 11.5). This correlation suggests that the basic mechanism for oxygen-induced loss of viability and for free radical formation may be similar. The difference between them is that for the former equilibrium between carrier X and oxygen is rapidly attained whereas for the latter equilibrium between oxygen acceptor (Y) and oxygen is slowly attained. That compounds which protect against viability loss also reduce free radical formation could indicate that species X and Y are not too dissimilar. On the other hand, protection through a diffusion control (Section 10.4) of oxygen would lead to the same result without species X or Y being similar.

The above explanation cannot be totally correct. This is because species YO_2 as given in equation 11.7 should reversibly convert back to species Y and oxygen when the sample is evacuated, i.e. equation 11.7 predicts that oxygen-induced free radical content of the sample should decrease if it were to be exposed to a hard vacuum. Given the value of $0.014\,h^{-1}$ (see above) for the velocity constant for the backward reaction, on evacuation, the free radical concentration in about 50 h should reduce to one half of its value compared to when evacuation began. Heckly and Dimmick (1968) show that the free radical signal intensity is not reduced by pumping at 10^{-5} mm Hg for 48 h. The formation of oxygen-induced free radicals therefore would not seem to be simply through an equilibrium process but rather through an irreversible one. One possible reaction mechanism given previously (Cox, 1973b) is when species YO_2 (equation 11.7) reacts irreversibly with another species D, i.e.

$$YO_2 + D \rightarrow DO_2 + Y \tag{11.9}$$

with DO_2 representing the irreversibly formed free radicals. However, for such a mechanism it was shown (Cox, 1973b) that given sufficient reaction time the same final free radical concentration would accrue for any given oxygen concentration greater than zero. In practice this does not seem to occur, rather the situation is more like that described by equation 11.7 except for the non-reversibility (through evacuation) of the free radical formation. Consequently, at present a complete explanation for the kinetics of oxygen-induced free radical formation in *S. marcescens* 8UK is lacking. This difficulty may be due in part to the phenomenon being more involved than previously indicated. Heckly (personal communication) using a more refined technique has observed that the electron paramagnetic resonance (EPR) spectrum of oxygen-induced free radicals in freeze-dried bacteria (unlike that in animal tissues) broadens with time due to the accumulation of a second free radical. The g value of the free radicals developing first is about 2.0038 while that of the much more slowly developing second free radicals is about 2.0057. Whether this represents a change of the first free radical species into the second or whether the former decays is not certain. But, given the first possibility then the previously discussed difficulty concerning non-reversibility through evacuation may be due to a mechanism such as,

$$A + O_2 \rightarrow AO_2 \tag{11.10}$$

$$AO_2 + Y \rightleftharpoons YO_2 + A \tag{11.11}$$

where AO_2, YO_2 represent free radical species.

From the point of view of oxygen-induced loss of viability the readily detectable oxygen-induced free radicals would not seem to be involved (Figure 11.5). However, free radical involvement cannot be totally excluded at present since it may be more a question of technique and EPR spectrometer sensitivity rather than that oxygen-induced free radicals do not rapidly form.

11.5 CONCLUSIONS

Oxygen inactivates some bacteria and algae when desiccated, unlike spores, phages and viruses. Oxygen susceptibility usually increases with degree of desiccation, increasing oxygen concentration and time whether desiccation is through aerosolization, freeze-drying or drying on surfaces. At least for *E. coli* B, it seems that cell division is the process which is inactivated and leads to viability loss. Analyses of survival curves by means of a kinetic model support this possibility but argue against free radical involvement.

REFERENCES

Anderson, J. D. and Dark, F. A. (1967). *J. Gen. Microbiol.*, **46**, 95–105.
Benbough, J. E. (1967). *J. Gen. Microbiol.*, **47**, 325–333.
Benbough, J. E., Hambleton, P., Martin, M. L. and Strange, R. E. (1972). *J. Gen. Microbiol.*, **72**, 511–520.
Cox, C. S. (1970). *Appl. Microbiol.*, **19**, 604–607.
Cox, C. S. (1971). *Appl. Microbiol.*, **21**, 482–486.
Cox, C. S. (1973a). In *Fourth International Symposium on Aerobiology*, J. F. Ph. Hers and K. C. Winkler (Eds), Oosthoek, Utrecht, The Netherlands, pp. 108–110.
Cox, C. S. (1973b). In *Freeze-drying of Biological Materials*, Institut International du Froid, Paris, pp. 55–59.
Cox, C. S. and Baldwin, F. (1964). *Nature (Lond.)*, **202**, 1135.
Cox, C. S. and Baldwin, F. (1966). *J. Gen. Microbiol.*, **44**, 15–22.
Cox, C. S. and Baldwin, F. (1967). *J. Gen. Microbiol.*, **49**, 115–117.
Cox, C. S. and Heckly, R. J. (1973). *Can. J. Microbiol.*, **19**, 189–194.
Cox, C. S., Baxter, J. and Maidment, B. J. (1973). *J. Gen. Microbiol.*, **75**, 179–185.
Cox, C. S., Bondurant, M. C. and Hatch, M. T. (1971a). *J. Hyg. (Camb.)*, **69**, 661–672.
Cox, C. S., Bondurant, M. C. and Hatch, M. T. (1971b). *Bact. Rev.*, A134.
Cox, C. S., Gagen, S. J. and Baxter, J. (1974). *Can. J. Microbiol.*, **20**, 1529–1534.
Dewald, R. R. (1966). *Appl. Microbiol.*, **14**, 568–572.
Dimmick, R. L., Heckly, R. J. and Hollis, D. P. (1961). *Nature (Lond.)*, **192**, 776–777.
Goldberg, L. J. and Ford, I. (1973). In *Fourth International Symposium on Aerobiology*, J. F. Ph. Hers and K. C. Winkler (Eds), Oosthoek, Utrecht, The Netherlands, pp. 86–89.
Heckly, R. J. and Dimmick, R. L. (1967). *Nature (Lond.)*, **216**, 1003–1004.
Heckly, R. J. and Dimmick, R. L. (1968). *Appl. Microbiol.*, **16**, 1081–1085.
Heckly, R. J. and Dimmick, R. L. and Windle, J. J. (1963). *J. Bacteriol.*, **85**, 961–966.
Hess, G. E. (1965). *Appl. Microbiol.*, **13**, 781–787.
Israeli, E. (1973). In *Fourth International Symposium on Aerobiology*, J. F. Ph. Hers and K. C. Winkler (Eds), Oosthoek, Utrecht, The Netherlands, pp. 110–113.
Lion, M. B., Kirby-Smith, J. S. and Randolph, M. C. (1961). *Nature (Lond.)*, **192**, 34–36.
Schwartz, H. M. (1971). *Cryobiology*, **8**, 255–264.
Strange, R. E. and Cox, C. S. (1976). *Symp. Soc. Gen. Microbiol.*, **26**, 111–154.
Webb, S. J., Dumasia, M. D. and Singh Bhorjee, J. (1965). *Can. J. Microbiol.*, **11**, 141–149.

Chapter 12

The open air factor

12.1 INTRODUCTION

Unlike the airborne transmission of disease in enclosed spaces such as rooms, that occurring in outside air is subject to an additional and important environmental parameter. As recently as 20 years ago the existence of this parameter was neither known nor suspected. It was largely the development of the microthread technique by May and Druett (1968) which led to its discovery. This technique which is described in more detail in Section 2.7 employs fine spider web wound onto small stainless steel frames. Loading such frames into a 'sow' through which fine aerosols are passed, provides them in a 'captive' form. Such captive aerosols are not subject to physical loss nor dispersion by normal air movements. Comparing the survivals of microorganisms (as captive aerosols) in the laboratory with those in outside air Druett and May (1968) observed that outside air often was much more toxic than inside air under the same conditions of photoactivity, RH and temperature. This finding was later confirmed by other investigators (Hood, 1971, 1973; Benbough and Hood, 1971; Southey and Harper, 1971). The nature of this Open Air Factor (Druett and May, 1969) or more simply OAF initially was difficult to envisage. For example, why should outside air when drawn along a narrow pipe lose its OAF property but when drawn with a fairly high velocity along a wide pipe retain it? Why should the level of its effect have depended also upon the composition and cleanliness of the pipe whereas drawing air through a particulate filter removed its OAF property? Several possibilities were suggested to account for such behaviour and these included rapid fluctuations of RH (known to occur in outside air), penetrating radiation, toxic particulates, pollutants and ozone. Each of these was shown experimentally not to be individually responsible although when the ozone level in outside air was high the toxicity of outside air often was high. But in the laboratory synthetically generated ozone at levels of a few parts per million (ppm), which correspond to those observed in outside air, failed to induce the rate of loss of viability observed in open air. It was found also that the biological decay rate under the latter conditions varied from day to day and with wind direction for vegetative bacteria but to a much lesser degree for viruses, while bacterial spores were unaffected. In addition, when the level of air pollution was high (e.g. when there was an inversion layer) the biological decay rate often was high. But in the laboratory as for ozone none of the pollutants was sufficiently germicidal to be identified with OAF. On the other hand, if small amounts (ppm) of ozone were allowed to react with automobile exhaust, petrol vapour or olefins held in large chambers an OAF-like product was obtained (Druett and May, 1969; May *et al.*, 1969; Druett, 1970, 1973; May, 1972; Harper, 1973; Hood, 1973; de Mik and de Groot, 1973; Nash, 1973). It seemed that OAF was most likely to be highly reactive products of ozone and olefin reactions. These products, due to their short half-lives (Druett, 1973), would produce OAF-like effects such as that of ventilation through pipes, loss of activity in particulate filters and the association with levels of ozone and pollution. However, this was a presumptive identification of OAF and attempts to detect and identify it in outside air by chemical means and by long path infrared studies failed (Druett, 1973; Cox *et al.*, 1973; Nash, 1973; Cox, 1979). It was not until 1973 that much more direct evidence was obtained for the identity of OAF. This is given in Section 12.3, while more specific details of OAF action will provide the remainder of this section.

As for other stresses different species of microorganisms have different sensitivities to OAF. Spores of *Bacillus subtilis* var *niger* and *B. anthracis* in the dark are not affected by OAF and *Micrococcus radiodurans* also is very resistant (May *et al.*, 1969). Foot-and-mouth disease (FMD) virus likewise is resistant under all conditions of photoactivity (Donaldson and Ferris, 1975) as is swine vesicular disease (SVD) virus (Donaldson and Ferris, 1974). Other viruses such as vaccinia and Semliki Forest like coliphage φX174, T1 and T7 are much more sensitive (May *et al.*, 1969; Benbough and Hood, 1971; de Mik *et al.*, 1977). Bacteria such as *E. coli*, *S. marcescens*, *F. tularensis*, *Br. suis*, group C streptococcus, *Micrococcus albus* and *Erwinia* strains show comparable sensitivities (May *et al.*, 1969; Southey and Harper, 1971; Dark and Nash, 1970; Graham *et al.*, 1979). The very toxic nature of OAF probably is best illustrated by example. Semliki Forest virus when stored at high RH in chambers (i.e. no OAF) loses viability slowly (Section 10.3). After about 5 min in the airborne state in a closed system, Benbough and Hood (1971) report a viability (or infectivity) of about 80% whereas under open air conditions (in the dark and at the same RH and temperature) the corresponding viability was only about 3% (Figure 12.1), i.e. the presence of OAF caused about a 25-fold decrease in infectivity. *E. amylovora* under corresponding conditions, except that the holding time was 120 min, has a viability of 100% and about 1%, respectively, i.e. a 100-fold decrease in viability (Southey and Harper, 1971). Other *Erwinia* strains are even more sensitive (Graham *et al.*, 1979). Obviously OAF can cause a very detrimental response upon survival and infectivity of both plant and animal pathogens when airborne and consequently their transmission by the airborne route. It may even be argued that OAF is beneficial for plants and animals.

In all the work on OAF a consistent observation was the very variable extent to which OAF affected bacterial viability. Therefore the survival of *E. coli*

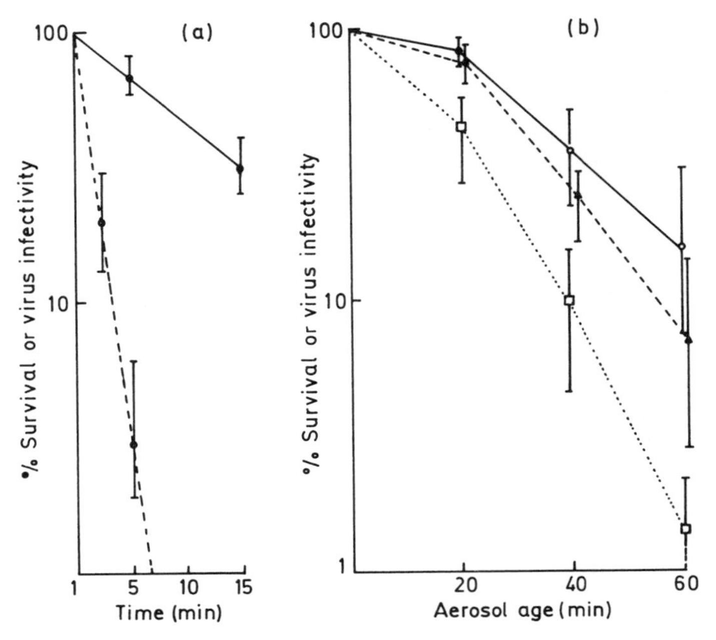

Figure 12.1 (a) Inactivation of Semliki Forest virus held on microthreads in enclosed air (●) and in open air (○). (b) Inactivation of aerosolized Semliki Forest virus in a well-ventilated sphere. Particle size (○) greater than 6 μm diameter; (▲) 3–6 μm diameter; (□) particles less than 3 μm diameter. Each point is the average of 20 experiments carried out on different days at temperatures from 2°–12 °C and relative humidities from 75–96%. (From Benbough and Hood, *J. Hyg. (Camb.)*, **69**, 619.) Reproduced by permission of the authors.

commune (i.e. MRE 162) was measured alongside that for other microorganisms so that some estimate could be made of the day-to-day variation in OAF levels and the relative sensitivities of different microorganisms. Likewise the action of OAF in different parts of the world then could be compared; as mentioned above no analytical technique has been found so far which is sufficiently specific and sensitive for the measurement of OAF concentration. Through the use of the test organism *E. coli* commune OAF has been detected in various parts of Europe and North America (Druett, 1973; Harper, 1973; Benbough and Hood, 1971; de Mik and de Groot, 1977; Graham *et al.*, 1979).

Another common finding is that the concentration of OAF at a given location depends upon wind direction, factors influencing the upward mixing of air masses, the time of day, etc., and that OAF toxicity is greatest in air masses which have crossed urban and industrial areas. In addition, during daylight hours, it is thought that the germicidal action of photoactivated molecules may be superimposed. In contrast to these findings for *E. coli* commune Benbough and Hood (1971) observed that the large day-to-day variations in *E. coli* biological decay rate did not occur for coliphage T1 and T7

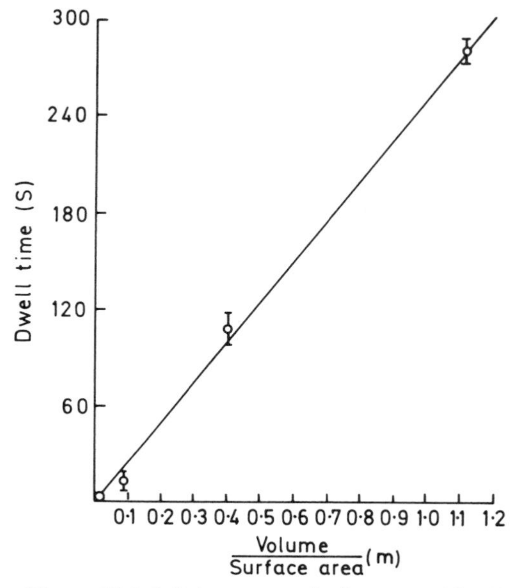

Figure 12.2 Minimum ventilation rates (dwell time) to preserve the open air factor (OAF) as a function of vessel volume/surface area ratio. (From Hood (1974), *J. Hyg. Camb.*, **72**, 53.) Reproduced by permission of the author.

and Semliki Forest virus. These workers suggest that this may be because of a *constant viricidal* activity of open air, i.e. a pollutant whose concentration is constant regardless of meteorological conditions. Alternatively, viruses may not be responsive to changes in OAF concentration, i.e. unlike *E. coli* commune, viruses are affected equally by very low and very high concentrations of OAF. This possibility seems more plausible, as explained in the next section which deals with the kinetics of OAF action.

As an alternative method to that of using microthreads for studying OAF, Hood (1971, 1973) has described an indoor system. While OAF usually is quickly lost when contained, if the holding vessel is rapidly ventilated with outside air OAF levels can be maintained. Hood (1974) showed that a good correlation exists between the minimum rates of ventilation required to preserve OAF levels and the ratio of vessel volume to surface area (Figure 12.2). One possible disadvantage with Hood's method in certain applications is that the microbial aerosols become highly diluted (e.g. 90% loss every 10 min for a 7 m diameter sphere). On the other hand, it does allow studies with true rather than captive aerosols which can be of great importance due to the sometimes severe artefacts produced by the microthread (Benbough and Hood, 1971). In addition, the ventilation method indicates that Semiliki Forest Virus (Figure 12.1b) and T1 coliphage in the presence of OAF survived better in large than in small particles and provided data from which a diffusion coefficient of 0.0752 cm^2/s, and a molecular weight range of 50–150, were calculated for OAF (Hood, 1974). Such values are consistent with those for the initial complexes of ozone and olefin reactions. That such reactions produce highly germicidal entities was demonstrated by Nash (1973), Dark and Nash (1970) and de Mik *et al.* (1977). In these studies one of the most significant correlations was that between the structure of the olefin and its activity. Ring olefins were the most effective for producing OAF-like products by reaction with ozone. Such observations fit with other aspects of the ozone and olefin reaction as well as the well-established principle that good air disinfectants work best at about 80% RH and must have a low vapour pressure (Nash, 1951, 1962; Dark and Nash, 1970). A possible mechanism is that attack by ozone splits the olefin double bond one end becoming a ketone or aldehyde and the other a peroxide 'zwitterion'.

(a) $R_1CH{=}CHR_2 + O_3 \rightarrow R_1CHO + R_2CH^{(+)}{-}O{-}O^{(-)}$

(b) (ring) $CH{=}CH + O_3 \rightarrow$ (ring) $CHO \ \ CH^{(+)}{-}O{-}O^{(-)}$

Comparing these two possibilities, one an open chain and the other a ring, the active fragment from the former will be smaller and have a higher vapour pressure than the latter (Dark and Nash, 1970). De Mik *et al.* (1977) discuss an alternative mechanism which involves a biradical, i.e.

(a) $R_1CH{=}CHR_2 + O_3 \rightarrow$ (ozonide ring: $R_1CH{-}O{-}O{-}O{-}CHR_2$) $\rightarrow R_1CH{-}O \ \ R_2CH{-}O{-}O$

but it is not really known which products are germicidal. As pointed out by Nash (Dark and Nash, 1970) it is not necessary to postulate that such products have exceptional germicidal activity. Rather, their physical properties and conditions of formation are such that they condense extremely effectively on aerosol particles, or other nearby surfaces.

12.2 KINETIC MODEL

Loss of viability caused by OAF is the only method presently available for estimating its concentration. Consequently, a relationship between viability, OAF concentration and time was required if meaningful estimates of OAF concentration were to be obtained from biological decay rate. This need is also apparent from a study of published work where decay rates based on simple exponential decay (equation 10.1) often are given but without any real account being taken of the OAF concentration prevailing at the time of the experiments. Furthermore, as will be discussed in section 12.3, the most direct evidence as to the nature of OAF could not have been obtained without estimates for prevailing OAF concentrations.

The only known method for deriving OAF concentrations is that of Cox *et al.* (1973). It involves a postulated mechanism for OAF killing action in which OAF combines reversibly with a carrier Y, i.e.

$$\text{OAF} + Y \underset{k_-}{\overset{k_+}{\rightleftharpoons}} Y\text{OAF} \tag{12.1}$$

with this complex then reacting irreversibly with an acceptor B, i.e.

$$Y\text{OAF} + \text{B} \xrightarrow{k} \text{BOAF} + Y. \tag{12.2}$$

The mechanism, essentially, is the same as that for oxygen-induced damage (Chapter 11) except that the equilibrium (equation 12.1) can be established slowly. This difference permits a marked change in the shape of the decay curves predicted by the two death mechanisms, viz. it provides for an 'induction period' before the onset of OAF-induced viable decay which is not observed in oxygen-induced viable decay. Such a slow attainment of equilibrium could correspond to a fairly slow build-up of OAF on aerosol particles through condensation as indicated in section 12.1. The differential equations representing the mechanism cannot be analytically integrated. However, numerical integration showed that provided the rate at which the reaction (12.2) proceeds is very slow or constant (i.e. steady state) an approximate analytical solution is possible. This solution, when combined with the equation derived from probability theory (Section 10.7) for the relationship between % viability and concentration of the critical species B, namely,

$$\ln V = K_1[B] - K_1[B]_0 + \ln 100 \tag{12.3}$$

gives,

$$\ln V = K_1[B]_0\left\{exp\left[\left(-\alpha t - \frac{\alpha}{\beta}\,(\mathrm{e}^{-\beta t} - 1)\right]\right)\right\} - K_1[B]_0 + \ln 100 \tag{12.4}$$

where,
$$\alpha = \frac{k[Y]_0 k_+[\text{OAF}]}{k_+[\text{OAF}] + k_-}$$

$$\beta = k_+[\text{OAF}] + k_-$$

$[B]_0$ = initial concentration of acceptor B,

$[Y]_0$ = initial concentration of carrier Y,

$[\text{OAF}]$ = concentration of OAF, assumed to remain constant,

V = % viability,

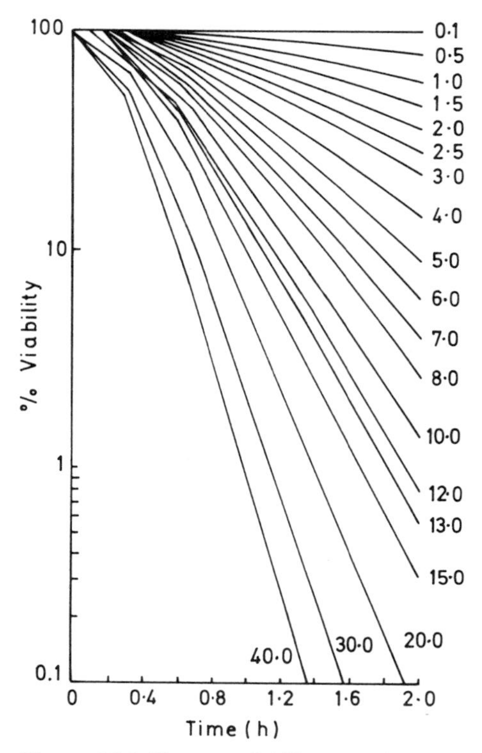

Figure 12.3 Percent viability as a function of OAF concentration and time as calculated by equation 12.4 (see text).

$$K_1,\ k_+,\ k_-,\ k = \text{constants}$$

$$t = \text{time (h)}.$$

The procedure used the method of Rosenbrock (1960) to fit equation 12.4 to a typical OAF-induced viable decay curve of *E. coli* commune corrected for any viable loss occurring under similar conditions but in the absence of OAF. The prevailing concentration was set to an arbitrary value of 10. The following values for the constants were obtained:

$$K_1[B]_0 = 16.44 \text{ (unitless)},$$

$$k[Y]_0 = 1.37 \text{ (h}^{-1}\text{)},$$

$$k_+ = 0.02 \text{ (arbitrary units)},$$

$$k_- = 0.88 \text{ (arbitrary units)}.$$

Using these values for the constants a family of decay curves (Figure 12.3) was calculated from equation 12.4 for other OAF concentrations. Comparison of experimental decay curves (obtained with *E. coli* commune grown, etc., under

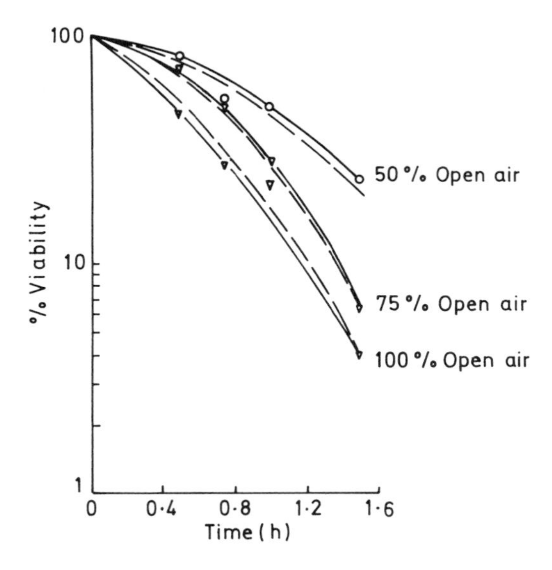

Figure 12.4 Comparison of experimental and theoretical viability decay curves induced by OAF as a function of different OAF dilutions: ——— experimental curves, – – – – theoretical curves (Figure 12.3). Percent open air values of 100, 75 and 50 have OAF values of 10.7 (100%), 8.2 (77%) and 4.8 (45%), respectively. Average experimental variation for replicate experiments = 25% of viability value and = 10% of percent open air value for values less than 100%.

similar conditions) with those in Figure 12.3 enabled a value for OAF concentration to be assigned to each decay curve. Hence, viable decay curves could be converted to the single parameter of relative OAF concentration. By exposing *E. coli* commune (held on microthreads) to open air and to different concentrations of it (obtained by dilution with OAF-free air) it was shown that the method applied in practice (Figure 12.4).

Even though OAF probably occurs in the parts per hundred million (pphm) range and no analytical method is available for independently estimating OAF concentrations (Druett and May, 1969; Cox *et al.*, 1973) approximate values for OAF concentrations can be derived. In this context assigning the units of pphm to the OAF concentrations given in Figure 12.3 may not be grossly in error. The approach described above also may find application in other cases where loss of viability (or infectivity) etc. is the only available method having the required sensitivity to detect the presence of toxic species. Applied to OAF it indicates that the usual day-to-day variation in OAF concentration is about eight-fold.

The kinetic model also provides an explanation for the apparently different bactericidal and viricidal actions of OAF. From equation 12.1 when equilibrium is established then,

$$\frac{[YOAF]}{[Y]_0} \simeq \frac{[OAF]}{k_-/k_+ + [OAF]} \tag{12.5}$$

The proportion of carrier Y which is in the form of $YOAF$ for a given value of OAF concentration depends upon the ratio of k_-/k_+. If the value of k_-/k_+ is much smaller than the given value for OAF concentration the proportion of Y as YOAF will be close to 1. Alternatively, if the value of k_-/k_+ is very much greater than the given value for OAF concentration the proportion of Y as $YOAF$ will be much less than 1. If the value for OAF concentration were now to be increased, say 10-fold, it would make little difference in the former case to the fraction $[YOAF]/[Y]_0$ because previously it was close to 1 (i.e. the maximum value). In the latter case, though, the fraction becomes greater than before. Because loss of viability is related to the fraction $[YOAF]/[Y]_0$ in the former case loss of viability would appear not to be greatly affected by OAF concentration while for the latter case it would be. This is the situation found experimentally for OAF viricidal and bactericidal actions, respectively. The above argument is exactly analogous to that used to account for enzymes becoming saturated by their substrates (i.e. Michalis-Menten kinetics) and to account for Langmuir (i.e. monolayer) adsorption or condensation (see above). Since the ratio of k_-/k_+ for viruses and bacteria may be different the implication is that carrier Y (or the condensation surface) is different in viruses (or phages) and bacteria.

12.3 NATURE OF OAF

The evidence given in Section 12.1 suggested that products from reactions between ozone and olefins have properties which mimic those of natural OAF. De Mik and de Groot (1977a) obtained much more direct evidence for this equivalence. Using equation 12.4 and the values of the constants given above the 'mean viability' (i.e. the arithmetic mean of % viability after exposure times of 30, 60 and 120 min) was calculated at various OAF concentrations. Similar 'mean viabilities' were calculated from experimental viable decay curves. Comparison of the sets of values enabled relative OAF concentrations to be assigned to each experimental curve thus translating each survivor curve into the single parameter of relative OAF concentration. At the same time and location that survivor curves were determined de Mik and de Groot obtained values for concentrations of O_3, NO, NO_2, SO_2, C_3H_8, C_3H_6, C_2H_4, and C_2H_2. Correlation coefficients (with 95% confidence limits) between these concentrations and their corresponding relative OAF concentrations are given in Table 12.1. Best correlations for data obtained in the Delft area of The Netherlands were obtained between OAF and ozone and OAF and C_3H_6 (r =

Table 12.1. Correlation coefficients (r) with 95% confidence limits of OAF concentration and concentrations of different air pollutants measured at Delft (n = 14) Reproduced by permission of the authors.

Comparison	r	95% limits	
OAF/O_3	0.90	0.97	0.72
OAF/NO	−0.17	0.39	−0.64
OAF/NO_2	0.24	0.68	−0.33
OAF/SO_2	−0.13	0.43	−0.62
OAF/C_3H_8	0.07	0.60	−0.50
OAF/C_3H_6	0.61	0.86	0.11
OAF/C_2H_4	−0.27	0.30	−0.70
OAF/C_2H_2	−0.34	0.23	−0.74

0.90 and 0.67, respectively). De Mik and de Groot (1977a) showed also that the correlation between ozone and OAF activity is higher for data combined from studies in various locations when the most important hydrocarbon source is motor traffic (r = 0.83) rather than when it is industrial (r = 0.56). The negative values for the correlation coefficients for OAF/NO, OAF/SO_2, OAF/C_2H_4, and OAF/C_2H_2 (Table 12.1) could mean that the amount of OAF is decreased by NO, SO_2, C_2H_4 and C_2H_2. Appurtaining to this possibility is that Nash (1973) observed that 10 ppm SO_2 diminished the bactericidal activity of ozone + 2-pentene and of ozone + cyclohexene, while C_2H_4 reacts comparatively slowly with ozone. Future work related to a more precise definition of OAF should allow for pollutants which can reduce OAF activity.

12.4 CAUSES OF DEATH

Only phage φX174 and *E. coli* commune have been studied in this context. As discussed in Chapter 10 desiccation denatures the protein coat of this phage but under the conditions for their experiments de Mik *et al.* (1977) found that cyclohexene did not exacerbate this process. Alternatively, ozone increased such damage (de Mik and de Groot, 1977b). Among the amino acids the most susceptible are cysteine, tryptophan and methionine (Mudd *et al.*, 1969). Ozone has been found as well to modify pyrimidine bases in the nucleic acids of *E. coli* (Pratt *et al.*, 1969) and of purines (Christensen and Giese, 1954). Appropriately, de Mik and de Groot (1977b) observed that DNA when isolated from phage φX174 and then aerosolized showed a time-dependent degradation caused by ozone. But this damage to the isolated DNA could not be differentiated from that occurring in the presence of ozone + cyclohexene (see later). Although unlikely, damage attributed to ozone alone may have been caused by reaction with materials slowly desorbing from the walls of the aerosol containment chamber. As for isolated DNA that within the phage head

is inactivated by ozone but at a slower rate possibly through protection by the protein coat. Consequently, it appears that ozone first reacts with the phage φX174 protein coat causing irreversible denaturation and then more slowly damages the DNA irreversibly but without causing strand breakage.

Ozone + cyclohexene (i.e. synthetic OAF) reacts very much more rapidly than ozone alone with the phage φX174 protein coat, as well as the DNA, whether isolated or not, with the phage protein coat providing virtually no protection for the phage nucleic acid.

Results from de Mik (1976) showed that the aerosol survival at 80% RH of *E. coli* commune sprayed from 50% spent culture medium into clean air, or with 1000 pphm cyclohexene, or with 40 pphm ozone added, were similar. On the other hand, when both cyclohexene and ozone were present together viable decay was very much greater. DNA extracted with alkaline sodium dodecyl sarcosinate from cells stored in clean air showed slight strand breakage. The extent is similar to that found for the DNA extracted from these bacteria refluxed in a Collison atomizer (Chapter 2) and when collected from aerosols stored in air at 80% RH plus 1000 pphm cyclohexene or 40 pphm ozone. In these cases about 10% of the DNA had strand breakage. When stored in air at 80% RH containing both 1000 pphm cyclohexene and 40 pphm ozone recovered *E. coli* commune had a very marked increase in DNA strand breakage and the sedimentation rate of the extracted DNA progressively decreased with decreasing viability. At the lowest viability level (0.02%) virtually all of the DNA molecules showed strand breakage but whether the synthetic OAF caused strand breakage, or alkali-labile lesions, is not certain. The uncertainty arises because alkali (used in the DNA extraction process) exposes pre-existing single strand breaks by denaturing the DNA in addition to producing breaks in DNA molecules having alkali-labile regions (de Mik, 1976). Nonetheless, the apparent action of synthetic OAF on both phage φX174 and *E. coli* commune is to break DNA strands. Whether proteins of *E. coli* also are inactivated (like those of the phage) is not known, but due to the severe nature of OAF induced damage is not too surprising that *E. coli* B following exposure to OAF cannot fully support colony formation or the replication of coliphage T7 (Cox, unpublished data). While similar experiments to those of de Mik have not been performed with other microorganisms it seems likely that OAF sensitive microorganisms are damaged in an analogous manner. Possibly, the OAF resistance of bacterial spores, FMD and SVD viruses and *M. radiodurans* arises from their outer structures having special properties which prevent access of OAF to the nucleic acids contained within them or that resistant microorganisms have a much superior repair capability (Chapter 14), or even both.

12.5 CONCLUSIONS

The ephemoral nature of OAF is related to its ready condensation onto surfaces and to the mechanism of its formation which involves ozone–olefin reactions. Its toxic action on some bacteria and viruses results in inactivation of coat constituents and of nucleic acids. The effectiveness of OAF is due less to a high toxicity than to a readiness to condense onto aerosol particles, a property possibly reflected by the kinetics of its action.

REFERENCES

Benbough, J. E. and Hood, A. M. (1971). *J. Hyg. (Camb.)*, **69**, 619–626.

Christensen, E. and Giese, A. C. (1954). *Arch. Biochem. Biophys.*, **51**, 208–216.

Cox, C. S. (1979). In *Biometeorological Survey*, vol. 1, 1973–1978, part A, *Human Biometeorology*, S. W. Tromp and Janneke J. Bouma (Eds), Heyden, London, Philadelphia, pp. 156–161.

Cox, C. S., Hood, A. M. and Baxter, J. (1973). *Appl. Microbiol.*, **26**, 640–642.

Dark, F. A. and Nash, T. (1970). *J. Hyg. (Camb.)*, **68**, 245–252.

Donaldson, A. I. and Ferris, N. P. (1974). *Veterin. Record.*, **95**, 19–23.

Donaldson, A. I. and Ferris, N. P. (1975). *J. Hyg. (Camb.)*, **74**, 409–416.

Druett, H. A. (1970). In *Third International Symposium on Aerobiology*, I. H. Silver (Ed.), Academic Press, London, New York, p. 212.

Druett, H. A. (1973). In *Fourth International Symposium on Aerobiology*, J. F. Ph. Hers and K. C. Winkler (Eds), Oosthoek, Utrecht, The Netherlands, pp. 141–149.

Druett, H. A. and May, K. R. (1968). *Nature (Lond.)*, **220**, 395–396.

Druett, H. A. and May, K. R. (1969). *New Scientist*, **41**, 579–581.

Graham, D. C., Quinn, C. E., Sells, I. A. and Harrison, M. D. (1979). *J. Appl. Bact.*, **46**, 367–376.

Harper, G. J. (1973). In *Fourth International Symposium on Aerobiology*, J. F. Ph. Hers and K. C. Winkler (Eds) Oosthoek, Utrecht, The Netherlands, pp. 151–154.

Hood, A. M. (1971). *J. Hyg. (Camb.)*, **69**, 607–617.

Hood, A. M. (1973). In *Fourth International Symposium on Aerobiology*, J. F. Ph. Hers and K. C. Winkler (Eds), Oosthoek, Utrecht, The Netherlands, pp. 149–151.

Hood, A. M. (1974). *J. Hyg. (Camb.)*, **72**, 53–60.

May, K. R. (1972). In *Assessment of Airborne Particles, Fundamentals, Applications and Implications to Inhalation Therapy*, T. T. Mercer (Ed.), Thomas, Springfield, Illinois, pp. 480–494.

May, K. R. and Druett, H. A. (1968). *J. Gen. Microbiol.*, **51**, 353–366.

May, K. R., Druett, H. A. and Packman, L. P. (1969). *Nature (Lond.)*, **221**, 1146–1147.

de Mik, G. (1976). *The Open Air Factor*, Ph. D. Thesis, University of Utrecht, The Netherlands.

de Mik, G. and de Groot, I. (1973). In *Fourth International Symposium on Aerobiology*, J. F. Ph. Hers and K. C. Winkler (Eds), Oosthoek, Utrecht, The Netherlands, pp. 155–158.

de Mik, G. and de Groot, I. (1977a). *J. Hyg. (Camb.)*, **78**, 175–187.

de Mik, G. and de Groot, I. (1977b). *J. Hyg. (Camb.)*, **78**, 199–211.

de Mik, G., de Groot, I. and Gerbrandy, J. L. F. (1977). *J. Hyg. (Camb.)*, **78**, 189–198.

Mudd, J. B., Leavitt, R., Ongun, A. and McManus, T. T. (1969). *Atmos. Environ.*, **3**, 669–682.

Nash, T. (1951). *J. Hyg. (Camb.)*, **49**, 382–399.

Nash, T. (1962). *J. Hyg. (Camb.)*, **60**, 353–358.

Nash, T. (1973). In *Fourth International Symposium on Aerobiology*, J. F. Ph. Hers and K. C. Winkler (Eds), Oosthoek, Utrecht, The Netherlands, pp. 158–162.

Pratt, R., Nofre, C. L. and Cier, A. (1969). *Ann. Inst. Past.*, **114**, 595–608.

Rosenbrock, H. H. (1960). *Computer J.*, **3**, 175–184.

Southey, R. F. W. and Harper, G. J. (1971). *J. Appl. Bact.*, **34**, 547–556.

Chapter 13

Other environmental parameters

13.1 INTRODUCTION

In addition to the environmental parameters of Chapters 10, 11 and 12 there are several others which can influence the survival, infectivity and transmission of microorganisms when airborne. They include air movements, pressure fluctuations, air ions, radiation and pollutants. Some members in the last category were considered in Chapter 12 and were shown to be extremely detrimental for some microbial species. Parameters to be considered here likewise can be important but usually to a lesser degree than the Open Air Factor (Chapter 12). General review articles which cover them are: Anderson and Cox, 1967; Akers, 1969, 1973; Serat, 1969; Hatch and Wolochow, 1969; Krueger *et al.*, 1969; Bridges, 1976; Krinsky, 1976; Morita, 1976; Hugo, 1976; Strange and Cox, 1976; Donaldson, 1978; Cox, 1979; Tromp, 1979; Spendlove and Fannin, 1982. Not all of these articles deal specifically with airborne microorganisms but as shown earlier many similarities exist for microorganisms and isolated biochemical moieties stressed by aerosolization and by other means.

13.2 AIR MOVEMENTS

The Aerobiological Pathway involves air movements not only in terms of aerial transport as described in Chapter 8 but also through a direct action on the survival and infectivity of airborne microorganisms. One example is when OAF is present and high rates of chamber ventilation are required to preserve its toxicity, as discussed in the previous chapter. Likewise, in outside air high wind speeds are more detrimental than slow ones possibly through increased collision frequency between OAF molecules and microorganisms. Changing wind speeds produce pressure changes and RH fluctuations that are relatively rapid with a frequency of about 10 Hz but Harper (1973) reported little influence of them on the survival of *E. coli* held on microthreads at high RH. Abrupt changes in RH, though, according to Hatch significantly decrease the survival of both bacteria and viruses particularly when stored at low RH (e.g. Hatch and Wolochow, 1969).

A very different aspect of air movements is the possibility of them sustaining colonies of bacteria in the airborne state. Such possibilities are not only related to the aerosol spread of infection on Earth but also to the contamination by space craft of the atmospheres of planets such as Jupiter and Saturn and their satellites. Before 1975 there was no direct evidence that metabolic processes could occur in airborne microorganisms. Since then Dimmick and co-workers (Dimmick *et al.*, 1975, 1977, 1979; Straat *et al.*, 1977) have shown that aerosolized *Serratia marcescens* stored at 95% RH and 30 °C produce $^{14}CO_2$ from ^{14}C-labelled glucose, incorporate ^{3}H-thymidine into DNA and divide for at least two generations. For these experiments *S. marcescens* was aerosolized under conditions most conducive for survival and cell division. It seems unlikely that division could continue much beyond two divisions because of inhibition through the accumulation of waste products in the aerosol droplets, and depletion of nutrients. But given circumstances in which wastes can be removed and nutrients replenished, growth could continue much further especially if collisions of aerosol droplets also occurred.

While division of bacteria in aerosol droplets can be demonstrated in the laboratory it is unlikely to occur to any great extent in the atmosphere of Earth. This is because of the presence of pollutants and radiation. As discussed in Chapter 12 and Section 13.5, these are likely to induce loss of viability at a rate greater than the growth rate. The net effect, therefore, would be for the microbial population to die. However, the atmospheres of other planets may be sufficiently different from the atmosphere of Earth that viable populations of microorganisms are maintained and, therefore, transmitted as aerosols.

13.3 PRESSURE FLUCTUATIONS

The best demonstration of the effects of pressure fluctuations is that of Druett (1973). Rapid expansion and recompression of air under conditions which cause condensation of water vapour on, and re-evaporation from, aerosol particles, has a marked germicidal action upon bacteria such as *E. coli*, *Brucella suis* and *Staphylococcus aureus* contained within them. Semliki Forest virus also is very sensitive to this stress which at least for *E. coli* commune causes an excessive denaturation of its surface structures. Recovered *E. coli* when observed under the electron microscope show cell wall damage ranging from loss of detailed structure to complete disruption. Following collection of aerosols exposed to pressure fluctuations damage to *E. coli* B cell walls is expressed, as well, by a greatly increased sensitivity to lysozyme. Consistent with the above and also results given in Chapter 10 is the fact that NaCl added to the spray fluid causes the same germicidal action to take place, but without the need for water vapour to saturate. In contrast, addition of glucose or glycerol to the spray fluid while not preventing surface damage does not produce the same enhancement as NaCl. Possibly, then

NaCl action is due to its toxicity rather than through enhanced condensation of water vapour. Such would be more consistent with the known action of solutes on the stability of biological macromolecules (Chapter 10).

13.4 AIR IONS

When alpha, beta or gamma radiation displaces an electron from a gas molecule, the molecule is left with a positive charge and under normal conditions the displaced electron is immediately captured by another molecule which then becomes negatively charged. Since the rate of collision of molecular ions with surrounding molecules is about 10^{19}/s transfer of charge is readily accomplished. As a result positive charges tend to reside on molecules possessing lowest ionization potential while negative charges tend to reside on molecules with highest ionization potential. When water molecules abound clusters of them form around charged molecules. The small air ions thereby produced vary in composition which through incorporation of trace gases can change with time.

Similar sequences result from lightning, shearing of water (e.g. water falls) and the rapid flow of air masses over land, etc. Under ideal conditions air ions may exist for several minutes, decaying by combination with ions of opposite charge, with uncharged condensation nuclei, and with surfaces. In clean air over land small air ions total about 1200–1500/ml and the ratio of small positive to small negative air ions is roughly 1.2 : 1. In polluted air small air ion concentrations are much lower and are inversely related to the degree of air pollution.

Small air ions can have broad biological impact (Krueger and Reed, 1976). For aerosols they increase the rate of physical decay through increased electrostatically induced agglomeration, while rate of loss of viability is increased by $5 \times 10^4 - 1 \times 10^5$ negative ions/ml (Krueger *et al.*, 1969). Phillips *et al.* (1964) found that for *Serratia marcescens* apparent loss of viability (in particles less than 5 μm in diameter) is doubled by positive ions and approximately tripled by negative ions, for ion densities of about 9×10^5/ml. However, positive ions affected only physical decay whereas negative ions increased both physical and biological losses. Similar antimicrobial activity is observed for microorganisms when exposed on plates or in small drops (Krueger *et al.*, 1957), with negative ions being more effective than positive ions. When radioactivity is used to discharge aerosol particles, or for measuring physical decay (Section 2.11), or when corona discharges and high electric fields are required in sampling (Section 3.9), increased physical and biological decay may result (Morris *et al.*, 1961) through free radical action (Krueger and Reed, 1976).

13.5 RADIATION

Microorganisms and isolated biochemical moieties exposed to radiation in aerosols and other states are likely to be affected by similar mechanisms. The reviews by Bridges (1976) and Krinsky (1976) may be pertinent, therefore, but as pointed out in articles dealing with microbial aerosols (Anderson and Cox, 1967; Hatch and Wolochow, 1969; Akers, 1973; Donaldson, 1978) the sensitivity of microorganisms as aerosols to radiation also is a function of degree of desiccation and of oxygen tension.

Electromagnetic radiation ranging from very short to very long wavelengths can affect microorganisms as can gamma-rays, alpha-particles, etc. The lethal action of gamma-rays is well demonstrated by its use for sterilizing laboratory equipment such as petri dishes or even ointments (Oie and Fystro, 1975). Nonetheless, resistant spores and vegetative bacteria have been described (Parisi and Antoine, 1975). X-rays, like gamma-rays, also inactivate microorganisms. Their maximum action for *E. coli* aerosols seems to occur at 70–80% RH (Webb, 1965a, b; Webb and Dumasia, 1964) whereas *S. marcescens* is least affected at this RH (Tetteh and Cormack, 1968). These workers found also that the presence of oxygen enhanced X-ray action and that aerosols were more susceptible than films. Water activity similarly influences sensitivity to ultraviolet radiation (Bridges, 1976; Krinsky, 1976; Webb, 1963; Won and Ross, 1969), as does oxygen (Bridges, 1976). In the aerosol state *F. tularensis*, *E. coli*, *S. marcescens*, *M. pneumoniae*, the RNA phage F2 and Rous sarcoma virus are all inactivated by UV, but the extent depends upon intensity, wavelength and RH (Beebe, 1959; Webb, 1961, 1963; Webb and Tai, 1968; Webb and Walker, 1968; Wright and Bailey, 1969). In addition, photoreactivation at certain wavelengths has been observed (Webb, 1961; Bridges, 1976). As for gamma-rays, UV-resistant mutants are known (Bridges, 1976; Krinsky, 1976). Visible light, like short wavelengths, can be lethal (Krinsky, 1976) and lasers offer intense sources (e.g. Takahashi *et al.*, 1975). Aerosolized *F. tularensis*, *F. pestis*, *M. pneumoniae* and *S. marcescens* are inactivated by such light with extent depending upon RH, particle size, and spray fluid (Beebe and Pirsch, 1958; Goodlow and Leonard, 1961; Webb, 1963, 1965b; Wright and Bailey, 1969). Lethality also is a function of oxygen tension, photosensitization, and quenching phenomena. Even so, some viruses (e.g. foot-and-mouth disease virus and poliovirus) are very resistant (Donaldson, 1978) although VEE is very sensitive especially when sodium fluorescein is used as a tracer (Berendt and Dorsey, 1971). Infrared and microwaves are thought to be less important, except through their heating action (Hatch and Wolochow, 1969). On the other hand Webb and Booth (1969) have found that certain microwave frequencies markedly alter the normal growth rate of *E. coli*, as well as their ability to perform certain biochemical functions.

Dry disseminated bacteria also are affected by radiation although to a lesser degree than when wet disseminated (Beebe, 1959; Goodlow and Leonard, 1961). It seems likely that other dry disseminated microorganisms respond similarly. Differences in sensitivity between bacterial strains to radiation damage is very much related to their repair capability. Such, in turn, is a function of their genetic make-up, as described in Chapter 14.

Causes of viability and infectivity loss induced by radiation depend upon its wavelength (or energy), the storage RH and whether oxygen or other oxidizing gases are present. Highly energetic radiation such as gamma- and X-rays produce air ions (Section 13.4) as well as acting more directly upon aerosolized microbes. Toxicity due to the former, therefore, is superimposed on the latter. Webb and co-workers suggest that damage caused by X-irradiation results from its direct action on the water molecules associated with airborne microbes. While, the added action of oxygen may be through photoactivated oxidation. At least for *E. coli* ionizing radiation inactivates in a different way to ultraviolet (UV) light, possibly involving free radicals. Ionizing radiation produces random single- and double-strand breaks in the bacterial DNA whereas UV light being absorbed most strongly by DNA at about 250–260 nm activates dimer formation between adjacent pyrimidines. Such dimerization sometimes may be reversed by exposure to other wavelengths. Another radiation-induced product is cytosine hydrate, which occurs predominantly in single-stranded regions of DNA. Processes such as prophage induction have been observed as well as other forms of DNA damage, the formation of long bacterial filaments, and a loss of ability to utilize oxygen (Bridges, 1976). Such biochemical abnormalities appear to be essentially the same as those resulting from desiccation and exposure to oxygen or OAF (Chapters 10, 11 and 12). Some of the co-factors for enzymes involved in respiration are sensitive to both near UV and visible light and when damaged inhibition of respiration results (Krinsky, 1976). Epel (1973) indicates that visible light can inhibit one component of cytochrome a_3 of bacteria, yeasts and animal mitochondria. Likewise, other haem proteins, such as catalase, can be destroyed by visible light (Mitchell and Anderson, 1965). Many other enzymes also are inactivated by radiation, which in addition may cause increased membrane permeability. On the other hand, the carotenoid pigments of microorganisms serve to protect them through quenching against harmful photosensitized oxidations (Krinsky, 1976).

That radiation (like desiccation) can induce damage in nucleic acids, enzymes and membranes implies that exposure increases mutation frequency and that repair mechanisms may exist. Both have been observed and there is extensive literature dealing with these phenomena (see, for example, Anderson and Cox, 1967; Bridges, 1976; Krinsky, 1976; Hatch *et al.*, 1973). Similarly, a wide range of isolated variants with differing repair capabilities have been isolated. This capability which depends upon genetic composition is discussed in Chapter 14.

13.6 POLLUTANTS

The pollutant most effective in causing damage to aerosolized microorganisms is that termed the 'Open Air Factor' or OAF. It is thought to arise through reactions between ozone and olefins (Chapter 12). The review articles by Strange and Cox (1976), Donaldson (1978) and Spendlove and Fannin (1982) cover, in general terms, work on the action of pollutants upon airborne microorganisms. Serat (1969) and Biersteker (1979) provide a broader summary of known pollutants and their effects upon living organisms while Goulden (1978) discusses methods for measuring concentrations of atmospheric pollutants.

Pollutants, apart from OAF, which have received the most attention are ozone (see Chapter 12), NO_2 and SO_2. Toxicity of NO_2 and SO_2 is greatest at high RH for aerosols of *Rhizobium melioti* (Won and Ross, 1969); NO_2 similarly increases the rate of loss of viability of VEE virus (Ehrlich and Miller, 1972) as well as Flavobacterium, a comparatively aerostable microorganism (Chatigny *et al.*, 1973). According to Lighthart *et al.* (1971) SO_2 is most toxic to aerosolized *S. marcescens* at low RH; however, due to the oxygen sensitivity of this bacterium (Chapter 11) the true effect of SO_2 may have been masked in their experiments.

The experiments described above used pollutant concentrations much greater than those known to occur in outside air. At environmental levels O_3, oxides of nitrogen, SO_2, HCHO, CO, Hcl, HF, C_2H_2, C_2H_4, C_3H_8 and C_3H_8 are not particularly toxic, especially when compared to OAF (Chapter 12) (Druett, 1973; de Mik and de Groot, 1973). Ozone as discussed in Chapter 12 causes loss of viability and infectivity through damage to DNA and proteins. Ehrlich and Miller (1972) suggest that NO_2 may react with water molecules to form nitric and nitrous acids. This possibility is supported by the work of Chatigny *et al.* (1973) which implicates hydrogen ions and pH in NO_2 toxicity. Possibly, then, NO_2 action is through denaturation of moieties which are acid labile with SO_2 acting in a similar manner, at least for *R. melioti*. However, under certain conditions SO_2 may enhance aerosol survival as discussed in Chapter 12. Also, as pointed out in that chapter and by Chatigny *et al.* (1973) artefacts are likely to arise when pollutants are confined in holding vessels. Nonetheless, from the point of view of the airborne transmission of disease OAF probably is the most important pollutant species due to the special conditions of its formation and its vapour pressure (Chapter 12).

13.7 CONCLUSIONS

Air movements, RH and pressure changes, air ions, radiation and pollutants all tend to decrease the survival and infectivity of airborne microorganisms, but extent depends on species and, very importantly, whether the induced damage is repaired.

REFERENCES

Akers, T. G. (1969). In *An Introduction to Experimental Aerobiology*, R. L. Dimmick and Ann B. Akers (Eds), Wiley Interscience, New York, London, pp. 296–339.
Akers, T. G. (1973). In *Fourth International Symposium on Aerobiology*, J. F. Ph. Hers and K. C. Winkler (Eds), Oosthoek, Utrecht, The Netherlands, pp. 73–81.
Anderson, J. D. and Cox, C. S. (1967). *Symp. Soc. Gen. Microbiol.*, **17**, 203–226.

Beebe, J. M. (1959). *J. Bact.*, **78**, 18–24.
Beebe, J. M. and Pirsch, G. W. (1958). *Appl. Microbiol.*, **6**, 127–138.
Berendt, R. F. and Dorsey, E. L. (1971). *Appl. Microbiol.*, **21**, 447–450.
Biersteker, K. (1979). In *Biometeorological Survey*, vol. 1, 1973–1978, part A, *Human Biometeorology*, S. W. Tromp and Janneke J. Bouma (Eds), Heyden Press, London, Philadelphia, pp. 73–67.
Bridges, B. A. (1976). *Symp. Soc. Gen. Microbiol.*, **26**, 183–208.
Chatigny, M. A., Wolochow, H., Lief, W. R. and Herbert, J. (1973). In *Fourth International Symposium on Aerobiology*, J. F. Ph. Hers and K. C. Winkler (Eds), Oosthoek, Utrecht, The Netherlands, pp. 94–97.
Cox, C. S. (1979). In *Biometeorological Survey*, vol. 1, 1973–1978, part A, *Human Biometeorology*, S. W. Tromp and Janneke J. Bouma (Eds), Heyden Press, London, Philadelphia, pp. 156–161.
Dimmick, R. L., Straat, P. A. Wolochow, H., Levin, G. V., Chatigny, M. A. and Schrot, J. R. (1975). *J. Aerosol. Sci.*, **6**, 387–393.
Dimmick, R. L., Chatigny, M. A., Wolochow, H. and Straat, P. (1977). *Cospar Life Sci. Space Res.*, **15**, 41–45.
Dimmick, R. L., Wolochow, H. and Chatigny, M. A. (1979). *Appl. Environ. Microbiol.*, **37**, 924–927.
Donaldson, A. I. (1978). *Vet. Bull.*, **48**, 83–94.
Druett, H. A. (1973). In *Fourth International Symposium on Aerobiology*, J. F. Ph. Hers and K. C. Winkler (Eds), Oosthoek, Utrecht, The Netherlands, pp. 90–94.
Ehrlich, R. and Miller, S. (1972). *Appl. Microbiol.*, **23**, 481–484.
Epel, B. L. (1973). In *Photophysiology*, vol. 8, A. L. Giese (Ed.), Academic Press, New York, pp. 209–229.
Goodlow, R. H. and Leonard, F. A. (1961). *Bact. Rev.*, **25**, 182–187.
Goulden, P. D. (1978). *Environmental Pollution Analysis. Heyden International Topics in Science*, Heyden Press, London, Philadelphia.
Harper, G. J. (1973). In *Fourth International Symposium on Aerobiology*, J. F. Ph. Hers and K. C. Winkler (Eds), Oosthoek, Utrecht, The Netherlands, pp. 151–154.
Hatch, M. T. and Wolochow, H. (1969). In *An Introduction to Experimental Aerobiology*, R. L. Dimmick and Ann B. Akers (Eds), Wiley Interscience, New York, London, pp. 267–295.
Hatch, M. T., Bondurant, M. C. and Ehresmann, D. W. (1973). In *Fourth International Symposium on Aerobiology*, J. F. Ph. Hers and K. C. Winkler (Eds), Oosthoek, Utrecht, The Netherlands, pp. 103–105.
Hugo, W. B. (1976). *Symp. Soc. Gen. Microbiol.*, **26**, 383–413.
Krinsky, N. I. (1976). *Symp. Soc. Gen. Microbiol.*, **26**, 209–239.
Krueger, A. P. and Reed, E. J. (1976). *Science.*, **193**, 1209–1213.
Krueger, A. P., Kotaka, S. and Andriese, P. C. (1969). In *An Introduction to Experimental Aerobiology*, R. L. Dimmick and Ann B. Akers (Eds), Wiley Interscience, New York, London, pp. 100–112.
Krueger, A. P., Smith, R. F. and Go, I. G. (1957). *J. Gen. Physiol.*, **41**, 359–381.
Lighthart, B., Hiatt, V. E. and Rossano, A. T. (1971). *J. Air Poll. Contr. Assoc.*, **21**, 639–642.
de Mik, G. and de Groot I. (1973). In *Fourth International Symposium on Aerobiology*, J. F. Ph. Hers and K. C. Winkler (Eds), Oosthoek, Utrecht, The Netherlands, pp. 155–158.
Mitchell, R. L. and Anderson, I. C. (1965). *Crop Sci.*, **5**, 588–591.
Morita, R. Y. (1976). *Symp. Soc. Gen. Microbiol.*, **26**, 279–298.
Morris, E. J., Darlow, H. M., Peel, J. F. H. and Wright, W. C. (1961). *J. Hyg. (Camb.).*, **59**, 487–496.
Oie, S. H. and Fystro, D. (1975). *Appl. Microbiol.*, **30**, 514–518.

Parisi, A. N. and Antoine, A. D. (1975). *Appl. Microbiol.*, **29**, 34–39.
Phillips, G., Harris, G. J. and Jones, M. W. (1964). *Int. J. Biometeor.*, **8**, 27–37.
Serat, W. F. (1969). In *An Introduction to Experimental Aerobiology*, R. L. Dimmick and Ann B. Akers (Eds), Wiley Interscience, New York, London, pp. 113–123.
Spendlove, J. C. and Fannin, K. F. (1982). In *Methods in Environmental Virology*, C. P. Gerba and S. M. Goyal (Eds), Marcel Dekker, New York, Basel, pp. 261–329.
Straat, P. A., Wolochow, H., Dimmick, R. L. and Chatigny, M. A. (1977). *Appl. Environ. Microbiol.*, **34**, 292–296.
Strange, R. E. and Cox, C. S. (1976). *Symp. Soc. Gen. Microbiol.*, **26**, 111–154.
Takahashi, P. K., Toups, H. J., Grienberg, D. B., Dimopoullos, G. T. and Rusoff, L. L. (1975). *Appl. Microbiol.*, **29**, 63–67.
Tetteh, G. K. and Cormack, D. V. (1968). *Rad. Res.*, **34**, 532–543.
Tromp, S. W. (1979). In *Biometeorological Survey*, vol. 1, 1973–1978, part A, *Human Biometeorology*, S. W. Tromp and Janneke J. Bouma (Eds), Heyden, London, Philadelphia, pp. 186–190.
Webb, S. J. (1961), *Can. J. Microbiol.*, **7**, 607–619.
Webb, S. J. (1963), *J. Appl. Bact.*, **26**, 307–313.
Webb, S. J. (1965a). *Bound Water in Biological Integrity*, Thomas, Springfield, Illinois.
Webb, S. J. (1965b). In *First International Symposium on Aerobiology*, R. L. Dimmick (Ed.), Naval Biological Laboratory, Naval Supply Center, Oakland, California, pp. 369–379.
Webb, S. J. and Booth, A. D. (1969). *Nature (Lond.).*, **222**, 1199–1200.
Webb, S. J. and Dumasia, M. D. (1964). *Can. J. Microbiol.*, **10**, 877–885.
Webb, S. J. and Tai, C. C. (1968). *Can. J. Microbiol.*, **14**, 727–735.
Webb, S. J. and Walker, J. L. (1968). *Can. J. Microbiol.*, **14**, 565–572.
Won, W. D. and Ross, H. (1969). *Appl. Microbiol.*, **18**, 555–557.
Wright, D. N. and Bailey, G. D. (1969). *Can. J. Microbiol.*, **15**, 1449–1452.

Chapter 14

Repair

14.1 INTRODUCTION

When microorganisms are stressed by aerosolization, freeze-drying, cold shock, freezing and thawing, osmotic shock, radiation, or mild heat stress, some may die others are injured and the remainder apparently are unaffected. The injured and 'dead' microbes are of particular interest since they may fail to grow under conditions adequate for growth of unstressed species but given suitable environments are capable of replication. For example, aerosolized or freeze-dried bacteria often need pampering with metabolic supplements before they revert to their normal growth characteristics. The occurrence of phenomena such as these is because of a capability for the repair of damage induced by a variety of applied stresses.

14.2 REPAIR OF SURFACE STRUCTURES

Dehydration–rehydration stress causes damage to microbial surface structures and membranes (Chapter 10). One manner in which this damage is expressed is sensitivity of microorganisms to hydrolytic enzymes. Hambleton (1970) studied such sensitivities for five strains of bacteria following their collection from aerosols. These bacteria being gram-negative have bacterial walls comprising an inner rigid layer of mucopolysaccharide and a more flexible lipoprotein–lipopolysaccharide component (Salton, 1957; Weidel *et al.*, 1960). The latter overlays the mucopolysaccharide which normally is thereby protected from the action of, for example, lysozyme. Treatments which damage this outer lipoprotein–lipopolysaccharide layer, if sufficiently severe, result in sensitization to lysozyme.

E. coli (strains B, Jepp and commune), *S. marcescens* 8UK and *Aerobacter aerogenes* H prior to aerosolization are resistant to hydrolytic enzymes (Hambleton, 1970). However, *E. coli* B organisms recovered from aerosols aged 1 s at 75% RH are sensitive to lysozyme, ribonuclease, deoxyribonuclease and trypsin. *E. coli* Jepp is less affected whereas the more aerostable commune strain is not sensitized by aerosolization. Response of recovered *E. coli* strains to hydrolytic enzymes follows their ability to survive. *A. aerogenes* survives similarly to *S. marcescens* and likewise following aerosol recovery is unaffected by ribonuclease, deoxyribonuclease and trypsin. However, unlike *S. marcescens*, *A. aerogenes* becomes fairly sensitive to lysozyme. Raffinose, dextran, glucose, glycerol and sodium glutamate added to the distilled water spray fluid all enhance survival and concomitantly decrease sensitivity of recovered cells to the hydrolytic enzymes. Raffinose is the most effective additive (Hambleton, 1970).

Hambleton and Benbough (1973) using lysozyme conjugated with the fluorescent dye, lissamine rhodamine B, confirmed that this enzyme binds to aerosolized *E. coli* B unlike unaerosolized controls. Similarly, electrophoretic mobility studies demonstrate an altered surface chemistry while immunological studies indicate the release of two surface antigens caused through aerosolization.

Events described above fit with those described in Chapter 10 and also provide a basis for estimating repair of cell wall damage. *E. coli* B (recovered from aerosols aged 1 s at 75%) when incubated at 37 °C for 1 h in the presence of Mg^{2+}, Fe^{3+} or Zn^{2+} become more resistant to subsequent lysozyme treatments. Mn^{2+} is less effective whereas Ca^{2+} has no effect (Hambleton, 1971). However, complete repair takes place only in the presence of a carbon energy source as well. Chloramphenicol, an inhibitor of protein synthesis in bacteria, has no influence on this repair whereas actinomycin D (an inhibitor of RNA synthesis) only slightly impairs it. Likewise, penicillin G which interrupts cell wall mucopeptide synthesis is ineffective. But 2-4-dinitrophenol severely inhibits repair showing that production of energy by these bacteria is essential for the repair of dehydration–rehydration induced damage. Since this stress when prolonged leads as well to a failure of energy production (Chapter 10), there is no natural way in which *E. coli* strains can repair and divide following prolonged desiccation stress. In terms of the definition of viability (i.e. being able to divide and form colonies on a nutrient medium) such bacteria must be classified as dead.

Whereas storage of aerosols for a few seconds causes membranes to become leaky yet repairable, storage for longer periods leads to additional inactivation of membranes of dehydrated microorganisms and Mackey (1984) attributes such viability loss to amino-carbonyl reactions between cell membrane protein molecules and cell membrane reducing sugar molecules. As described in Chapter 10 such reactions are enhanced by removal of protecting water molecules during drying while added sugar molecules protect possibly by reversibly competing for cell membrane protein reaction sites. Oxidation also is said to lead to membrane damage and to single strand breaks in DNA.

Streptococcus lactis and *faecalis* following dehydration both show an absolute peptide requirement for repair and growth whereas for other bacteria minimal media recovery (i.e. complex media inhibit repair) occurs. As minimal media recovery probably is related to DNA repair it is important to note that complex media when autoclaved or heated to boiling then can contain mutagens. At these temperatures chemical reactions between media constituents produce mutagens of low molecular weight (G. Adams, personal communication). Whereas unstressed bacteria are unlikely to be affected by these mutagens, owing to the controlled semi-permeability of their membranes, dehydrated bacteria having damaged membranes are likely to mutate. Hence if complex media need to be employed, their sterilization by filtration rather than by heating would seem expedient.

Another difficulty when handling stressed populations of bacteria is that injured members can have different biochemical requirements. Yet another aspect of repair is the time factor in that one repair mechanism can take a few minutes whereas another may need as long as 6 h for completion (Mackey, 1984).

Vibrio metschnikovii, *E. coli* K12, *Lactobacillus bulgaricus* and *Streptococcus* strains following freezing or freeze-drying become in an analogous way sensitive to NaCl toxicity (Morichi and Irie, 1973; Morichi *et al.*, 1973; Morichi *et al.*, 1967). Incubation of these bacteria, following reconstitution, with specific peptides enables them to repair outer structures so that sensitivity to NaCl then disappears. L phase Group A *Streptococcus haemolyticus* seems especially sensitive to drying, suffering marked changes in morphology and physiology (Nei and Malkawa, 1973). Red blood cell membranes also are extremely sensitive (Anderson and Nei, 1973).

These findings not only help to explain why growth conditions can affect observed viabilities (Chapter 2) of aerosolized microorganisms but may be pertinent to their pathogenicity as well. Since damage to microbial surfaces and membranes following dehydration and rehydration seems a general phenomenon, pathogenic microorganisms probably are affected similarly. Unless their landing site in the host (see Chapter 9) is one in which repair to their surfaces can take place they are unlikely to become invasive due to their enhanced susceptibility to host enzymes. That vapour phase rehydration or rehumidification (Chapter 2) can reduce this sensitivity (Maltman and Webb, 1971) is of special significance in this context.

14.3 REPAIR OF TRANSPORT ACTIVITY

A further result for *E. coli* strains of desiccation is damage to some of their active transport systems (Chapter 10; Benbough and Hambleton, 1973; Benbough *et al.*, 1972). Storage for 1 s at different relative humidities diminishes the active transport of α-methyl glucoside and galactosides. The additional loss of control of K^+ ions (Anderson and Dark, 1967), which are essential cofactors in the phosphoenol pyruvate-phosphotransferase system, may have contributed (Benbough and Hambleton, 1973). Certainly, the loss of other materials is involved because transport activity is partially restored by exudate from aerosolized bacteria. One component of the exudate from *E. coli* B binds significant amounts of α-methyl glucoside. However, complete restoration of the active transport system has not been artificially achieved; possibly, integration of binding moieties into the membrane itself must occur for complete activity.

Damage to active transport systems is not in itself immediately lethal. Provided further damage to airborne *E. coli* (e.g. prolonged storage) is absent or prevented the natural repair of active transport systems therefore must take place.

14.4 REPAIR OF RADIATION DAMAGE

Radiation damage is expressed in various ways depending upon radiation wavelength, intensity, RH and oxygen concentration (Chapter 13). Sensitivity is a function also of the repair capability of the microorganisms being studied. Repair-deficient, radiation-sensitive strains as well as radiation-resistant variants of wild-types exist. One of the more profitable approaches for the elucidation of radiation damage, or any other damage, is through the isolation and study of mutants. Degree of sensitivity then can be correlated with particular genes or groups of genes. The subject is a large one as exemplified by the review articles by Bridges (1976), Krinsky (1976), Hanawalt and Setlow (1975), Town *et al.* (1973) and Lehmann (1976).

UV, unlike ionizing radiation, induces in DNA a cyclobutane pyrimidine dimer in which neighbouring cytosine or thymine moieties are joined in a cyclobutane ring. Many bacteria possess a highly specific enzyme capable of splitting these dimers, thereby restoring the DNA base sequence. The enzyme has a strict requirement for light (Bridges, 1976), i.e. photoreactivation. Excision-repair involves the excision of UV-induced pyrimidine dimers by a DNA polymerase or a specific exonuclease. The resulting single-strand gap is filled by the action of a DNA polymerase using the complementary DNA strand as the template. Polynucleotide ligase reconnects. Ionizing radiation can produce analogous single-strand breaks and these too are subject to enolase action for converting them into small gaps which very rapidly are repaired by DNA polymerase and polynucleotide ligase (Bridges, 1976; Webb, 1964). In contrast to DNA, RNA in *E. coli* strains and in phage F2 is not affected by radiation levels which cause loss of viability (Webb and Walker, 1968).

Bacteria deficient in excision repair following irradiation with X-rays or UV light especially may be unable to give rise to recombinants. Desiccation alone can be similarly effective (Webb, 1968). However, since repair mechanisms can be error prone an increased mutation frequency commonly is associated with radiation damage (Webb, 1963a, b; Bridges, 1976) and with desiccation damage (Webb, 1961, 1963a, b, 1964, 1965; Webb *et al.*, 1963, 1964, 1965). Likewise repair inhibitors such as caffeine can increase mutation frequency and sensitivity to radiation as can sensitizers (Krinsky, 1976; Webb and Walker, 1968).

14.5 REPAIR-DEFICIENT MUTANTS

Repair-deficient mutants have been used to probe the nature of damage caused by aerosolization in the absence of radiation. The mutant *E. coli* B_{s-1} lacking host cell reactivation (Hcr) and excision repair (Exr) capability, is much more sensitive to dehydration–rehydration stress than *E. coli* B and *E. coli* B/r (Cox *et al.*, 1971). A similar relationship holds for oxygen-induced damage. However, work with the *E. coli* deficient strains 26x, 26xA3 (Cox *et al.*, 1971) and 26xA2, $WP2_s$, $B_{S2/r}1$ (Hatch *et al.*, 1973) showed that aerosol sensitivity of *E. coli* B_{s-1} is not because of a lack of Hcr Exr repair capability (or, conversely, that the relative aerosol stability of *E. coli* B/r is not due to this strain having these repair capabilities). Rather, as shown by Hatch, other genetic factors are involved and it is these which lead to aerosol stability (Hatch *et al.*, 1973). Such findings contradict the suggestion by Webb (1969) that oxygen toxicity (Chapter 11) is through inhibition of *E. coli* repair mechanisms.

14.6 EFFECTS DUE TO REPAIR

Due to the capability of repair of a wide range of aerosol-induced abnormalities it is now not so surprising that early reports on aerosol stability rarely agreed (Chapter 2). Findings that certain collecting fluids give higher viabilities or infectivities than others or that one strain is more sensitive than others, or that one strain is more sensitive than another when assayed by method A but not by method B, is understandable in terms of repair phenomena. Similarly, difficulties associated with obtaining a consistent response to a given applied stress (Chapter 2) (especially when tested at long intervals) could be due to repair mechanisms. Furthermore, whether or not such repair occurs could be crucial for infection by aerosols. As indicated in Chapter 15 unless studies with rehumidified aerosols are included estimates for dosages may be grossly in error.

14.7 CONCLUSIONS

The outcome of an applied stress depends not only on the extent of induced damage but also on the ability of microorganisms to repair that damage. In turn, ability to repair depends on genetic make-up of the microorganism as well as its chemical environment which, depending on composition, can switch repair mechanisms off as well as on. The question of repair mechanisms is complex and has far reaching implications for the Aerobiological Pathway. Repair is an evolving subject requiring much more study (Andrew and Russell, 1984).

REFERENCES

Anderson, J. D. and Dark, F. A. (1967). *J. Gen. Microbiol.*, **45**, 95–105.

Anderson, J. O. and Nei, T. (1973). In *Freeze-drying of Biological Materials*, Institut International du Froid, Paris, pp. 109–117.
Andrew, M. H. E. and Russell, A. D. (Eds) (1984). *The Revival of Injured Microbes*, Academic Press, London.
Benbough, J. E. and Hambleton, P. (1973). In *Fourth International Symposium on Aerobiology*, J. F. Ph. Hers and K. C. Winkler (Eds), Oosthoek, Utrecht, The Netherlands, pp. 135–137.
Benbough, J. E., Hambleton, P., Martin, K. L. and Strange, R. E. (1972). *J. Gen. Microbiol.*, **72**, 511–520.
Bridges, B. A. (1976). *Symp. Soc. Gen. Microbiol.*, **26**, 183–208.
Cox, C. S., Bondurant, M. C. and Hatch, M. T. (1971). *J. Hyg. (Camb.).*, **69**, 661–672.
Hambleton, P. (1970). *J. Gen. Microbiol.*, **61**, 197–204.
Hambleton, P. (1971). *J. Gen. Microbiol.*, **69**, 81–88.
Hambleton and Benbough (1973). In *Fourth International Symposium on Aerobiology*, J. F. Ph. Hers and K. C. Winkler (Eds), Oosthoek, Utrecht, The Netherlands, pp. 131–134.
Hanawalt, P. C. and Setlow, R. B. (1975). *Molecular Mechanisms for the Repair of DNA*, Plenum Press, New York.
Hatch, M. T., Bondurant, M. C. and Ehresmann, D. W. (1973). In *Fourth International Symposium on Aerobiology*, J. F. Ph. Hers and K. C. Winkler (Eds), Oosthoek, Utrecht, The Netherlands, pp. 103–105.
Krinsky, N. I. (1976). *Symp. Soc. Gen. Microbiol.*, **26**, 209–239.
Lehmann, A. R. (1976). In *Effects of Ionizing Radiation on Nucleic Acids*, Springer, Berlin.
Mackey, E. Y. (1984). In *The Revival of Injured Microbes*, M. H. E. Andrew and A. D. Russell (Eds), Academic Press, London, pp. 45–75.
Maltman, J. R. and Webb, S. J. (1971). *Can. J. Microbiol.*, **17**, 1443–1450.
Morichi, T. and Irie, R. (1973). *Cryobiol.*, **10**, 393–399.
Morichi, T., Irie, R., Yano, N. and Kembo, H. (1967). *Agr. Biol. Chem.*, **31**, 137–141.
Morichi, T., Okamoto, T. and Irie, R. (1973). In *Freeze-drying of Biological Materials*, Institut International du Froid, Paris, pp. 47–53.
Nei, T. and Maekawa, S. (1973). In *Freeze-drying of Biological Materials*, Institut International du Froid, Paris, pp. 89–95.
Salton, M. R. J. (1957). *Bact. Rev.*, **21**, 82–99.
Town, C. D., Smith, K. C. and Kaplan, H. S. (1973). *Curr. Top. Rad. Res. Quat.*, **8**, 351–399.
Webb, S. J. (1961). *Can. J. Microbiol.*, **7**, 621–632.
Webb, S. J. (1963a). *J. Appl. Bact.*, **26**, 307–313.
Webb, S. J. (1963b). *Nature (Lond.)*, **198**, 785–787.
Webb, S. J. (1964). *Nature (Lond.)*, **203**, 374–377.
Webb, S. J. (1965). *Bound Water in Biological Integrity*, Thomas, Springfield, Illinois.
Webb, S. J. (1968). *Nature (Lond.)*, **217**, 1231–1234.
Webb, S. J. (1969). *J. Gen. Microbiol.*, **58**, 317–326.
Webb, S. J. and Walker, J. L. (1968). *Can. J. Microbiol.*, **14**, 565–572.
Webb, S. J., Bather, R. and Hodges, R. W. (1963). *Can. J. Microbiol.*, **9**, 87–92.
Webb, S. J., Cormack, P. V. and Morrison, H. G. (1964). *Nature (Lond.)*, **201**, 1103–1105.
Webb, S. J., Dumasia, M. D. and Singh Bhorjee, J. (1965). *Can. J. Microbiol.*, **11**, 141–149.
Weidel, W., Frank, H. and Martin, H. H. (1960). *J. Gen. Microbiol.*, **22**, 158–166.

Chapter 15

Infectivity

15.1 INTRODUCTION

It has not always been accepted that disease can spread by the airborne route. For example, Chapin (1912) stated 'Most diseases are not likely to be dustborne and they are sprayborne only for two or three feet, a phenomenon which after all resembles contact infection more than it does aerial infection as ordinarily understood.' Chope and Smillie (1936) shared this view which arose partially because of the absence of suitable techniques for making infectivity studies. Even so, Lurie (1930) had reported the aerogenic transmission of tuberculosis among guinea pigs. Also, Wells (1933, 1934) had advanced the theory that droplet nuclei could transmit infectious organisms and had developed some techniques for the quantitative sampling of microbial aerosols. The importance of the aerosol route is now recognized (Conference on Airborne Infection, 1961), and formed the major topic for the *Fourth International Symposium on Aerobiology* (Hers and Winkler, 1973).

Survival of microorganisms is a prerequisite for their infectivity. Consequently, all parameters known to influence survival must affect infectivity as well. In addition, there are factors which modify infectivity *per se* and these are considered below.

15.2 AEROSOL PARTICLE SIZE AND HOST SUSCEPTIBILITY

The infectivity of a given microorganism is markedly dependent upon aerosol particle size (Figure 15.1). As described more fully in Chapter 9 this dependence is not only because survival is affected but also because the part of the respiratory tract where an aerosol particle is deposited depends upon its size. In general, aerosol particles about 0.5 μm diameter are deposited relatively inefficiently in the lung. Below 0.5 μm deposition efficiency increases through diffusion processes while from 0.5 μm to about 3 μm impaction forces raise it. Above about 3 to 5 μm lung penetration of aerosols decreases due to their inertial impaction in the upper respiratory system. In studies with microbial aerosols, aerosol particles are hygroscopic and increase in size during inhalation through condensation and sorption of

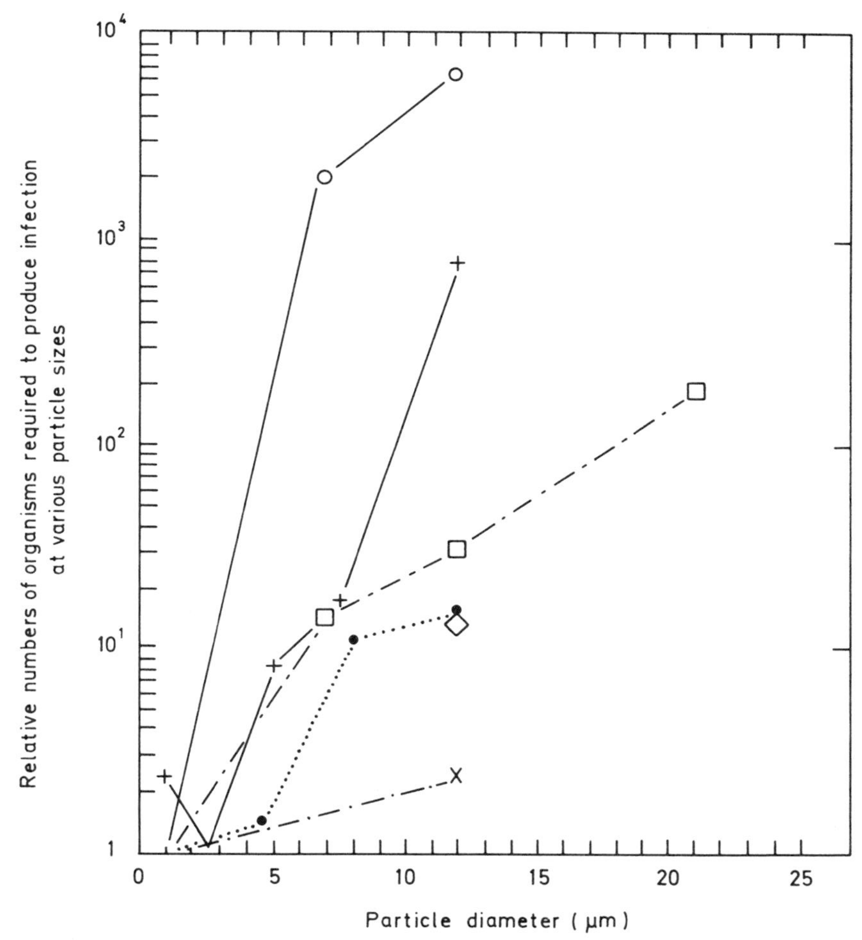

Figure 15.1 Relative numbers of microorganisms required to produce infection at various particle sizes: ○: *F. tularensis* in guinea pigs, □: *F. tularensis* in rhesus monkeys, +: *Br. suis* in guinea pigs, ●: *B. anthracis* in guinea pigs, ◇: *B. anthracis* in rhesus monkeys, X: *F. pestis* in guinea pigs. (From Druett (1967), *17th Symp. Soc. Gen. Microbiol.*) Reproduced by permission of Cambridge University Press.

water molecules. Points of aerosol deposition therefore are not simple functions of their size before inhalation but are related to particle hygroscopicity as well (Knight, 1973). In addition microbes may be carried on skin scales or other large particles (May and Pomeroy, 1973; Blowers *et al.*, 1973; Ayliffe *et al.*, 1973).

As shown in Figure 15.1 the number of microorganisms required to initate disease (e.g. LD_{50}) is a function of particle size. LD_{50} is a function also of the host and as pointed out by Pappagianis (1969) it is often difficult to translate information from measurements in animals to man. One factor is the difference

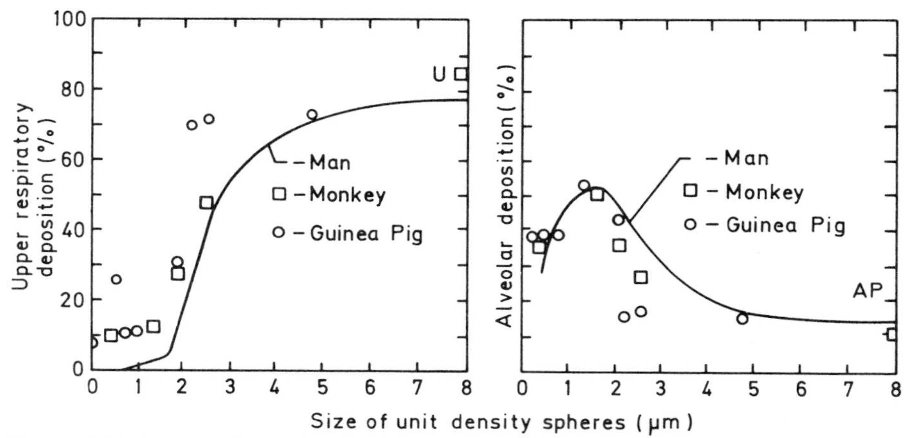

Figure 15.2 Percent deposition versus particle size of inhaled particles in the upper respiratory tract and in the lungs of the guinea pig and monkey compared with man. (From Pappagianis (1969), in *An Introduction to Experimental Aerobiology*, R. L. Dimmick and Ann B. Akers (Eds), Wiley-Interscience.) Reproduced by permission of Academic Press, New York.

in relative dimensions of respiratory passages but according to Hatch and co-workers equalizing factors operate when deposition in man is compared with that of smaller animals (Figure 15.2). The location of the lesion and character of the infection are associated too with the infecting agent. Often upper respiratory infections are associated with viruses whereas larger microorganisms (e.g. bacteria, rickettsiae, mycoplasmas) tend to induce infections in the lower respiratory tract. Hoorn (1973), Hoorn and Tyrrell (1966), Denny (1973) and Reed and Nolan (1973) have used the interesting approach of studying growth of respiratory viruses in cells derived from various tissues of the respiratory tract and have demonstrated some tissue specificity. Other studies have compared nasal instillations and aerosols (Tyrrell, 1965; Knight, 1973).

One difficulty in determining the relative importance of landing sites is that once deposited the infectious agent may not remain there. Cilia action and secretions tend to move deposited particles (Tyrrell, 1973), but some infective agents can interfere with these movements (Camner *et al.*, 1973; Liu, 1973; Morein, 1973; Taylor-Robinson, 1973). On the other hand, specialist cell systems such as alveolar macrophages kill and remove invasive microorganisms (Rylander, 1973). However, effectiveness of these processes can be reduced by tobacco smoke, air pollutants, barbiturate, hypoxia, alcohol, endotoxins etc. (Thomson *et al.*, 1973; Henry *et al.*, 1973; Goldstein *et al.*, 1973; Loosli *et al.*, 1973; Woolsgrove *et al.*, 1973) and mixed infections (Jakab and Green, 1973; Gromyko, 1973). Hypothermia likewise can increase host susceptibility (Won and Ross, 1971, 1973; Won *et al.*, 1976). While host diet had little effect on the pathology of airborne influenza A virus infection in mice (Loosli *et al.*, 1973) poor diet seems likely to generally increase host susceptibility whereas the administration of synthetic polycarboxylates enhances host resistance (Biliau *et al.*, 1973). Another factor affecting infectivity is plaque-forming ability, at least for strains of encephalomyocarditis virus, and also whether the infectious agent is free nucleic acid (Akers *et al.*, 1966).

A rather different type of infectious process is that due to phages. These pathogens invade bacteria and show a fairly specific host requirement. A symposium in 1968 deals with the entry and control of such foreign nucleic acids while the infectious process also is included in Chapters 10, 11 and 12.

In addition to antibodies present in blood, respiratory system surfaces are covered with mucus containing a range of immunoglobulins differing in spectrum. As discussed by Heremans (1973) the major immunoglobulin in respiratory secretions is a dimer of IgA which is associated with two other polypeptide chains. Other immunoglobulins are present as well. All of these plus interferon and circulating antibodies help to increase host resistance (Allison, 1973; Ogra, 1973; Gardner *et al.*, 1973; Artenstein and Brandt, 1973; Shore *et al.*, 1973). Likewise, immunization through the airborne or other routes can be effective (Hottle, 1969; Bartlema and Fontagnes, 1973; Terskikh and Gusman, 1973; Fournier *et al.*, 1973; Waldman *et al.*, 1973; Eigelsbach and Hornick, 1973; Liem *et al.*, 1973; Thomas, 1973; Hearn *et al.*, 1973; Bogaerts and Durville-van der Oord, 1973; Danilov, 1973; Akers *et al.*, 1966).

15.3 VIRULENCE

Infection via the aerosol route can result from a single infectious microorganism such as *Francisella tularensis* (Furcolow, 1961; Hood, 1961; McCrumb, 1961; Tigertt *et al.*, 1961). On the other hand, Moloney murine sarcoma virus and leukaemia virus aerosols do not appear to produce macroscopic lesions in lungs or other tissues in mice (Hinshaw *et al.*, 1976). Alternatively, respiratory infection with anthrax spores requires a large number of them (Druett *et al.*, 1953). Apart from reasons of host specificity, particle size, etc. (Section 15.2), aerosol age and the matrix in which the airborne microorganisms are imbedded can affect virulence. As shown by Schlamm (1960), Hood (1961) and Sawyer *et al.* (1966) infectivity of aerosolized microorganisms can decline faster than does viability. One reason for this is that some spray fluids include infectivity suppressors. Lester as early as 1948, had shown that the presence of NaCl in spray fluids was responsible for greatly diminished infectivity of aerosolized influenza A virus (PR8 strain). However, Lester's work did not seem to receive much attention. Later, Hood (1961) studying *F. tularensis* observed loss of virulence in aerosols due to spray fluids incorporating chloride ions, and components of yeast extract and casein hydrolysate. Changes induced by inclusion of these substances in spray fluids were elucidated by Hood in 1977. Fully virulent *F. tularensis* bacteria are encapsulated, whereas avirulent ones from aged aerosols are decapsulated. Similarly, *F. tularensis* bacteria previously decapsulated by suspension in hypertonic saline, while maintaining viability, are avirulent for guinea pigs when subsequently challenged intraperitoneally. The capsular and cell wall materials differ in lipid and carbohydrate composition and when isolated the former neither contains a lethal toxin nor induces an immunological response in mice or guinea pigs. The ease of removal of the capsule may explain reports that *F. tularensis* Schu 4 is not encapsulated (Hesselbrock and Foshay, 1945; Eigelsbach *et al.*, 1946; Shepard *et al.*, 1954) in contrast to those of Hood (1977) and Pavlova *et al.* (1967).

Capsules at least for *F. tularensis* seem essential for virulence and may provide essential protection from host defense mechanisms. In this respect the high concentration of the α-OH fatty acid in the capsule possibly is responsible for the extreme resistance of fully-virulent bacteria to intra-cellular (mononuclear cells) bactericidal factors in non-immune hosts (Hood, 1977). An analogous argument may apply also to viruses since their virulence can be modified by passage (Bradish *et al.*, 1971; Chapter 2). But any possible role of rehumidification in the loss of capsular material does not seem to have been investigated. On the other hand, provided that suitable precautions are taken virulence like viability can be observed following many years of storage in the dry state (Heckly and Blank, 1980).

15.4 NATURAL RESISTANCE

Natural resistance to inhaled microorganisms has at least the two facets of pulmonary clearance and of antimicrobial activity. Respiratory tract clearance by the mucociliary apparatus and by phagocytosis was considered in Chapter 9.

According to Rylander (1973) *E. coli* in the airways and lungs of guinea pigs declines rapidly. In the airways themselves barely any viable *E. coli* can be recovered 2 h after exposure, whereas for the lungs the viable count was reduced to about 15% 5 h after exposure. For comparison, for killed *E. coli* when inhaled about 50% of these particle still remained in the lungs at a comparable time. Results such as these, and similar ones for other animals and man, indicate that the viable count declines faster than the bacterial count. The effect is because of antimicrobial agents (contained in mucus) which mainly consists of IgA immunoglobulin antibodies. Consequently natural resistance results from both innate and acquired mechanisms. Their relative importance requires elucidation but is known to be dependent on air quality. Ozone at the level of about a part per million decreases pulmonary bactericidal activity. At slightly high concentrations nitrogen dioxide has a similar effect (Goldstein *et al.*, 1973). Antiviral activity also is reduced by ozone and nitrogen dioxide, as well as by sulphur dioxide (Loosli *et al.*, 1973). Most likely reduced antimicrobial activity caused by these pollutants is not through an inactivation of antibodies but rather through defective pulmonary alveolar macrophages (Goldstein *et al.*, 1973).

15.5 IMMUNOLOGICAL FACTORS

Antibodies present in the blood (i.e. circulating immunoglobulins) mainly comprise IgG, IgA and IgM whereas the immunoglobulins of mucus mainly comprise a special form of IgA (Heremans, 1973) having a molecular weight considerably higher than serum IgA. This IgA is a dimer form of circulating IgA plus two additional polypeptide chains. In humans it is derived most probably from local biosynthesis rather than from the circulation, that is secretory IgA occurs only at the site that has received the immunogenic stimulus. Studies with fluorescently labelled antibodies show that the source of locally produced IgA lies in the plasma cells populating the connective tissues of the mucosae and glands. The mechanisms of transfer of the IgA across the epithelial layer though is not understood but may involve an active process.

IgA complexes to other proteins especially the epithelial glycoproteins and forms part of the mucus layer. It operates probably like other antibodies by binding to specific sites of the corresponding antigens. But it is not clear whether it confers specific antigen-binding sites to the mucus, whether it sterically hinders specific functions of the antigen or whether it promotes phagocytosis.

Other immunoglobulins such as IgG, IgM and IgE also are found in normal bronchial washings but are at much lower concentrations than IgA. Their roles in respiratory secretions would seem to be of less importance than that of the secretory IgA.

15.6 RESPIRATORY IMMUNIZATION

This way of immunization was favoured in the past as being a natural way of introducing antigen into the host and that mass vaccination could be achieved by exposing large numbers of people simultaneously to the immunizing aerosol. However, this method suffers from the disadvantage that it is difficult to control the dose received. Therefore, for live vaccines the risk could be high. On the other hand, as described above, secretory antibody differs from the corresponding circulating antibody. Thus it may be expected that respiratory immunization provides a higher degree of protection than that conferred by the corresponding antibody induced through subcutaneous vaccination for airborne diseases. For diseases in which the respiratory systems is not involved, e.g. tetanus, intestinal infections, the use of respiratory immunization seems questionable by the same token. Another factor is that side-effects caused by immunization are least by the aerosol route (Waldman *et al.*, 1973; Liem *et al.*, 1973).

Even though aerosol vaccination has been shown to be immunologically and epidemiologically effective for certain diseases, it is not widely adopted. Yet when the vaccine is supplied in small aerosol cans (Thomas, 1973) the dose is controllable and given to recipients on an individual basis and without any discomfort. Operation of the valve on the aerosol can delivers a metered volume of 50 μl which can be repeated 100 times. A small adapter at the nozzle of the aerosol can enables aerosol vaccines to be simply administered via the nose or the mouth. Respiratory immunization carried out in this manner seems likely to overcome potential disadvantages of this technique.

15.7 MICROBIAL SURVIVAL IN ANIMALS

An important aspect of the disease process is the survival of microorganisms within the host. Pathogens invasive by the respiratory tract first must survive on mucous surfaces. Some such as whooping cough begin the disease process solely on the mucous surface while others such as streptococcal infections penetrate to the underlying surface epithelium. In both cases the first requirement is an ability to adhere to the mucous surface thereby avoiding mucocilliary action. The next requirement is to be unaffected by pH and the bactericidal or bacteriostatic action of host secretions. Furthermore pathogens must successfully compete with the indigenous microflora as well as avoid the action of the large number of phagocytes associated with the mucous layer. Hence during the initial phases of the disease process survival of pathogenic microorganisms in the host depends on avoiding the non-specific host defence mechanisms.

A few days after infection the immune reactions of the host increase thereby enhancing phagocytic activity and providing antibodies operating specifically against the invading pathogen. These antibodies directly neutralize bacterial aggressins (compounds that inhibit host defence mechanisms), sensitize cell lysis and block surface components by binding to specific antigens. Bacterial aggressins include capsular polysaccharides and polyamino acids as well as endotoxins.

Provided the pathogen can survive within phagocytes in spite of their antimicrobial activity through hydrogen peroxide, lysozyme and other enzymes, it is protected from antibodies, etc. As phagocytes themselves are short lived intracellular pathogens are liberated on their death and can infect surrounding tissues which may be distant from the portal of entry. Such spread of infection has been observed in brucellosis, in plague and in staphylococcal infections (Smith, 1976).

In vivo a dynamic equilibrium exists between the multiplication of microbes able to resist host defence mechanisms and those which are susceptible. An increase in their number leads to a progression of disease while a decrease results in their elimination.

15.8 EXPERIMENTAL PATHOGENICITY

For infections initiated by the subcutaneous or intraperitoneal routes, for example, an inoculum quantity initiating the disease can be determined relatively easily compared to the aerosol route. As already discussed particle size affects the outcome of an aerosol challenge. In the past the general procedure had been to generate a fine aerosol with a baffled aerosol generator (e.g. Collison, Wells atomizers). In this way dried aerosol particles about 1–2 μm diameter were generated and for bacteria represented usually one bacterium per particle. For viruses and other minute microbes each particle contained a number of microorganisms that depended on their concentration in the spray fluid.

Even though the generated aerosols were polydisperse it was assumed that the particles deposited mainly in one region of the host's lungs. With the availability today of monodisperse aerosol generators (e.g. vibrating orifice) problems due to polydispersity are avoidable.

Having generated a preferably monodisperse pathogenic aerosol, test animals are exposed to it for a known time, dependent on aerosol concentration and required inhaled dose. The challenge concentration of infective aerosol particles usually is determined by collecting an aerosol sample by impinger (or other suitable sampler) operating at a known flow rate and for a known time interval. This procedure is presumed to provide an accurate measure of the challenge airborne pathogen concentration but, as discussed in other chapters of this book, the observed colony or plaque assay counts (respectively for bacteria and viruses) depend on the collecting fluid and whether or not the aerosol is rehumidified prior to sampling. Since rehumidification of the aerosol occurs naturally in the host a similar process logically should be applied to the artificially collected reference sample. But even then there is the question of microbial repair following particle deposition. Given that it occurs in the host but not in the reference sample, for instance, a higher than true value will be assigned to the LD_{50} value (i.e. the lethal dose required to affect 50% of the host animals).

Supposing that the question of repair is resolved there is another one concerning the proportion of the inhaled microbes that deposit in the host respiratory tract. A figure of about one-third seems an acceptable general estimate, but infections in the intestinal tract caused by swallowing aerosol particles may still occur and thereby confuse the respiratory nature of the infection.

Taking the case when the above problems are solved and there are no infections due to swallowing, the animal exposure or dose is expressed in terms of an artificial number of colony forming units or plaque forming units. For example, when tissue cultures are employed in the assay of the reference sample, the LD_{50} is expressed in tissue culture infective doses ($TCID_{50}$).

Because of the difficulties mentioned above, quoted values for microbial pathogenicity should be treated with caution. However, to provide a more accurate estimate of pathogenicity account needs to be taken of the host, the aerosol particle size and its distribution, the technique used to determine the aerosol challenge level, whether or not injured microbes repair, their point of deposition and whether an intestinal infection occurs together with a respiratory one.

15.9 CONCLUSIONS

Since a microorganism must be viable to be infective all factors known to affect survival also must influence infectivity. In addition, there are factors which modify only infectivity and include chemical composition of the aerosol particles, while particle size controls the landing site in the host. This landing site and the efficacy of host defence mechanisms play essential roles in the outcome when pathogenic microorganisms enter host respiratory systems. Measurements of infectious doses are fraught with difficulties and quoted values need to be treated with caution.

REFERENCES

Akers, T. G., Bond, S. B., Papke, C. and Leif, W. R. (1966). *J. Immunol.*, **97**, 379–385.

Allison, A. C. (1973). In *Fourth International Symposium on Aerobiology*, J. F. Ph. Hers and K. C. Winkler (Eds), Oosthoek, Utrecht, The Netherlands, pp. 269–279.

Artenstein, M. S. and Brandt, B. L. (1973). In *Fourth International Symposium on Aerobiology*, J. F. Ph. Hers and K. C. Winkler (Eds), Oosthoek, Utrecht, The Netherlands, pp. 285–290.

Ayliffe, G. A. J., Babb, J. R. and Collins, B. J. (1973). In *Fourth International Symposium on Aerobiology*, J. F. Ph. Hers and K. C. Winkler (Eds), Oosthoek, Utrecht, The Netherlands, pp. 435–437.

Bartlema, H. C. and Fontagnes, R. (1973). In *Fourth International Symposium on Aerobiology*, J. F. Ph. Hers and K. C. Winkler (Eds), Oosthoek, Utrecht, The Netherlands, pp. 295–305.

Biliau, A., Karelse, M. P. C. and Claes, P. (1973). In *Fourth International Symposium on Aerobiology*, J. F. Ph. Hers and K. C. Winkler (Eds), Oosthoek, Utrecht, The Netherlands, pp. 253–257.

Blowers, R., Hill, J. and Howell, A, (1973). In *Fourth International Symposium on Aerobiology*, J. F. Ph. Hers and K. C. Winkler (Eds), Oosthoek, Utrecht, The Netherlands, pp. 432–434.

Bogaerts, W. J. C. and Durville-van der Oord, B. J. (1973). In *Fourth International Symposium on Aerobiology*, J. F. Ph. Hers and K. C. Winkler (Eds), Oosthoek, Utrecht, The Netherlands, pp. 341–345.

Bradish, C. J., Allner, K. and Meber, H. B. (1971) *J. Gen. Virol.*, **12**, 141–160.

Camner, P., Jarstrand, C. and Philipson, K. (1973). In *Fourth International Symposium on Aerobiology*, J. F. Ph. Hers and K. C. Winkler (Eds), Oosthoek, Utrecht, The Netherlands, pp. 236–328.

Chapin, C. V. (1912). *The Sources and Modes of Infection*, 2nd edn, Wiley, New York.

Chope, H. D. and Smillie, W. G. (1936). *J. Ind. Hyg. and Toxicol.*, **18**, 780–792.

Conference on Airborne Infection (1961), *Bact. Rev.*, **25**, 173–377.

Danilov, A. I. (1973). In *Fourth International Symposium on Aerobiology*, J. F. Ph. Hers and K. C. Winkler (Eds), Oosthoek, Utrecht, The Netherlands, pp. 344–345.

Denny, F. W. (1973). In *Fourth International Symposium on Aerobiology*, J. F. Ph. Hers and K. C. Winkler (Eds), Oosthoek, Utrecht, The Netherlands, pp. 186–189.

Druett, H. A. (1967). *Symp. Soc. Gen. Microbiol.*, **17**, 165–202.

Druett, H. A., Henderson, D. W., Packman, L. and Peacock, S. (1953), *J. Hyg. (Camb.)*, **51**, 359–371.

Eigelsbach, T. and Hornick, R. B. (1973). In *Fourth International Symposium on Aerobiology*, J. F. Ph. Hers and K. C. Winkler (Eds), Oosthoek, Utrecht, The Netherlands, pp. 318–322.

Eigelsbach, H. T., Chambers, L. A. and Coriell, L. (1946). *J. Bact.*, **52**, 179–185.

Fournier, J. M., Chomel, J. J. and Fontagnes, R. (1973). In *Fourth International Symposium on Aerobiology*, J. F. Ph. Hers and K. C. Winkler (Eds), Oosthoek, Utrecht, The Netherlands, pp. 312–317.

Furcolow, M. L. (1961). *Bact. Rev.*, **25**, 301–309.
Gardner, P. S., Mcquillin, J. and Scott, R. (1973). In *Fourth International Symposium on Aerobiology*, J. F. Ph. Hers and K. C. Winkler (Eds), Oosthoek, Utrecht, The Netherlands, pp. 282–284.
Goldstein, E., Warshauer, D., Hoeprich, P. D. and Eagle, M. C. (1973). In *Fourth International Symposium on Aerobiology*, J. F. Ph. Hers and K. C. Winkler (Eds), Oosthoek, Utrecht, The Netherlands, pp. 220–224.
Gromyko, A. I. (1973). In *Fourth International Symposium on Aerobiology*, J. F. Ph. Hers and K. C. Winkler (Eds), Oosthoek, Utrecht, The Netherlands, pp. 246–248.
Hearn, H. J., Bradish, C. J. and Allner, K. (1973). In *Fourth International Symposium on Aerobiology*, J. F. Ph. Hers and K. C. Winkler (Eds), Oosthoek, Utrecht, The Netherlands, pp. 338–341.
Heckly, R. J. and Blank, H. (1980). *Appl. Environ. Microbiol.*, **39**, 541–543.
Henry, M. C., Aranyi, C. and Ehrlich, R. (1973). In *Fourth International Symposium on Aerobiology*, J. F. Ph. Hers and K. C. Winkler (Eds), Oosthoek, Utrecht, The Netherlands, pp. 216–219.
Heremans, J. F. (1973). In *Fourth International Symposium on Aerobiology*, J. F. Ph. Hers and K. C. Winkler (Eds), Oosthoek, Utrecht, The Netherlands, pp. 261–269.
Hers, J. F. Ph. and Winkler, K. C. (Eds) (1973). *Fourth International Symposium on Aerobiology*, Oosthoek, Utrecht, The Netherlands.
Hesselbrock, W. and Foshay, L. (1945). *J. Bact.*, **49**, 209.
Hinshaw, V. S., Schaffer, F. L. and Chatigny, M. A. (1976), *J. Natnl. Cancer Inst.*, **57**, 775–778.
Hood, A. M. (1961). *J. Hyg. (Camb.)*, **59**, 497–504.
Hood, A. M. (1977). *J. Hyg. Camb.)*, **79**, 47–60.
Hoorn, B. (1973). In *Fourth International Symposium on Aerobiology*, J. F. Ph. Hers and K. C. Winkler (Eds), Oosthoek, Utrecht, The Netherlands, pp. 192–193.
Hoorn, B. and Tyrrell, D. A. J. (1966). *Amer. Rev. Resp. Dis.*, **93**, 156–161.
Hottle, G. A. (1969). *An Introduction to Experimental Aerobiology*, R. L. Dimmick and Ann B. Akers (Eds), Wiley Interscience, New York. Ch. 15.
Jakab, G. J. and Green, G. K. (1973). In *Fourth International Symposium on Aerobiology*, J. F. Ph. Hers and K. C. Winkler (Eds), Oosthoek, Utrecht, The Netherlands, pp. 214–216.
Knight, V. (1973). In *Fourth International Symposium on Aerobiology*, J. F. Ph. Hers and K. C. Winkler (Eds), Oosthoek, Utrecht, The Netherlands, pp. 175–182.
Lester, W. (1948). *J. Exp. Med.*, **88**, 361.
Liem, K. S., Marcus, E. A., Jacobs, J. and van Strik, R. (1973). In *Fourth International Symposium on Aerobiology*, J. F. Ph. Hers and K. C. Winkler (Eds), Oosthoek, Utrecht, The Netherlands, pp. 322–327.
Liu, C. (1973). In *Fourth International Symposium on Aerobiology*, J. F. Ph. Hers and K. C. Winkler (Eds), Oosthoek, Utrecht, The Netherlands, pp. 238–241.
Loosli, C. G., Buckley, R. D., Hwang-Kow, S. Y., Hertweck, M. S., Hardy, J. D. and Serebrin, R. (1973). In *Fourth International Symposium on Aerobiology*, J. F. Ph. Hers and K. C. Winkler (Eds), Oosthoek, Utrecht, The Netherlands, pp. 225–231.
Lurie, N. B. (1930). *J. Exp. Med.*, **51**, 743–751.
McCrumb, F. R. (1961). *Bact. Rev.*, **25**, 262–267.
May, K. R. and Pomeroy, N. P. (1973). In *Fourth International Symposium on Aerobiology*, J. F. Ph. Hers and K. C. Winkler (Eds), Oosthoek, Utrecht, The Netherlands, pp. 426–432.
Morein, B. (1973). In *Fourth International Symposium on Aerobiology*, J. F. Ph. Hers and K. C. Winkler (Eds), Oosthoek, Utrecht, The Netherlands, pp. 241–245.
Ogra, P. L. (1973). In *Fourth International Symposium on Aerobiology*, J. F. Ph. Hers and K. C. Winkler (Eds), Oosthoek, Utrecht, The Netherlands, pp. 280–282.

Pappagianis, D. (1969). *An Introduction to Experimental Aerobiology*, R. L. Dimmick and Ann B. Akers (Eds), Wiley Interscience, New York. Ch. 16.
Pavlova, I. B., Meshcheryakova, I. S. and Emelyanova, O. S. (1967). *J. Hyg. Epid. Microbiol. Immuno. (Praha)*, **11**, 320–327.
Reed, S. E. and Nolan, P. S. (1973). In *Fourth International Symposium on Aerobiology*, J. F. Ph. Hers and K. C. Winkler (Eds), Oosthoek, Utrecht, The Netherlands, pp. 193–196.
Romig, W. R. and Thorne, C. B. (Eds) (1968). *Symposium on Entry and Control of Foreign Nucleic Acids*, *Bact. Rev.*, **32**, 291–399.
Rylander, R. (1973). In *Fourth International Symposium on Aerobiology*, J. F. Ph. Hers and K. C. Winkler (Eds), Oosthoek, Utrecht, The Netherlands, pp. 201–208.
Sawyer, W. D., Jemski, J. V., Hogge, A. L., Eigelsbach, H. T., Elwood, K., Wolfe, E. K., Dangerfield, H. G., Gochenour, W. S. and Crozier, D. (1966). *J. Bact.*, **91**, 2180–2184.
Schlamm, N. A. (1960). *J. Bact.*, **80**, 818–822.
Shepard, C. C., Ribi, E. and Larson, C. (1954). *J. Immunol.*, **75**, 7–14.
Shore, S., Potter, C. W. and Stuart-Harris, C. H. (1973). In *Fourth International Symposium on Aerobiology*, J. F. Ph. Hers and K. C. Winkler (Eds), Oosthoek, Utrecht, The Netherlands, pp. 290–294.
Smith, H. (1976). *Symp. Soc. Gen. Microbiol.*, **26**, 299–326.
Taylor-Robinson, D. (1973). In *Fourth International Symposium on Aerobiology*, J. F. Ph. Hers and K. C. Winkler (Eds), Oosthoek, Utrecht, The Netherlands, pp. 196–200.
Terskikh, I. and Gusman, B. S. (1973). In *Fourth International Symposium on Aerobiology*, J. F. Ph. Hers and K. C. Winkler (Eds), Oosthoek, Utrecht, The Netherlands, pp. 305–312.
Thomas, G. (1973). In *Fourth International Symposium on Aerobiology*, J. F. Ph. Hers and K. C. Winkler (Eds), Oosthoek, Utrecht, The Netherlands, pp. 328–338.
Thomson, R. G., Gilka, F., Lillie, L. E. and Savan, M. (1973). In *Fourth International Symposium on Aerobiology*, J. F. Ph. Hers and K. C. Winkler (Eds), Oosthoek, Utrecht, The Netherlands, pp. 209–211.
Tigertt, W. D., Benenson, A. S. and Gochenour, W. S. (1961), *Bact. Rev.*, **25**, 285–293.
Tyrrell, D. A. J. (1965). *Common Colds and Related Diseases*, Williams and Wilkins, Baltimore, Maryland.
Tyrrell, D. A. J. (1973). In *Fourth International Symposium on Aerobiology*, J. F. Ph. Hers and K. C. Winkler (Eds), Oosthoek, Utrecht, The Netherlands, pp. 183–185.
Tyrrell, D. A. J. (1976), *Symp. Soc. Gen. Microbiol.*, **17**, 286–306.
Waldman, R., Fox, E., Mauceri, A. and Dorfman, A. (1973). In *Fourth International Symposium on Aerobiology*, J. F. Ph. Hers and K. C. Winkler (Eds), Oosthoek, Utrecht, The Netherlands, pp. 317–318.
Wells, W. F. (1933). *Amer. J. Publ. Hlth.*, **23**, 58–59.
Wells, W. F. (1934). *Amer. J. Hyg.*, **20**, 611–618.
Won, W. D. and Ross, H. C. (1971). *Aerospace Med.*, **42**, 642–645.
Won, W. D. and Ross, H. C. (1973). In *Fourth International Symposium on Aerobiology*, J. F. Ph. Hers and K. C. Winkler (Eds), Oosthoek, Utrecht, The Netherlands, pp. 211–214.
Won, W. D., Ross, H. C. and Dieg, E. F. (1976). *Aviat. Space Environ. Med.*, **47**, 704–707.
Woolsgrove, B., Buch, S. A. and Binns, R. (1973). In *Fourth International Symposium on Aerobiology*, J. F. Ph. Hers and K. C. Winkler (Eds), Oosthoek, Utrecht, The Netherlands, pp. 232–236.

Chapter 16

Catastrophe theory

16.1 INTRODUCTION

To mathematically model natural events such as the breaking of waves on the seashore or the sudden aggression shown by a dog was not possible until fairly recently. The main difficulty was the discontinuous nature of such events. In 1972 René Thom of the Institute des Hautes Études Scientific published his book entitled *Stabilité Structurelle et Morphogénèse* while D. H. Fowler (1975) provides an English translation. Thom represented discontinuous events in terms of equilibria in topographical form. Topology is involved because the underlying forces in nature can be represented by smooth surfaces of equilibrium.

In classical treatments, representation is as though chemical reactions proceed continuously whereas close examination suggests this to be an approximate solution. For example, in terms of individual molecules, collisions are not continuous events rather they are discontinuous. However, overall reactions involving large numbers of molecules may appear to behave in a continuous fashion because discontinuities become smoothed. When the number of reacting entities is comparatively small, discontinuous representation is preferable.

Such is the case when loss of viability or infectivity is considered, since relatively small numbers of cells are entailed. At the level of the individual cell, excluding repair for the moment, it is either alive or dead. Loss of viability or infectivity therefore is a discontinuous process with the sudden change between these two states being termed a catastrophe. In that sense it is analogous to the sudden breaking of a wave or to the sudden aggression shown by a dog. Consequently, the same mathematical treatment is applicable to catastrophes of wide diversity.

Applications of Thom's catastrophe theory range widely from, for example, the crash of a stock market to anorexia nervosa (Zeeman, 1976). All involve a sudden behavioural change or catastrophe. The bases of catastrophe theory are equilibria with behaviour being related to the potential energy of the system (Woodcock and Davis, 1978). The system's potential energy in turn is related to factors governing that equilibrium. These factors are called control parameters. The simplest case is when the system's potential energy is

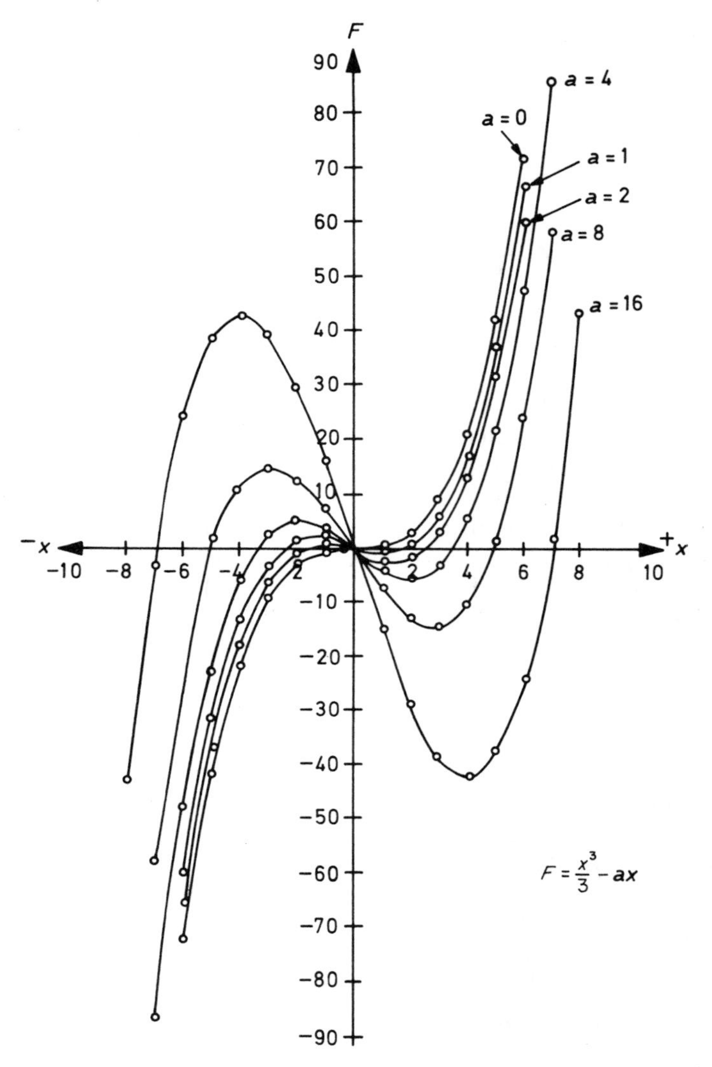

Figure 16.1 Potential energy of a system as a function of the control parameter (a).

affected by only one control parameter. Figure 16.1 shows how the potential energy changes with the value of that control parameter. For one range of values for the control parameter, the potential energy curve has a local minimum and a maximum and the system spontaneously tends to move to the minimum. The local minimum represents a stable equilibrium and the local maximum a metastable one. Provided that the potential energy of the system is at the minimum, stability results. A change in the value of the control parameter may cause the system potential energy to move to the maximum, i.e.

to a metastable state. A small perturbation one way puts the system back to the stable equilibrium while a small perturbation in the other direction to beyond some value of the control parameter causes a sudden fall in potential energy, i.e. a catastrophe occurs. In terms of response the system shows a dramatic change in behaviour (or state) and it is no longer in equilibrium (Figure 16.1). In the present context the stable equilibrium represents the viable or infective state and the catastrophic drop in potential energy, which normally is non-reversible, leads to the inactivated or non-equilibrium state.

Zeeman (1976) gives one possible form for the potential energy function F, of a system (like that shown in Figure 16.1) with one control parameter as,

$$F = \frac{x^3}{3} - ax \tag{16.1}$$

where, x represents the system's behaviour, and a represents the control parameter. In terms of viability (or infectivity) x corresponds to the viable (or infective) fraction, i.e.

$$F = \frac{(V/100)^3}{3} - a(V/100) \tag{16.2}$$

where, V = % viability. The control parameter a is the difference between the concentration of the crucial moiety upon which viability (or infectivity) depends and on some minimum values (possibly zero).

Hence, in terms of denaturation kinetics (Section 10.7), the control parameter is $([B] - [B]_{min})$ and,

$$F = \frac{\left(\frac{V}{100}\right)^3}{3} - K([B] - [B]_{min})\left(\frac{V}{100}\right) \tag{16.3}$$

where K is a proportionality constant.

For stable equilibrium, $(dF/dV) = 0$, i.e.

$$\frac{\mathrm{df}}{\mathrm{d}\left(\frac{V}{100}\right)} = \left(\frac{V}{100}\right)^2 - K([B] - [B]_{min}) = 0 \tag{16.4}$$

$$\text{Hence, } V = 100[K([B] - [B]_{min})]^{\frac{1}{2}} \tag{16.5}$$

For illustrative purposes, putting $K = 1$ and $[B]_{min} = 0.1$, the curve relating % viability to $[B]$ (Figure 16.2) can be derived for values of $[B]$ from 0.1 to 1.1. It demonstrates how as the value of $[B]$ decreases % viability correspondingly decreases reaching zero when $[B]$ equals $[B]_{min}$.

This case with only one control parameter is referred to as the simple fold model (Fowler, 1975) (Figure 16.3). More control parameters are possible with mathematical expressions for up to 25 control parameters having been

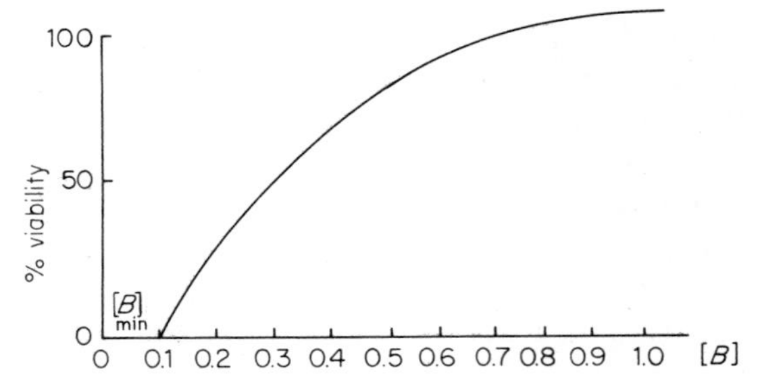

Figure 16.2 Form of the relationship between % viability and control parameter B.

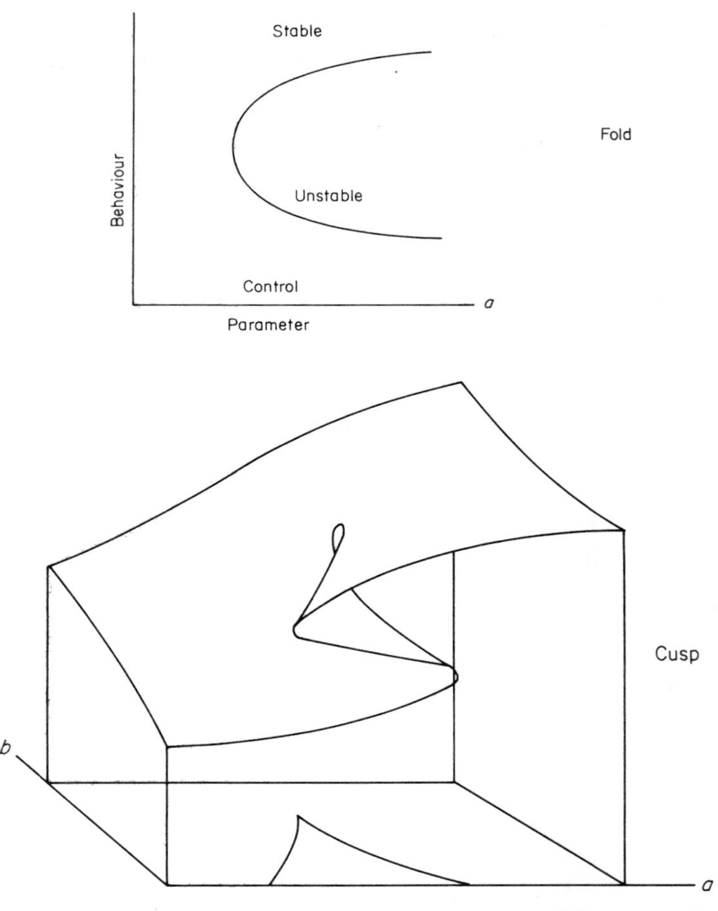

Figure 16.3 The behaviour surface of the simple fold model (1 control parameter, a) and the cusp model (2 control parameters, a, b).

derived (Zeeman, 1976). Probably, the most widely applied catastrophe model though is that with two control parameters, i.e. the cusp model. The name, like that of the simple fold, derives from the topographical shape of the equilibrium potential energy surface defining behaviour in terms of the control parameter (Figure 16.3). For more than two control parameters the topographical form of the equilibrium potential energy surface is difficult to envisage since more than three dimensions are required.

For the kinetic models in Chapters 10 to 12 the relationship (equation 10.10) between % viability and the concentration of the crucial moiety using probability theory is:

$$\ln V = K_1[B]_0 - K_1[B]_0 + \ln 100 \quad (16.6)$$

where, V = % viability

$[B]$ = concentration of crucial moiety at time t

$[B]_0$ = concentration of crucial moiety at time zero

K_1 = probability constant

The analogous rèlationship derived from catastrophe theory (equation 16.5) is,

$$V = 100[K([B] - [B]_{min})]^{\frac{1}{2}} \quad (16.7)$$

or,

$$\left(\frac{V}{100}\right) = [K([B] - [B]_{min})]^{\frac{1}{2}} \quad (16.8)$$

Rewriting equation 16.6 gives:

$$\left(\frac{V}{100}\right) = exp[K_1([B] - [B]_0)] \quad (16.9)$$

The similarity of these two relationships is quite marked when the viable fraction is close to unity. However, they differ fundamentally when the viable fraction is close to zero. That derived from catastrophe theory (equation 16.8) indicates that the viable fraction may be zero whereas that derived from probability theory equation (16.9) indicates a finite lower limit equal to $exp(-K_1[B]_0)$. While the limiting value can closely approach zero it can never be exactly zero. This possible difficulty was noted previously (Cox, 1976). Since the catastrophe theory derivation does not exhibit a finite lower limit and because it is based on a discontinuous process, it may be preferable. Consequently, equations corresponding to those in Chapters 10 to 12 will be given below in terms of catastrophe theory (equation 16.7).

16.2 CATASTROPHE THEORY AND DENATURATION KINETICS

In Chapter 10 it was shown that first order denaturation kinetics accounted extremely well for loss of viability or infectivity induced by desiccation. The equation corresponding to equation 10.12 but based on catastrophe theory is,

$$V = 100[K[B]_0\, e^{-kt} - K[B]_{min}]^{\frac{1}{2}} \quad (16.10)$$

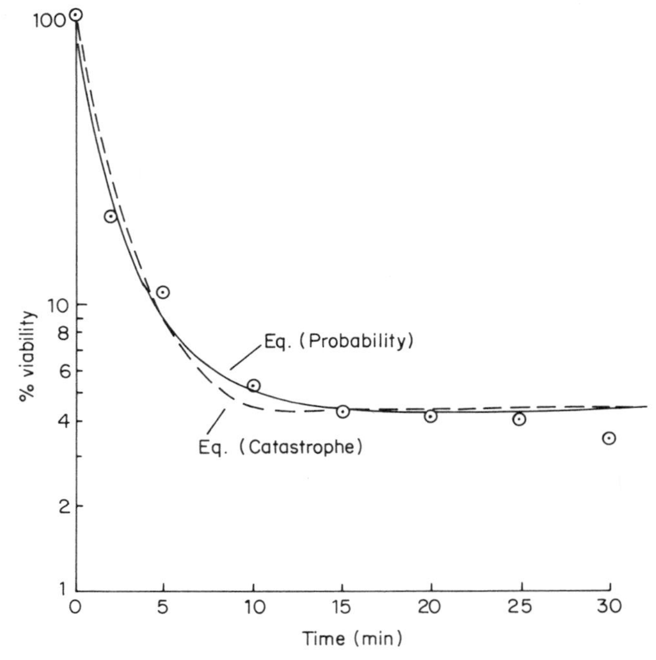

Figure 16.4 Comparison of the viability-time curve for *E. coli* Jepp wet disseminated into nitrogen. Points experimental data; ——— calculated by equation 10.12, – – – – calculated by equation 16.10.

where k is the first order denaturation velocity constant. Treating the first order denaturation process itself in terms of catastrophe theory leads to complications outside the scope of this book. Their extent may be gauged from the work of Calo and Chang (1980) on the kinetics of chemical reactions. Consequently, the traditional approach as used in Chapters 10 to 12 will be used to represent chemical reactions involved in the mechanisms discussed here.

As an example of the application of equation 16.10, it was found using the method of Rosenbrock (1960) to fit the data given in Figure 10.14 not quite as well as equation 10.12 (Figure 16.4). In other examples though the degree of fit of equation 16.10 can be slightly better than that of equation 10.12. Values for constants derived by applying equation 16.10 to data for *E. coli* B and SFV are given in Figures 16.5 and 16.6 (Figure 16.5 may be compared to Figure 10.18.) The value of $K[B]_0$ (Figure 16.5) is essentially independent of RH below 70%, but with slight deviations (<1%) above it. Whether these deviations are significant is not yet known. For SFV, the value of $K[B]_0$ is essentially independent of RH (Figure 16.6).

The marked increase in the value for the first order denaturation constant k in the range 75–85% RH for *E. coli* B (Figure 16.5) indicates that species B is particularly unstable in this RH region. Possible reasons for such instability are

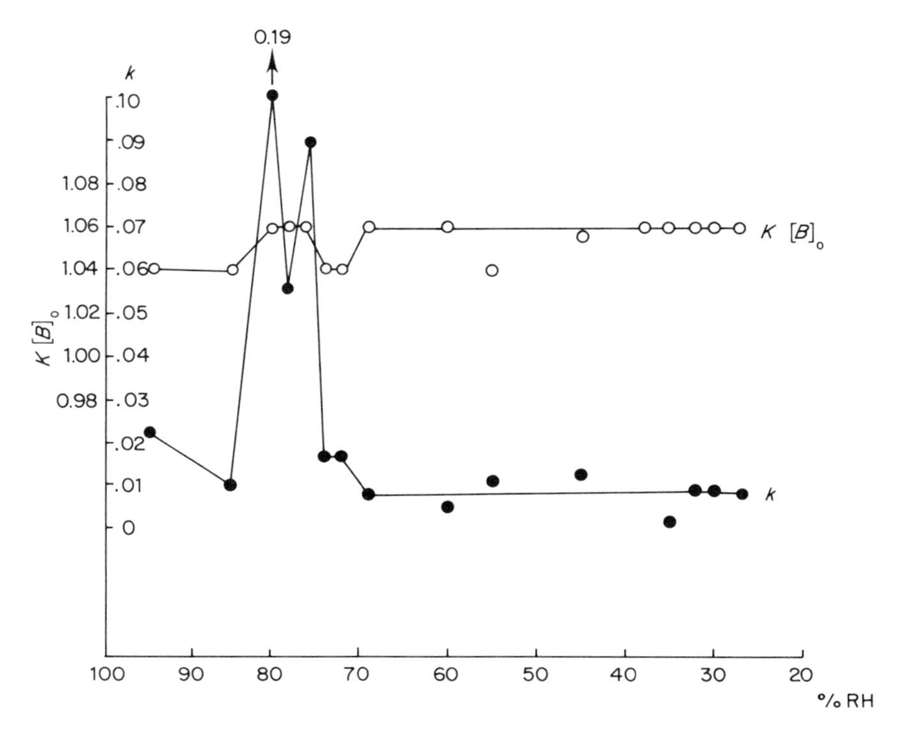

Figure 16.5 Constants derived by fitting equation 16.10 to experimental viability-time curves for *E. coli* B aerosolized from suspension in 0.13 M raffinose into nitrogen.

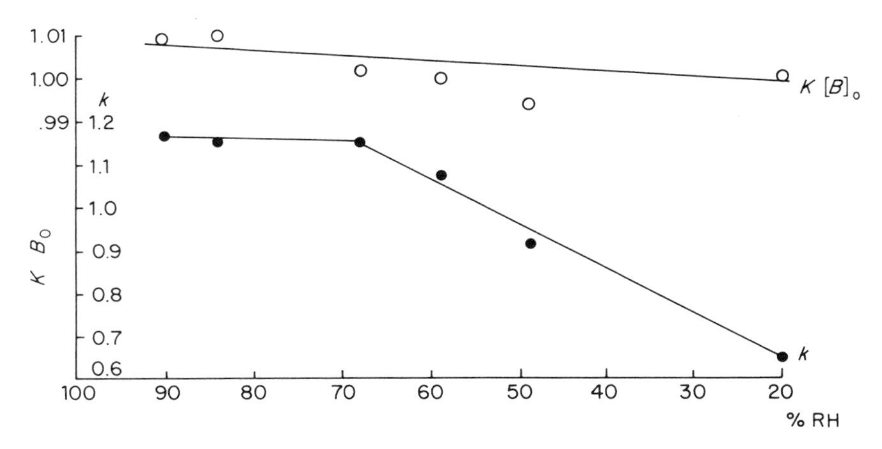

Figure 16.6 Values of the constants derived by fitting equation 16.10 to experimental viability-time curves for Semliki Forest virus aerosolized from suspension in medium 199 + 10% (vol/vol) calf serum.

discussed in Section 10.4. For SFV, in contrast, the value of k depends less critically upon RH. Even so both of these microorganisms are most stable (i.e. lowest values for the first order denaturation constant) at low RH.

From equation 16.10, it could be argued that the constant $K[B]_{min}$ is redundant, in that (from the boundary condition at $t = 0$, $V = 100\%$) it is given by:

$$K[B]_{min} = K[B]_0 - 1 \tag{16.11}$$

However, it was not evaluated as such for two reasons:
1. the microbial population may not be homogeneous, i.e. a resistant subpopulation may be contained within it (Cox, 1976);
2. owing to errors in normalization procedures (Section 2.11) the viability at $t = 0$ may not actually be 100%.

Consequently, separate evaluation of values of $K[B]_{min}$ provides an estimate for the relative magnitudes of the minimum value of B, the extent of resistance within the stressed population and possible normalization errors. $K[B]_{min}$, therefore, is equivalent to the following expression:

$$K[B]_{min} = K[B]'_{min} - K[B]_{res} + \Delta V \tag{16.12}$$

where, $K[B]'_{min}$ = the true value of $K[B]_{min}$,

$K[B]_{res}$ = the value representing the resistant proportion of the population

and, ΔV = a correction term to allow for normalization errors.

Derived values of $K[B]_{min}$ (equation 16.10) for *E. coli* B and SFV (corresponding to Figures 16.5 and 16.6) are given in Table 16.1. Due to the complex nature of the term $K[B]_{min}$ (equation 16.12) its wide variation as a function of RH is not too surprising. Values for ($K[B]_0 - 1$) in Table 16.1 correspond to 'ideal' values as calculated from equation 16.11. Compared with these values negative deviations of $K[B]_{min}$ imply population resistance and/or normalization errors whereas positive deviations imply normalization errors.

Biological decay may arise through denaturation mechanisms having higher than first order. For example, instead of monomolecular reactions bimolecular or higher order reactions may be involved as discussed in Section 10.7. Such possibilities are discussed in the following section which considers the additional parameter of temperature.

16.3 CATASTROPHE THEORY APPLIED TO ANALYSIS OF THE ROLE OF TEMPERATURE

The method of analysis is to fit an equation describing loss of viability (or infectivity) to experimental biological decay data obtained at several different temperatures. Derived values for the constants are then analysed as a

Table 16.1. Derived values of $K[B]_{min}$ and $(K[B]_0 - 1)$

Organism	% Relative humidity	$K[B]_{min}$	$K[B]_0 - 1$
SFV in air	20	-4.1×10^{-2}	-1.0×10^{-3}
	49	-7.9×10^{-3}	-6.0×10^{-3}
	59	-8.3×10^{-4}	-1.0×10^{-4}
	68	-1.0×10^{-4}	$+1.3 \times 10^{-2}$
	84	-1.0×10^{-6}	$+1.0 \times 10^{-2}$
	90	-2.5×10^{-4}	$+9.8 \times 10^{-3}$
E. coli B/ 0.3M raffinose in N_2	27	-7.5×10^{-4}	+0.062
	30	+0.075	+0.057
	32	+0.011	+0.064
	35	+0.107	+0.065
	45	+0.041	+0.057
	55	+0.109	+0.043
	60	-1.15×10^{-3}	+0.064
	69	+0.071	+0.058
	72	−0.018	+0.041
	74	+0.063	+0.043
	76	−0.10	+0.060
	78	$+1.12 \times 10^{-4}$	+0.060
	80	−0.232	+0.061
	85	+0.122	+0.044
	95	−0.043	+0.042

function of temperature as for other chemical reactions, e.g. by the Arrhenius equation.

Ehrlich and co-workers provide experimental data for loss of viability (or infectivity) of *Flavobacterium* sp., *F. tularensis* and VEE virus as a function of aerosol age and temperature (Ehrlich *et al.*, 1970; Ehrlich and Miller, 1971, 1973). However, the first order denaturation equations 10.12 (probability theory) and 16.10 (catastrophe theory) do not fit these data, especially those obtained at elevated temperatures. Therefore, the corresponding equation for the second order denaturation process, viz.

$$B + B \rightarrow 2B \tag{16.13}$$

was compared with the experimental data.

The equation, using catastrophe theory, is,

$$V = 100\left[\frac{K[B]_0}{1 + K[B]_0 kt} - K[B]_{min}\right]^{\frac{1}{2}} \tag{16.14}$$

where the constants $K[B]_0$, $K[B]_{min}$, and k are as previously defined.

Equation 16.14 fits the data (see above) of Ehrlich and co-workers quite well as exemplified by Figure 16.7 (lines calculated, points experimental data).

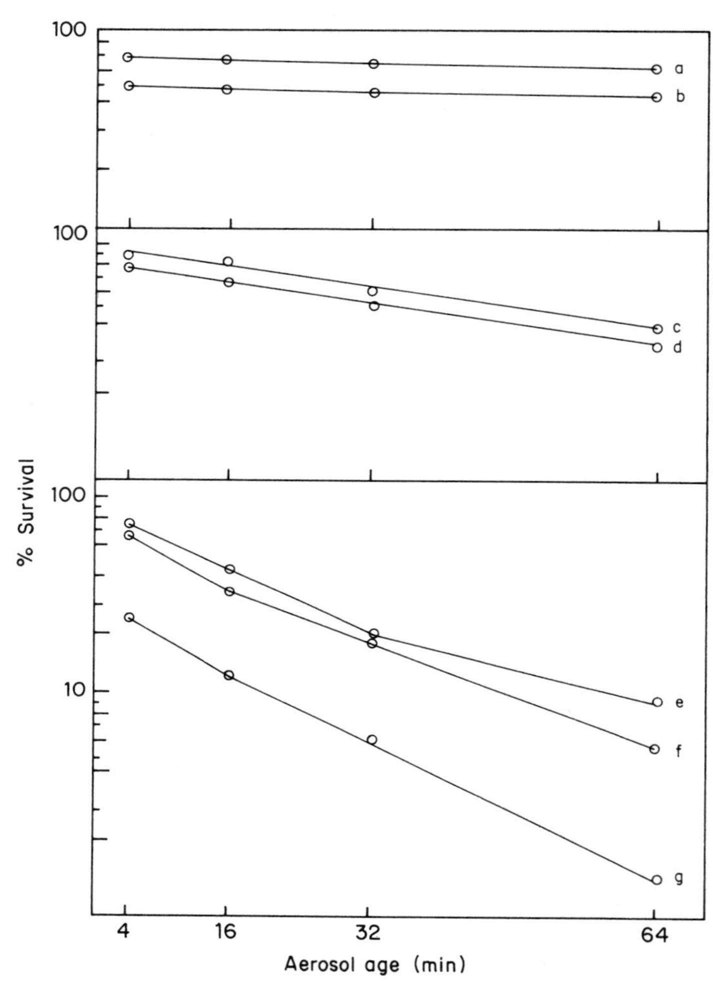

Figure 16.7 Effect of temperature on the aerosol survival of *Flavobacterium* sp. Points are experimental data, lines are calculated from equation 16.14 for a second order denaturation process. (Experimental data taken from Ehrlich *et al.* (1970), *Appl. Microbiol.*, **20**, 884.)

Curve	Temperature (°C)
a	18
b	40
c	24
d	2
e	29
f	38
g	49

Values of derived constants were analysed as a function of temperature in the following way. The Arrhenius equation relating the rate (or velocity) constant of a chemical reaction to temperature is,

$$k = A\,exp(-E/RT) \tag{16.15}$$

where, k = rate (or velocity) constant,

A = molecular collision frequency factor,

E = reaction activation energy (calories/mole),

R = gas constant (moles/calorie/degree),

and, T = absolute temperature.

Rewriting equation 16.15 gives,

$$\ln k = \ln A - E/RT \tag{16.16}$$

Therefore, the natural logarithm of values of k when plotted against the reciprocal of the corresponding absolute temperatures, should yield a straight line. Its negative slope is proportional to the activation energy of the reaction, whereas the ln k axis intercept value is equal to the natural logarithm of the frequency factor, A. The activation energy (E) is related to the enthalpy of activation (ΔH^*) by,

$$E = \Delta H^* + RT \tag{16.17}$$

Values for $K[B]_0$ are analysed in a somewhat analogous way. The standard free energy change, $\Delta F^\circ_{T,P}$, at constant temperature and pressure, is related to temperature by,

$$\Delta F^\circ_{T,P} = -RT \ln K_{\text{equil}} \tag{16.18}$$

However,

$$\Delta F^\circ_{T,P} = \Delta H_{T,P} - T\,\Delta S_{T,P} \tag{16.19}$$

where, $\Delta H_{T,P}$ = enthalpy change, $\Delta S_{T,P}$ = entropy change } at a constant temperature and pressure

and, $$\Delta F^\circ_{\text{P}} = \Delta H_{\text{p}} + T\left(d\frac{\Delta F^\circ_{\text{p}}}{dT}\right) \tag{16.20}$$

(Gibbs-Helmholtz equation)

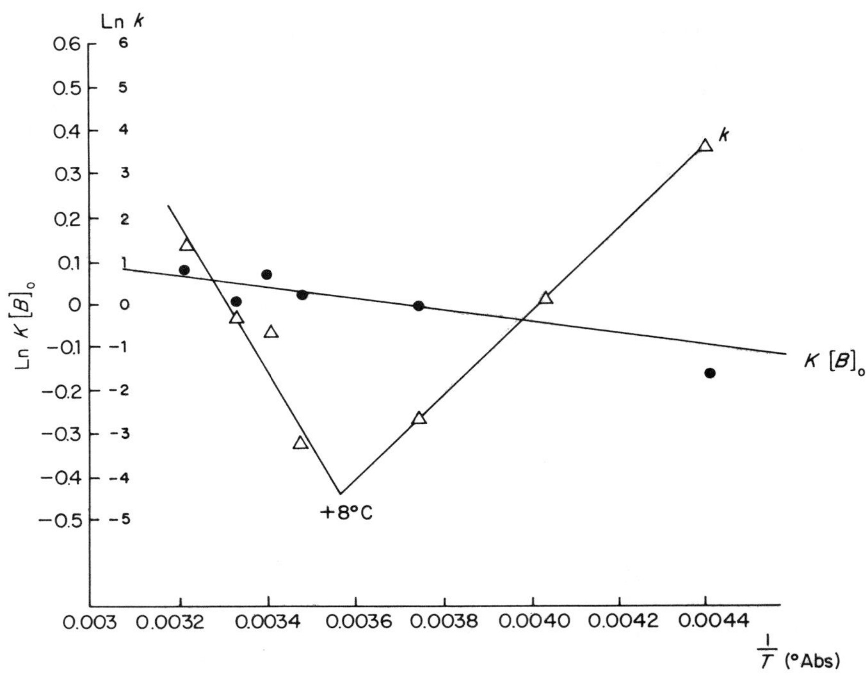

Figure 16.8 Values of the constants derived by fitting equation 16.14 (2nd order denaturation) to viability-time curves for aerosolized *Flavobacterium* sp. (Figure 16.7) as a function of reciprocal absolute temperature (see text).

Provided that $(d\Delta F^\circ_p/dT) = 0$ (i.e. the free energy change is independent of temperature), then,

$$\Delta F^\circ_P = \Delta H_p \tag{16.21}$$

Alternatively, if $(d\Delta H_p/dT) = 0$ (i.e. ΔH_p is independent of temperature) then integration of equation 16.20 gives,

$$\frac{\Delta F_2}{T_2} - \frac{\Delta F_1}{T_1} = \frac{\Delta H_p(T_2 - T_1)}{T_1 T_2} \tag{16.22}$$

From equations 16.18 and 16.22,

$$\ln\left(\frac{K_2}{K_1}\right) = \frac{-\Delta H_p}{R}\left(\frac{T_1 - T_2}{T_1 T_2}\right) \tag{16.23}$$

Hence, by analogy when $\ln K[B]_0$ values are plotted as a function of the reciprocal of the corresponding absolute temperatures a straight line should be obtained (provided ΔH_p is independent of temperature).

Natural logarithms of values of $K[B]_0$ and of k (derived by fitting equation 16.14 to the data of Ehrlich *et al.*) are plotted as a function of $1/T$ in Figures 16.8, 16.9 and 16.10. For *Flavobacterium* sp. (Figure 16.8) values of

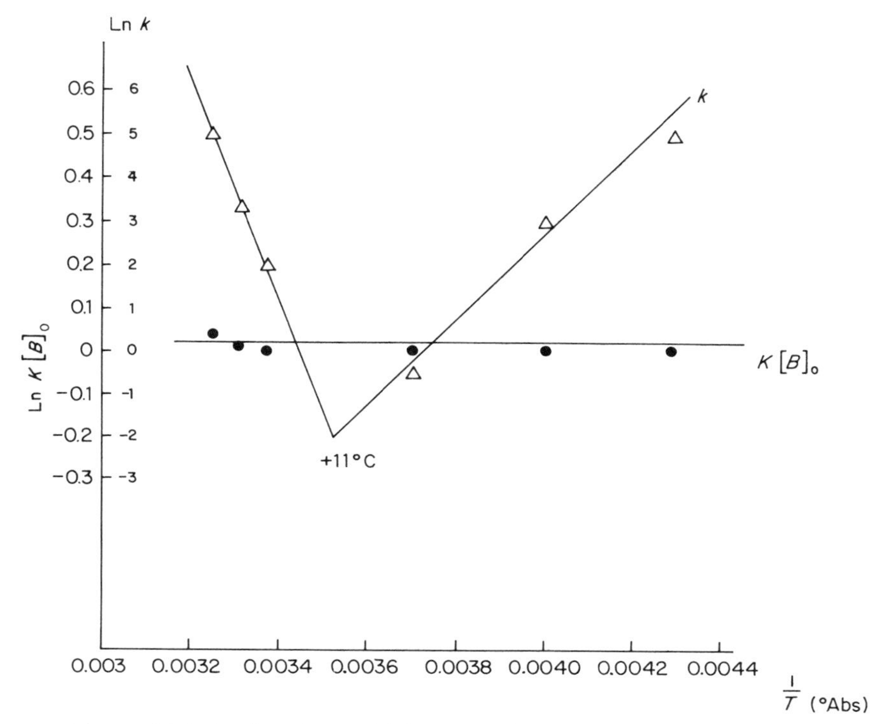

Figure 16.9 As for Figure 16.8 except for wet aerosolized *F. tularensis*. (Experimental data taken from Ehrlich and Miller (1973), *Appl. Microbiol.*, **25**, 369.)

ln $K[B]_0$ are a linear function of $1/T$ whereas values of ln k fall on two straight lines which intersect at about +8 °C. For *F. tularensis* (Figure 16.9) values of $K[B]_0$ are independent of temperature, whereas values of ln k fall on two straight lines, intersecting at +11 °C. With wet and dry disseminated VEE (Figure 16.10) temperature has little effect up to +24 °C. Above +24 °C, the transition temperature, the activation energy and enthalpy both suddenly increase.

Values for derived thermodynamic parameters are given in Table 16.2 for each of the three microorganisms. The pattern for each of them is similar in that there is a transition temperature above which the activation energy E is positive, and below which it is negative or zero. The enthalpy for all three microorganisms is quite small. The transition temperature for the two bacteria is the temperature at which they should show maximum aerostability, under the conditions of test. The transition temperature for VEE marks the temperature above which this virus becomes relatively unstable in the aerosol. In all cases, the transition temperatures are above 0 °C and, therefore, do not correspond to a change of state of solvent water.

Proteins and nucleic acids are known to change their structures with temperature (e.g. melting) and with RH (see Section 10.4), but such changes

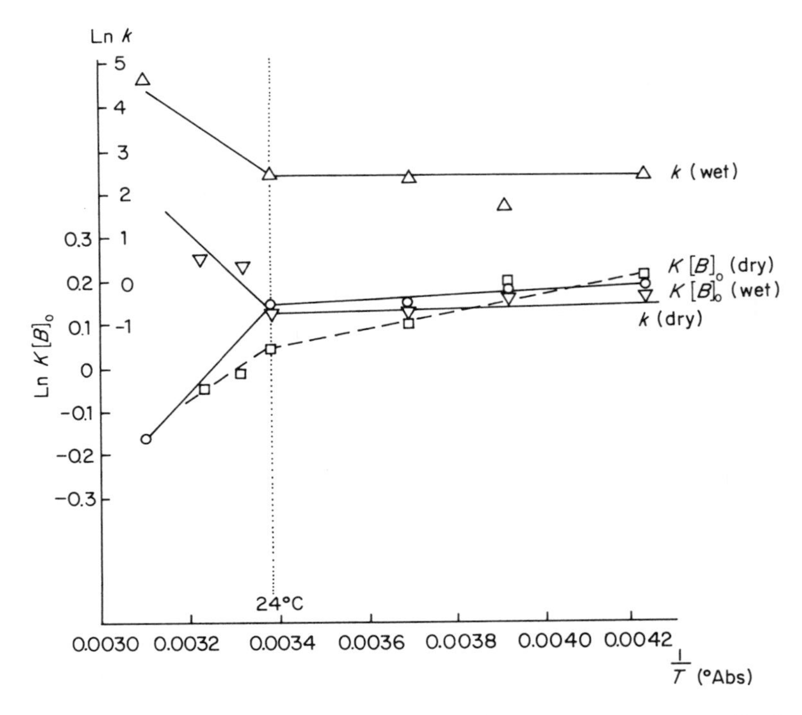

Figure 16.10 As for Figure 16.8 except for wet and dry aerosolized Venezuelan Equine Encephalitis virus. (Experimental data taken from Ehrlich and Miller (1971), *Appl. Microbiol.*, **22**, 194.)

occur with activation energies greater than 100 Kcal/mole and at much higher temperatures than those given in Table 16.2. These values correspond more closely with those for enzymes (Dixon *et al.*, 1979). In addition, hydrolysing enzymes such as trypsin, pepsin, lipase, fumarate hydratase, etc., exhibit changes in activation energy at similar transition temperatures as in Table 16.2, while fumarate hydratase inhibitors show activation energies which change sign above and below a transition temperature in the region of 20 °C (Dixon *et al.*, 1979). Explanations proposed to account for such behaviour include parallel and successive reactions, protein structural changes and phase changes in lipids (Dixon *et al.*, 1979; Sondergaard, 1979). Lipid phase changes have been noted also for synthetic phospholipids (Blume, 1979; Cho *et al.*, 1981) as well as in membranes of *Anacystis nidulans* (Furtado *et al.*, 1979), in *E. coli* membrane vesicles and in lipid–water and lipid–protein–water model systems (Dupont *et al.*, 1972).

For *A. nidulans* the transition temperature (in the range −15 °C to +30 °C) for the membrane lipid phase separation depends on the growth temperature, while the kinetics of the transition in *E. coli* membrane vesicles is similar to that for the lipid–protein–water system (Dupont *et al.*, 1972). Both the vesicles

Table 16.2. Activation energy and enthalpy of denaturation

Organism	Temperature range (°C)		Transition temperature (°C)	Activation energy (kcal/mole)	ΔHp (cal/mole)
Flavobacterium sp.	+8	+49	+8	+28.9	−345
	−40	+8		−21.0	−345
F. tularensis	+11	+35	+11	+43.1	0
	−40	+11		−22.3	0
VEE	+24	+49	+24	+17.6	+2500
(wet disseminated)	−40	+24		0.0	+170
VEE	+24	+49	+24	+17.6	+1450
(dry disseminated)	−40	+24		0.0	+340

and the model system demonstrate very marked hysteresis in the order–disorder transition of the paraffin chains of the lipid as a function of temperature. Lipids without protein present show only slight hysteresis which indicates that the presence of protein strongly hinders lipid diffusion and reaggregation (Dupont *et al.*, 1972). Whether similar phenomena are involved in desiccation damage in microorganisms is not known, but many parallels exist. At least for the microorganisms discussed above it seems more likely that temperature operates in terms of molecular or lipid phase changes rather than absolute humidity (Section 10.8). Further experimental data are required where both temperature and RH are varied. In addition, structural and lipid phase changes might be looked for by means of, for example, infrared microspectroscopy (Falk *et al.*, 1963; Baddiel *et al.*, 1972; Shiraishi *et al.*, 1977) or other physical techniques such as membrane probes, EPR, NMR, etc.

16.4 CATASTROPHE THEORY AND MORE COMPLEX DENATURATION KINETICS

In the previous sections it was shown how loss of viability or infectivity due to desiccation of *E. coli* and SFV could be accounted for by a first order denaturation process. Alternatively, for *Flavobacterium* sp., *F. tularensis* and VEE, a second order, rather than first order, denaturation process is much more consistent with the experimental data. The model given in equation 16.13 is that for such a process where molecule B dimerizes to give an inactive form 2B (see also Section 10.4). Another second order process would be when two different molecular species were to combine to give the inactive form, e.g. a reaction between a reducing sugar and an amino acid (Section 10.4), i.e.

$$B + C \xrightarrow{k} BC \tag{16.24}$$

Corresponding to this situation the equation for loss of viability (or infectivity) is,

$$V = 100(K[B]_0 - \phi - K[B]_{\min})^{\frac{1}{2}} \tag{16.25}$$

$$\text{where, } E = \left(\frac{1}{[B]_0} - \frac{1}{[C]_0}\right)$$

$$\beta = -\left(\frac{1}{[B]_0[C]_0}\right)$$

$$\sigma = (E^2 - 4\beta)$$

$$\alpha = \frac{(E - \sigma)}{(E + \sigma)} exp(\sigma \times [B]_0 \times [C]_0 \times k \times t)$$

$$\phi = \frac{(E \times (\alpha - 1) + \sigma(1 + \alpha))}{2 \times \beta \times (1 - \alpha)}$$

The experimental data analysed in Section 16.3 were analysed also with equation 16.25 but it does not fit the data significantly better than equation

16.14, especially when the greater number of constants of equation 16.25 is taken into account. At present, therefore, it seems unnecessary to postulate a denaturation process more complex than that involving dimerization (equation 16.14). Whether such a process occurs in reality may not be known until there are further studies along the lines outlined at the end of Section 16.3. Nonetheless the kinetics of viability loss due to desiccation are consistent which the chemical reactions discussed in Section 10.4.

16.5 CATASTROPHE THEORY AND OXYGEN-INDUCED LOSS OF VIABILITY

The equation based on catastrophe theory corresponding to that for oxygen-induced death (equation 11.4) is,

$$\left[V = 100\left[K[A]_0\right] \times exp\left(-\frac{k[X]_0[O_2]t}{K_x + [O_2]}\right) - K[A]_{min}\right]^{\frac{1}{2}} \qquad (16.26)$$

where, $[A]_0$ = concentration of crucial moiety at time = $t = 0$,

K = proportionality constant,

K_x = equilibrium constant between carrier X and oxygen,

$[X]_0$ = total carrier concentration,

k = velocity constant for reaction of species A with XO_2,

$[O_2]$ = oxygen concentration.

The data of Cox *et al.* analysed by equations 11.3 and 11.4 were analysed also using equation 16.26. As shown in Figure 16.11 both equations fit the data extremely well which reflects the small differences usually found between the probability and catastrophe approaches discussed in Section 16.1. Unfortunately, there are no known data corresponding to those analysed here, but made over a range of temperatures.

16.6 CATASTROPHE THEORY AND OAF-INDUCED LOSS OF VIABILITY

The equation corresponding to equation 12.4 is,

$$V = 100\left[K[B]_0\left\{exp\left[\left(-\alpha t - \frac{\alpha}{\beta}(e^{-\beta t} - 1)\right)\right]\right\} - K[B]_{min}\right]^{\frac{1}{2}} \qquad (16.27)$$

where, α, β are as previously defined in Chapter 12.

As presently there is no independent way to assess OAF concentration (see Chapter 12), equations 12.4 and 16.27 cannot be compared independently.

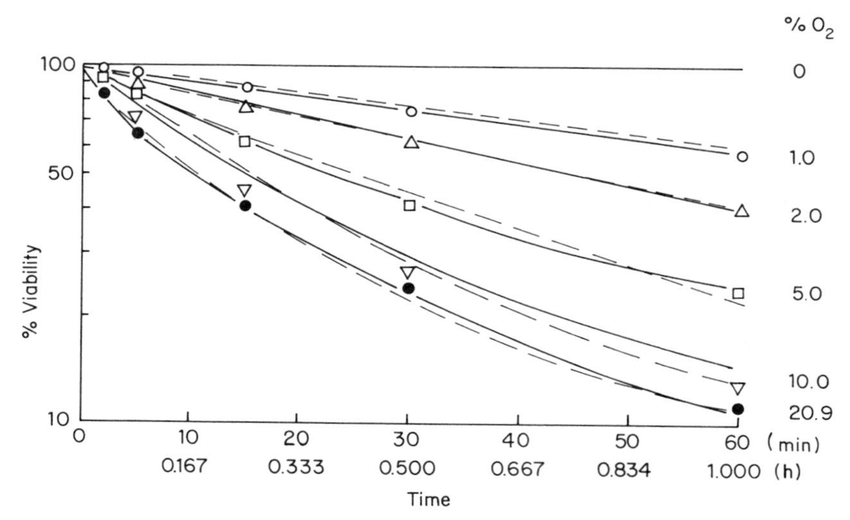

Figure 16.6 Values of the constants derived by fitting equation 16.10 to experimental viability-time curves for Semliki Forest virus aerosolized from suspension in medium 199 + 10% (vol/vol) calf serum.

16.7 CATASTROPHE THEORY AND REPAIR

Some microorganisms following their recovery from aerosols repair inflicted damage (Chapters 10, 13 and 14). Application of catastrophe theory enables repair phenomena to be represented mathematically.

Section 16.1 shows how behaviour can be described by catastrophe theory and in the examples given in Sections 16.3 to 16.6 one control parameter is invoked. Introduction of a second control parameter allows repair to be taken into account. One form of the potential energy function for this case (Zeeman, 1976) is,

$$F = \frac{x^4}{4} - ax - \frac{1}{2}bx^2 \tag{16.28}$$

where, x = behaviour,
a, b = control parameters.

For equilibrium, $(dF/dx) = 0$, therefore,

$$\frac{dF}{dx} = x^3 - a - bx = 0 \tag{16.29}$$

i.e. the cusp model (Section 16.1, Figure 16.3). Control parameter a is as previously defined, while control parameter b represents repair. In the simple fold model (one control parameter) once catastrophe occurs no way exists for the system to regain the stable or viable equilibrium state. In the cusp model (two control parameters) over certain ranges of values of control parameters a and b following the catastrophe the system can regain stable equilibrium through repair. In the present context (Figure 16.3) setting $b = 0$ as the value of a decreases the fraction of microorganisms in the viable state declines while the fraction in the non-viable state increases correspondingly. If now the repair parameter b is increased from zero a point is reached where non-viable cells convert back to viable ones, i.e. a point is reached in the cusp behaviour surface where the previously forbidden transition from non-equilibrium back to the viable equilibrium state becomes permissible (Figure 16.3). Consequently, catastrophe theory can take account of the repair phenomena demonstrated by stressed microorganisms (Chapter 14). Since Thom's catastrophe theory can encompass large numbers of control parameters, all the aerosol stresses could be represented simultaneously together with their corresponding repair mechanisms. The derived behaviour surface although multidimensional and difficult to envisage may provide explanation for complex synergism of compounded stresses or even the oscillatory behaviour described by Heckly and Dimatteo (1975).

16.8 CONCLUSIONS

Kinetic models derived on the basis of catastrophe theory can account for loss of viability induced by desiccation, by oxygen and by OAF. Calculated curves agree very well with experimental data. Degree of fit, particularly for denaturation induced by desiccation, is better than that for curves calculated from equations derived on the basis of probability (Chapter 10). Analysis of temperature effects suggests that desiccation results in irreversible lipid phase changes and/or sugar-protein complex formation in the membranes of bacteria and of viruses. Also *Flavobacterium* sp., *F. tularensis* and VEE have optimum survival temperatures of respectively, +8 °C, +11 °C, $< +24$ °C. Catastrophe theory which can account for discontinuous processes seems well suited to analysing loss of viability of desiccated microorganisms and to allowing for repair.

The suggested mechanism of loss of microbial viability through denaturation of lipoprotein and of lipoprotein-carbohydrate membrane complexes and their stabilization by raffinose and other sugars is analogous to that for nematodes surviving dehydration. Their survival is achieved through their synthesis of large quantities of trehalose, a sugar which combines with phospholipids of the cell to form a structure which is unaffected by loss of water (Wright, 1984). Hence, membrane lipoprotein-carbohydrate complex destabilization through water loss and its stabilization by sugars may be a fairly widespread phenomenon in nature.

REFERENCES

Baddiel, C. B., Breuer, M. M. and Stephens, R. (1972). *J. Coll. and Int. Sci.*, **40**, 429–436.
Blume, A. (1979). *Biochim. Biophys. Acta*, **557**, 32–44.
Calo, J. M. and Chang, H. C. (1980). *Chem. Eng. Sci.*, **35**, 264–272.
Cho, K. C., Choy, C. L. and Young, K. (1981). *Biochim. Biophys. Acta*, **663**, 14–21.
Cox, C. S. (1976). *Appl. Environ. Microbiol.*, **31**, 836–846.
Dixon, M., Webb, E. C., Thorne, C. J. R. and Tipton, K. F. (1979). *Enzymes*, 3rd edn, Academic Press, New York, San Francisco.
Dupont, Y., Gabriel, A., Chabre, M., Gulik-Krzywicki, T. and Schechter, E. (1972). *Nature (Lond.)*, **238**, 331–333.
Ehrlich, R. and Miller, S. (1971). *Appl. Microbiol.*, **22**, 194–198.
Ehrlich, R. and Miller, S. (1973). *Appl. Microbiol.*, **25**, 369–372.
Ehrlich, R., Miller, S. and Walker, R. L. (1970). *Appl. Microbiol.*, **20**, 884–887.
Falk, M., Hartman, K. A. Jr. and Lord, R. C. (1963). *J. Amer. Chem. Soc.*, **85**, 387–391.
Fowler, D. H. (1975). *Structural Stability and Morphogenesis*, W. A. Benjamin Inc., New York.
Furtado, D., Williams, W. P., Brain, A. R. P. and Quinn, P. J. (1979). *Biochim. Biophys. Acta*, **555**, 352–357.
Heckly, R. J. and Dimatteo, J. (1975). *Appl. Microbiol.*, **29**, 565–566.
Rosenbrock, H. H. (1960). *Computer J.*, **3**, 175–184.
Shiraishi, H., Hiltner, A. and Baer, E. (1977). *Biopolymers*, **16**, 2801–2806.
Sondergaard, L. (1979). *Biochim. Biophys. Acta*, **557**, 208–216.
Woodcock, A. and Davis, M. (1978). *Catastrophe Theory: The Landscape of Change*, Dutton, New York.
Wright, P. (1984). Science Report, *The Times*, 7 September.
Zeeman, E. C. (1976). *Scientific American*, **234**, 65–83.

Chapter 17

The aerobiological pathway in practice

17.1 INTRODUCTION

Knowledge of the survival and infectivity of microorganisms as aerosols is required in many areas of scientific endeavour including those of human health, animal and crop disease, microbiology, genetic engineering, industry, environmental studies, space exploration and the biological control of insects. This chapter relates some of the implications of earlier sections to these particular pursuits.

17.2 HUMAN HEALTH

While there are a number of studies concerned with survival and infectivity of microorganisms when wet disseminated very few cover dry dissemination. Yet, as shown in Chapters 10 and 16, wet and dry disseminated microbes can behave differently. Disturbing dust and making beds, etc., therefore may produce a different epidemiology to that from sneezing and coughing, etc. Furthermore, the analyses in Chapter 16 suggest that temperature could be more important for aerosol stability than hitherto considered. The combined action of temperature and RH is not altogether clear but could be critical in determining aerosol infectivity. Consequently, the indoor spread of disease by the airborne route may be mainly dependent on the combined effects of these two environmental parameters. In high flying aircraft, though, ozone through chemical reaction with vapours released from plastics, etc. could form a synthetic OAF and thereby become paramount for decreasing aerosol survival and infectivity of airborne microorganisms (Chapter 12). In closed operating theatres, etc., synthetic OAF could be of benefit in reducing aerial contagion but when ventilated quite large quantities of ozone may be needed. Even then for some microorganisms, unless the concentration is high, OAF sterilizing action is not rapid.

Whether disease develops in hosts exposed to an infective aerosol depends upon several factors including the relative levels of specific host antibodies and the infective dose. The number of microorganisms required to initiate disease even in non-immune hosts is greatly affected by the occurrence of infectivity suppressors (Chapter 15) in natural body fluids. This number is a function also of the landing site in the host and therefore of aerosol aerodynamic particle size as this parameter largely controls that landing site. The importance of landing site possibly arises in terms of tissue specificity relationships (Chapter 9) or of providing specific components for microbial repair (Chapter 14).

On inhalation, aerosol particles are exposed to very humid air thereby becoming hydrated and enlarged in size. This process of vapour phase rehydration (Chapter 2) can cause marked changes in viability (Chapter 10) but whether there are concomitant changes in infectivity is not known (Chapters 2 and 15). Consequently, current values for LD_{50}'s may bear little relationship to actual values (Chapters 2 and 15).

17.3 ANIMAL AND CROP DISEASE

The comments of Section 17.2 above in the main apply in an analogous manner to the airborne infection of farm animals and birds which when confined seem especially vulnerable. Even when in the open air farm animals can become infected by foot-and-mouth disease virus, for example. Nearly simultaneous outbreaks of this disease sometimes arise at farms which are physically isolated one from another and are many miles distant. Whether long-range transmissions occur by the aerosol route is not easy to prove, but the virus responsible for the disease is extremely resistant to aerosol stresses such as desiccation, radiation, oxygen, ozone and OAF (Chapter 12). Consequently, when carried in small aerosol particles the virus potentially could travel hundreds of kilometres and remain infective (Chapter 8). Swine vesicular disease and anthrax are two other animal pathogens having high aerosol stabilities. In contrast *Erwinia* strains are readily inactivated by OAF, but nonetheless potato blight may still be transmitted by the airborne route because of the close spacing of plants and the ensuing short-lived aerosol state.

Spore forming microorganisms, such as *Puccinia*, *Aspergillus*, *Alternaria*, *Cladosporium*, *Areospores*, *Ustilago*, are aerosol stable and well adapted for long-distance airborne transmission of crop diseases, although questions of particle take-off, deposition and landing remain (Chapter 9).

17.4 MICROBIOLOGY

Perhaps one of the more surprising observations is that tail-less phages, lipid-free and lipid-containing viruses and bacteria on suffering dehydration–rehydration stress (in the absence of radiation, oxygen, OAF, etc.) lose viability by the same basic mechanism. This mechanism is one of denaturation which through a first or second order process probably involving lipids, proteins and sugars causes inactivation of membrane constituents. Denaturation arises through destabilization induced by loss of solvent water molecules on dehydration (Chapter 10). Lipid-containing viruses and bacteria are least stable at mid to high RH and benefit from vapour phase rehydration. In contrast, lipid-free viruses are least stable at low RH and do not benefit from vapour phase rehydration. Presumably, lipid–protein–water systems become metastable at mid to high RH, whereas protein–water systems become destabilized at low RH (Chapters 10 and 16). Compounds protecting against viability loss induced by desiccation follow the same ranking as those for stabilizing isolated macromolecules (Chapter 10). While such compounds through chemical reactions (e.g. Maillard) protect against denaturation induced by aerosolization or by freeze-drying, mutation is not prevented. Rather, desiccation increases mutation frequency (Chapter 14). The alternative of freezing offers some advantages for cell preservation, but mutation continues in frozen samples. On the other hand, as denaturation usually causes volume increases, cell preservation possibly may be achieved by storage under high pressures (Chapters 10 and 16). Potentially, this storage method could concomitantly lower mutation frequency as some natural mutation processes have positive volume changes.

Another aspect of earlier chapters is that of the role of water in biological systems. Aerosolization provides a technique whereby the water content of microorganisms and of isolated biological materials can rapidly be changed. That isolated molecules and microbial structures suffer denaturation on their dehydration shows that one role of water in nature is stabilization of biopolymers and their complexes. Another is that for bacteria to metabolize and to divide, liquid water is required as a solvent and mobility matrix.

17.5 MICROBIAL CONTAMINATION

In the home, laboratory and industry problems are encountered due to microbial contamination. Compared to the airborne route spread of microorganisms by contact is slow. Moreover, the ease with which fine aerosol particles are carried by air currents permits particles to reach the most inaccessible of places. In the laboratory and industry the use of absolute filters reduce but rarely totally eliminate aerosol contamination problems. Maintaining environments sterile is further complicated if humans require access, since we continuously shed microorganisms from our bodies and clothes (Chapter 9). Depending upon the size, shape, density, etc. of these liberated particles they can disseminate as aerosols.

For space exploration, providing sterile spacecraft is extremely difficult. Consequently, space probes entering the atmosphere of another planet or landing on its surface may cause contamination by microorganisms from Earth. Even if conditions at the planet's surface are unequitable for the support of life the results of Dimmick and co-workers (Chapter 13) demonstrate that microbial colonies could become established in an aerosol form, given certain atmospheric conditions. This situation in our own atmosphere though appears unlikely due to the presence of oxygen, ozone, OAF, pollutants and radiation. All cause viability loss especially for log phase microorganisms (Chapters 10, 11 and 12). However, aerosol spread of microorganisms over the Earth's surface would appear inevitable given that the microbes survive the airborne state (Chapter 8). In this context spore forming microorganisms are well adapted (Section 2.11).

17.6 GENETIC ENGINEERING

Genetic engineering involves the transfer of a piece of DNA from one type of cell into another type of cell, often a particular strain of *Escherichia coli*. Following isolation of the *E. coli* bacterium having the required genetic complement the new *E. coli* variant is grown in fermenters to provide the desired microbial product, as for microorganisms isolated naturally. However, genetically engineered variants, by definition, are those to which previously we have not been exposed. It is therefore sensible to try to contain or at least to try to insure that such variants survive poorly in the environment.

As indicated in Chapter 6 perfect containment is a difficult goal especially when there is a reasonable probability that aerosol filters can become wet. Consequently, some reliance has to be placed on the poor capability of the variant to survive on surfaces and in the airborne state. As the contents of previous chapters show, to establish that aerosol survival capability is itself difficult due to the very large number of factors known to affect this property. In addition, natural mutation and selection will cause that capability to change with time. Also of concern are the findings of Hatch and co-workers (Chapter 14) that aerosol survival and repair capability are under genetic control. Yet genetic engineering implicitly involves changes to microbial genetic complement.

17.7 BIOLOGICAL CONTROL OF INSECTS AND PESTS

Disadvantages associated with the use of chemicals for control of insects and pests led to the currently expanding use of pathogenic microbes for this purpose. Prime examples are variants of *Bacillus thuringiensis* for the control of gypsy moth larvae and of mosquitoes, and nuclear polyhedrosis virus for controlling cotton bollworm (Chapter 7). Each year many hundreds of tons of preparations containing spores of *B. thuringiensis* are aerosolized for controlling insects. On a smaller scale similar preparations are available for the control of garden insects and pests.

Factors governing the effectiveness of airborne insect and pest pathogens are similar to those pertaining to infection of man and animals by airborne pathogens. For instance, particle size controls whether the infective aerosol particle can enter the breathing tubes of insects (Chapter 15) or whether the pathogen will be deposited on leaves (Chapter 9) to be eaten subsequently by developing larvae. Aerosol stability of such pathogens (Chapters 10 to 13 and 16) together with particle size, wind speed and direction (Chapter 8) will control the effectiveness and extent of downwind travel while the rate of aerosol generation will control dosages (Chapter 7). The specific insect or pest affected by such treatments is determined by the host specificity and pathogenicity of the particular pathogen which is aerosolized (Chapter 15). That the approach is a successful one is demonstrated by its increasing practice.

17.8 CONCLUSIONS

While the given examples indicate some implications of aerosol survival and infectivity, complete interpretation may require the detailed elucidation of repair and genetic control. These decisive mechanisms must be the most important of all in determining the fate of airborne microorganisms.

A glossary of terms

AEROBIOLOGY The study of phenomena associated with microorganisms, pollens, spores etc. when suspended in gases.
AEROSOL A collection of particles suspended in a gaseous medium.
AEROSOL ADDITIVES Additives or protecting agents incorporated into the SPRAY FLUID to modify the biological activity of microorganisms when AEROSOLIZED.
AEROSOL AGE The time interval between aerosol generation and sampling.
AEROSOL BEHAVIOUR The observed changes in properties of particles in aerosol form, e.g. concentration, size distribution, viability, etc.
AEROSOL CHAMBERS Containers for aerosols.
AEROSOL IMMUNIZATION Immunization by the aerosol route.
AEROSOL PARTICLES Some of the particles within an AEROSOL.
AEROSOL ROUTES AEROSOL pathways.
AEROSOL SAMPLERS Devices for recovering AEROSOL PARTICLES.
AEROSOL STABILITY Measure of the ability of aerosolized microorganisms to maintain viability or infectivity, or other defined biological activity.
AEROSOL STABILIZERS See AEROSOL ADDITIVES.
AEROSOLIZATION Act of generating AEROSOL PARTICLES.
AEROSOLIZED Operating the process of dispersal required for generating AEROSOL PARTICLES.
AGGLOMERATION Particles sticking together to form larger masses or agglomerates.
AGI-30 Standard glass impinger. AGI stands for 'all glass impinger'; 30 refers to the distance (mm) between the bottom of the critical orifice and the impinger base. Usual operating rate = 12.5 l/min.
AIRBORNE INFECTIONS Infections spread by the AEROSOL ROUTE.
AIRBORNE TRANSMISSIONS See AIRBORNE INFECTIONS.
AIRBORNE PARTICLES See AEROSOLS and AEROSOL PARTICLES.
AIR IONS Submicron (i.e. less than 1×10^{-4} cm) diameter molecular clusters carrying an electric charge.
ANTIBODIES Proteinaceous molecules occurring in the blood of hosts in response to their invasion by microorganisms or other biological entities.
ANTIGENS Materials eliciting the production of ANTIBODIES.

ATOMIZED AEROSOLIZED or sprayed to form a fine mist.
ATOMIZER A two-fluid aerosol generator.
ATOMIZER FLUID A liquid or liquid suspension which is to be ATOMIZED.
BIOLOGICAL DECAY Loss of biological function of a microbiological aerosol, as opposed to loss of aerosol particles due to PHYSICAL DECAY or loss.
BIOLOGICAL DECAY RATE Rate of BIOLOGICAL DECAY.
BIOPOLYMER Large natural molecule (for example protein, nucleic acid) built of amino acids, purines, pyrimidines, simple sugars, etc., as opposed to one that is man-made, e.g. polyethylene, polyurethane.
CAPILLARY IMPINGER See AGI-30.
CASCADE IMPACTOR An AEROSOL SAMPLER having two or more IMPACTORS in series.
CATASTROPHE THEORY A theory by René Thom to account for non-continuous events that occur in nature.
CLOUD A visible aerosol.
CLOUD CHAMBERS See AEROSOL CHAMBERS.
COAGULATION Similar to AGGLOMERATION.
COLLECTION EFFICIENCY Sampling efficiency of an AEROSOL SAMPLER usually measured relative to another sampler.
COLLISION ATOMIZER/SPRAY: A REFLUX ATOMIZER developed by W. E. Collison.
CONTAMINANTS Microorganisms or substances which can interfere in the course of an experiment.
CRITICAL ORIFICE An orifice through which gas flow is sonic, thereby regulating the gas flow rate.
DEATH RATE The rate of LOSS OF VIABILITY per unit time.
DECAY RATE The rate of BIOLOGICAL or PHYSICAL LOSS.
DENATURATION A mechanism causing loss of biological activity, described by,

$$\frac{-dx}{dt} = kx^n$$

where, x = concentration of biologically active species at time = t,

n = whole or fractional number (e.g. 1, $1\frac{1}{2}$, 2), its value defining the order of the reaction.

See also RENATURATION, INACTIVATION.
DIMER Product of linking two molecules.
DISSEMINATION Liquid (i.e. wet) or powder (i.e. dry) dispersal to form an AEROSOL.
DYNAMIC AEROSOLS Continuously flowing AEROSOLS.
FLASH EVAPORATOR Evaporator based on a rotating sphere containing a much smaller volume of the liquid to be evaporated so that a large liquid surface area is generated.
HAEM PROTEIN Protein which contains iron as an integral part of its structure, e.g. haemoglobin, catalase.
HETEROGENEOUS AEROSOLS AEROSOLS comprising non-uniform particles, e.g. non-uniform size or shape.

HOMOGENEOUS AEROSOLS AEROSOLS comprising uniform particles, e.g. shape, or AEROSOLS having uniform spatial distributions.
HYPERTONIC Solution with an osmotic pressure greater than that pertaining in cells (or sea water). Converse: HYPOTONIC.
IMPACTION Deposition of AEROSOL PARTICLES through inertial forces.
IMPINGEMENT IMPACTION onto a surface bathed with liquid.
INACTIVATION Loss of biological activity through ill-defined mechanisms.
INFECTIVITY Measure of ability to cause infection or disease, under a defined set of conditions.
INFECTIVITY SUPPRESSOR Spray fluid component resulting in loss of INFECTIVITY.
LOG PHASE Phase of growth when bacteria, mammalian cells, etc., double in number per unit of time.
LOSS OF VIABILITY Reduction in the proportion of a population of microorganisms able to replicate under a given set of growth conditions.
MOIETY Part of a larger assemblage, e.g. the HAEM of haemoglobin.
MONODISPERSE AEROSOLS Those comprising particles of uniform size.
NEBULIZER See ATOMIZER.
OAF Products of ozone + olefins reactions which have high chemical activity and induce high BIOLOGICAL DECAY RATES.
OLEFIN Series of hydrocarbons related to ethylene and having the general formula C_nH_{2n}.
OPEN AIR FACTOR See OAF.
PHYSICAL DECAY Removal of particles from an aerosol due to physical processes such as sedimentation, impaction, diffusion, electrostatic forces, etc.
POLYDISPERSE AEROSOLS Those comprising particles of non-uniform size.
PROTECTIVE (-ING) AGENTS See AEROSOL STABILIZERS.
REFLUX ATOMIZERS ATOMIZERS in which unaerosolized fluid, after passing through a two-fluid jet, returns by means of a baffle to a reservoir, e.g. COLLISON ATOMIZER, WELLS ATOMIZER.
REHUMIDIFICATION Vapour phase REHYDRATION.
REHYDRATION Replacement of water molecules lost through their evaporation, sublimation or other driving forces, e.g. osmotic pressure, temperature gradients.
RELATIVE HUMIDITY Ratio of actual water vapour pressure to the vapour pressure of liquid water at the same temperature.
RENATURATION Reverse of DENATURATION usually by non-metabolic reactions leading to a recovery of biological activity.
REPAIR Recovery of biological activity as a consequence usually of metabolic processes, e.g. enzymatic reactions.
ROTATING DRUMS Cylindrical AEROSOL containers slowly rotated on their axes to reduce PHYSICAL DECAY.

SERIAL PASSAGE Virus population maintained by sequential inoculation between cell cultures. A fresh cell culture is inoculated with the infective virus culture when it needs replenishing.
SHEAR FORCE Strain imposed on an object by a force, e.g. the bending of a tree caused by the force of the wind.
SHIPE IMPINGER Modified CAPILLARY IMPINGER.
SLIT SAMPLER AEROSOL SAMPLER impacting AEROSOL PARTICLES onto a rotating nutrient agar surface.
SLURRY A microbial suspension.
SPENT CULTURE FLUID Depleted medium following microbial growth.
SPRAY FLUIDS Liquids, containing or destined to contain microorganisms, which are ATOMIZED.
STABILIZING ADDITIVES See AEROSOL STABILIZERS.
STATIC AEROSOLS AEROSOLS which are not continuously generated and are held in a single aerosol chamber.
SURVIVAL See VIABILITY.
SURVIVAL CURVES Curves describing microbial SURVIVAL as a function of time or other parameters.
SUSPENDING FLUIDS See SPRAY FLUIDS.
TRACERS Observed numbers of VIABLE microorganisms recovered from their AEROSOLS decline with AEROSOL AGE due to the sum of their PHYSICAL DECAY and BIOLOGICAL DECAY. These two effects must be separated to obtain the true BIOLOGICAL DECAY RATE. Their separation is achieved by adding to SLURRIES materials which measure only the PHYSICAL DECAY of AEROSOLIZED microorganisms. Such materials are termed TRACERS, and, ideally, their incorporation in no way influences the outcome of an experiment, while their PHYSICAL DECAY RATE exactly matches that of the AEROSOLIZED microorganisms.
VIABILITY The proportion (usually expressed as a percentage) of a population which is VIABLE.
VIABLE The state in which microorganisms replicate under a given set of growth conditions.
WELLS ATOMIZER A REFLUX ATOMIZER developed by W. F. Wells.

Index